GENETICS AND BREEDING

Editor

Hartmann Pfenning

scitus academics

Genetics and Breeding
Edited by **Hartmann Pfenning**

This Book is Printed on Acid-Free Paper. ♾

ISBN: 9781681179254
Library of Congress Control Number: 2017948954

SCITUS Academics LLC,
616, Corporate Way, Suite 2, 4766,
Valley Cottage, NY 10989

www.scitusacademics.com

Notice

Printed in United States of America.

Preface

Animal breeding and genetics is concerned with maximizing desirable genetic traits, such as producing animals that have leaner meat. Animal geneticists have identified elements within genes that can enhance animal growth, health, and ability to utilize nutrients. These genetic advances can increase production while reducing environmental impacts. Animals and livestock contribute 40 percent of the global value of agricultural output and contribute to the livelihoods and food security of almost a billion people worldwide. Advances in animal breeding, genetics, and genomics are facilitating a more efficient industry. For example, the number of cattle has decreased over the past decade, yet the total production of beef and milk has increased. This was largely possible because genetic advancements led better animal feed efficiency, which is critical to improving livestock production and lowering costs for producers. Reproduction and early development in domestic animals are biological processes in which complex genetic events like formation of the germ cells, meiosis, syngamy, zygote formation, cleavage, midblastula transition, dose compensation of sex chromosomes, genetic imprinting, and multiple cell differentiation take place. Many of these processes have great impact in veterinary reproductive medicine and are influenced by the assisted reproductive techniques applied. Transformed environmental and metabolic conditions caused by intensive livestock farming influence reproductive success by altered gene expression via epigenetic changes. In this concept molecular markers give valuable information. The increasing availability of PCR-based molecular markers allows the detailed analyses and evaluation of genetic diversity in animals and also, the detection of genes influencing economically important traits.

This volume 'Genetics and Breeding' discusses ways and means to improve productivity through breeding are discussed and some basic issues requiring further research pointed out. It aims to provide a glimpse into the dynamic process of genetic disparity in animals by presenting the thoughts of scientists who are engaged in the generation of new idea and techniques used for the assessment of genetic variety, often from very different perspectives. This book is a comprehensive source of knowledge on the latest advances in the field of animal genetics and breeding. The book should prove useful to students, researchers, and experts in the area of genetics and breeding, and molecular biology.

Table of Contents

CHAPTER 1

Piebald Camels

Gabriele Volpato[1], Maurizio Dioli[2] and Antonello Di Nardo[3]

[1]*Center for Integrative Conservation Research, University of Georgia, Athens, GA, USA*

[2]*Independent Veterinarian, Alfaz del Pi, Spain.*

[3]*The Pirbright Institute, Pirbright, Woking, Surrey, UK.*

ABSTRACT

Animal breeds are the diverse outcome of the thousands-year-long process of livestock domestication. Many of these breeds are piebald, resulting from the artificial selection by pastoralists of animals bearing a genetic condition known as leucism, and selected for their productive, behavioural, or aesthetical traits. Piebald dromedary camels have not been studied or discussed before, and their same existence is often overlooked. Based on fieldwork in Western Sahara, direct observations across Northern and East Africa and the Middle East, and a literature review, we address the morphological and behavioural traits, geographical distribution, taxonomy, and material and cultural importance of piebald (painted) camels. They are a hundreds-year-old camel breed used for caravans, as mounts, and for aesthetical and cultural reasons across Sudan, Niger, Mali, Mauritania, Western Sahara, and Morocco. While they are increasingly bred out of a pastoral context for tourism and entertainment in the Canary Islands, mainland Europe, and the USA, in part of their original African range, piebald camels are under threat due to wars, droughts, and demise of pastoral livelihoods. More research is needed about these 'beautiful and dignified' animals.

Keywords: Livestock breeds, Leucism, Paint dromedary, camels, Pastoral nomads, Sahara

Have you ever seen a piebald mahri[1] *before?;*

Have you ever seen a camel with such grace, lightness, and stature?

Have you ever seen anything more beautiful and dignified?

(Al-Koni 2008)

1. INTRODUCTION

A breed is a group of domesticates that has specific characteristics or traits artificially selected by man and transmitted through inheritance (Driscoll et al. 2009; Larson and Fuller 2014). Breeds are the diverse outcome of the thousands-year-long process of livestock domestication (Francis 2015). They have been selected in accordance to productive, cultural, and aesthetical traits and are often a key element of pastoral livelihoods and identities. Breed diversity is nowadays endangered by processes of livestock intensification and cultural homogenization (Faye et al. 2004), and as such, its study is a fundamental step for breeds' conservation. Among the thousands of breeds from tens of domesticates existing in the world, several are piebald, i.e. spotted, painted, or patched of white and solid colour. Many of these piebald breeds are the result of a selective breeding for a genetic condition known as leucism.

Animals' colouring is the result of either presence or absence of the pigment melanin in the skin, hair, and eyes. Among the known conditions affecting melanin production and animal colouring are albinism, melanism, and leucism; the latter resulting from defects in pigment cell differentiation and/or migration from the neural crest to the skin, hair, or feathers during development (Cieslak et al. 2011; Wilkins et al. 2014). In its most common form, leucism results in irregular patches on the body surface expressing as white on an animal that otherwise has normal colouring and patterning (Rook et al. 1998; Cieslak et al. 2011); when this happens, the animal is 'pied' or 'piebald' (also 'paint', 'spotted', or 'speckled').[2] Some leucistic animals also exhibit coloration of the irises that matches the surrounding skin (blue eyes for pink skin, brown for dark).

Leucism and piebald colorations occur sporadically in the wild, as they usually reduce the individuals' selective fitness providing less colour protection from predators (Woolf 1991). But recessive conditions in domesticates have been propagated voluntarily or not by humans since the very beginning of the domestication process, and changes in size and colour are among the first effects (Epstein 1955; Trut et al. 2009; Ludwig et al. 2009). Voluntary propagation includes the selective breeding of specific phenotypes for their religious, ritual, social, or subsistence/economic value.

Involuntary propagation relates to the genetic and physiological links existing between tameness (a goal of early domesticators) and colour phenotypes (Wilkins et al. 2014; Cieslak et al. 2011; Price 2002) and to the effects of genetic drift in domesticates' isolated populations (Zeder 2012). There is a general agreement that a wide variation of colour phenotypes were brought about by domestication, 'but it is still largely unknown how color phenotypes were distributed in domesticated animals in earlier times, which roles they played for humans during these times and when they first occurred' (Cieslak et al. 2011).

The piebald leucistic phenotype is known in a variety of domesticates, including Tobiano and Appaloosa horses (Brooks and Bailey 2005; Brooks et al. 2007), Belgian Blue and Shorthorn cattle breeds (Seitz et al. 1999), goats, and sheep, as well as dogs, cats, and pigs (Cooper et al. 2005; Giuffra et al. 1999). Several piebald animals have evolved into valued breeds, where Holstein cows and Dalmatian dogs are among the most notable examples.

Some of the genetic mutations related to the piebald character bear pleiotropic effects such as deafness, developmental disorders of the eyes and night blindness, and osteopetrosis (Stritzel et al. 2009; Bellone 2010; Wilkins et al. 2014). Because the development of the optical system is highly dependent on the presence of melanin, depigmentation has an effect on the development of the visual system (Grandin and Deesing 1998; Wilkins et al. 2014). Congenital stationary night blindness has been reported in a variety of spotted and piebald animals (e.g. Appaloosa horses; Sandmeyer et al. 2007). Animal breeders throughout the world have since early times recognized that a lack of body and eye pigmentation may be accompanied by neurological defects. A relationship between depigmentation and congenital deafness has been found, among other domesticates (Webb and Cullen 2010), in dogs (e.g. Dalmatian dogs with extensive white depigmented areas are most likely to be deaf; Strain 1996) and in llamas and alpacas (Camelids), where it is associated with pale blue eyes (Gauly et al. 2005).

A great deal of research has been conducted on the genetics, behavioural, and productive aspects of piebald breeds among several species (e.g. horses, dogs). In spite of this, the same existence of piebald camels has been largely overlooked, even in the scientific literature on camels (Wardeh 2004). Little and scattered information is available on piebald (painted) dromedary camels, and to the best of our knowledge, no single article has ever been written about them. The only author that has addressed their existence is Bulliet (1975), while too often it has not been properly acknowledged. For example, in the most recent and otherwise comprehensive book on animal domestication (Francis 2015), while discussing the correlation between coat colour and camel domestication, the author states that 'In the United States, paint camels – white and brown, like pinto horses – have been produced, but paint camels are virtually absent in Arabia and North Africa.' Though it is true that there are no painted camels in Arabia and East Africa, we contend that they indeed exist in North Africa since hundreds of years, and that the USA is just the latest country where they have been introduced after their diffusion westward from central-eastern Sahara, where they probably originated.

This paper departs from and contributes to the study of human-animal relationships and domestication from an historical and anthropological perspective (Bulliet 2005; Francis 2015; Hurn 2012). Most research on camels revolves around their potential as milk and meat producers and associated conditions (e.g. diseases) (Farah and Fischer 2004), whereas relatively less studies have been carried out on the productive and cultural links between pastoral populations and camels. In general, camel breeds are not as differentiated and classified as in other livestock species and their classifications are often derived from names of ethnic groups or geographical breeding regions rather than based on phenotypic characters (Dioli 2016). Therefore, camel breed study could help their conservation and support camel pastoralists (Kakar et al. 2011). Studies on the cultural relevance of piebald livestock (e.g. cattle) have already proven to be useful in describing complex human-animal relationships particularly in regard to livestock management (e.g. in genetic selection) and cultural identity (e.g. founding myths, cultural identity, and social values) (Coote 1994). In this paper, we address piebald camels' physiological, genetic, and behavioural characteristics; their geographical distribution; taxonomy; and material and cultural roles among pastoral populations. Then, we discuss piebald camels' origin and diffusion through Africa and out of Africa, we address trends in piebald camels' husbandry, and we invite further research about and support to these 'beautiful and dignified' animals.

1.1 Study area

The area under study includes large parts of the Sahara desert where camels are bred, with a focus on Western Sahara and the Sahrawi pastoralists (Figure. 1, Table 1). The Sahara is bordered by the Atlantic Ocean on the western edge, the Atlas Mountains and the Mediterranean Sea to the north, the Red Sea on the east, and the Sudan and the valley of the Niger River on the south. Half of the Sahara receives less than 25 mm of rain per year, while the rest receives up to 100 mm per year. Most of the Sahara is characterized as rocky hamada, a type of desert landscape defined by barren rocky plateaus. Large areas are covered by sand and dunes. The central part has extremely limited vegetation, while the northern, southern, and western reaches of the desert, and the highlands and mountain areas (e.g. Aïr, Hoggar, Saharan Atlas, Tibesti Mountains), have sparse grassland

and desert shrub, with trees (mainly Acacia species) usually along the dry riverbeds (Julivert 2003). Across most of the Sahara, the dromedary camel (Camelus dromedarius L.) is the main livestock species. It was progressively introduced to large areas of the Sahara beginning about 2,500 years ago, and it provides pastoralists with milk, meat, and transport, as well as with a means of utilization of the local desert environment (Gauthier-Pilters and Dagg 1981).

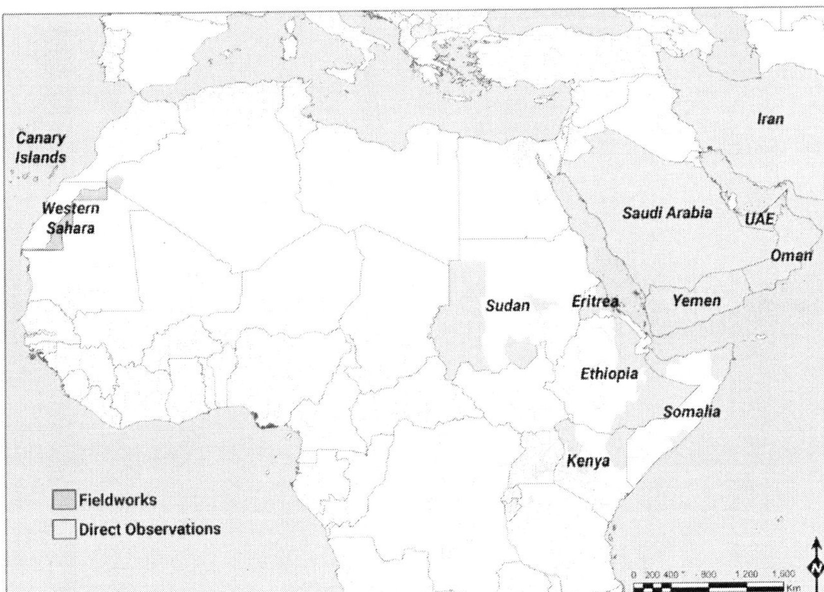

Figure 1 Map of the areas where fieldwork and direct observations have been conducted

In-depth fieldwork was conducted in inland Western Sahara, northern Mauritania, and the part of Algeria to the south and south-east of the Hamada of Tindouf, which are the customary nomadic territories of Sahrawi pastoralists. Across this area, the climate is continental: summer daytime temperatures pass 50 °C, while winter night-time temperatures may drop to 0 °C. Rainfalls are torrential, unpredictable, and patchy, with an average annual rainfall of 30 to 50 mm and recurrent droughts. Generally occurring from the end of the summer through autumn, these rains are driven by the extreme northerly penetration of the African Monsoon from the south or are associated with the Atlantic Westerlies (Brooks et al. 2005). Biogeographically, we can distinguish two main areas: Zemmur to the north and Tiris to the south.

Table 1 Location, duration, and time periods of fieldwork and direct observations

Time period	Duration	Country	Region	Type of study
1981 to 1991	10 years	Kenya	North west North east	Direct observations
1992 to 2013	3 years	Somalia	Middle Shebelle (Jowhar) Awdal (Borama) Bari (Bosaso) Woqooyi Galbeed (Berbera) Somaliland Puntland	Direct observations
1998 to 2001	3 years	Ethiopia	Afar Dire Dawa Somali (Jijiga, Ogaden)	Direct observations
2003 to 2006	3 years	Eritrea	Anseba (Keren) Gash-Barka (Barentu, Tesseney) Northern Red Sea (Nakfa, Massawa)	Direct observations
2004	Weeks	Western Sahara		Fieldwork
2006	Weeks	Kingdom of Saudi Arabia		Direct observations
2006 to 2007	Weeks	Canary Islands	Fuerteventura Lanzarote	Direct observations
2007 to 2009	1 year	Sudan	Darfur Khartoum (Omdurman) South Kurdufan (Nuba)	Direct observations
2008 to 2010	Months	Western Sahara		Fieldwork
2009 to 2010	3 months	Yemen		Direct observations
2010	Weeks	Iran		Direct observations
2010 to 2011	Months	United Arab Emirates		Direct observations
2012	2 weeks	Oman		Direct observations
2012	Months	Kenya	North west North east	Direct observations
2016	Weeks	Kingdom of Saudi Arabia		Direct observations

The first runs east-west between northern Western Sahara and northern Mauritania: it is characterized by gravel plains with occasional surface of sandstone and granite in its eastern and central parts and by higher relief and hilly terrain in its western part. All Zemmur, and especially its central and western areas, is drained by inactive or occasionally active river channels that flow west into the Saguia el-Hamra, a large ephemeral river. After the rains, Zemmur displays a savannah-like environment dominated by Acacia-Panicum vegetation, while flowering prairies may appear on flat gravel areas. The southern sector, known as Tiris, is more arid and characterized by flat sand and gravel plains from which characteristic black granite hills arise in either clusters or in isolation. In Tiris, there are no dry riverbeds, and hence, vegetation is mostly herbaceous and adventitious and includes large areas covered by halophytic plants (Soler et al. 1999). For a background on the Sahrawi refugees and nomads and on their camel husbandry, see Volpato and Howard (2014) and Caro Baroja (1955).

2. METHODS

The data analysed in this paper are drawn from three types of sources: (i) a review of scientific and grey literature about piebald camels and related topics, (ii) direct observations, and (iii) fieldwork carried out between 2008 and 2010 in the Sahrawi refugee camps of west Algeria and in Western Sahara.

Fieldwork included semi-structured interviews ($n = 32$) with Sahrawi camel owners (26 of them had piebald camels in their herds) about morphological, physiological, and behavioural characteristics of piebald camels and their uses and cultural values. Interviews were conducted in Hassaniya (the Arabic language with Berber substrate spoken by the Sahrawi), recorded and translated into Spanish by local research assistants. Interviews were recorded and transcribed with the help of the same research assistant to minimize translation errors and clarify information. Transcripts were then entered into NVivo qualitative data management software, and codes, concepts, and categories were generated during analysis. In every case, prior informed consent was obtained verbally before the interview was conducted, according to the ethical guidelines adopted by the American Anthropological Association (2009) and by the International Society of Ethnobiology (2006).

Besides fieldwork in Algeria and Western Sahara, direct observations on the presence (or absence) and distribution of piebald camels have been conducted by the authors in the Canary Islands (2006 to 2007), Sudan (2007 to 2009), Kenya (2012, 2015), Ethiopia (1998 to 2001), Eritrea (2003 to 2006), Somalia (2013), Yemen (2009 to 2010), Oman (2012), UAE

(2010 to 2011), Kingdom of Saudi Arabia (2006), and Iran (2010) (Table 1). The areas where fieldwork and direct observations have been conducted are represented in Figure 1 with a focus on Africa, where piebald camels are found. While fieldwork included interviews about piebald camels, direct observations are reports from the authors on piebald camels' presence in the areas where they have worked or conducted research and did not include interviews about piebald camels.

3. RESULTS AND DISCUSSION

3.1 Morphological, physiological, and behavioural characteristics

Piebald camels have a white and solid (black, brown, tawny, red, or grey) coloration that varies between individuals in the relative cover of white or solid and in the shape that the patches assume on the body. There is a high variation in the amount of white in the body, from individuals who are all solid-coloured but the snout to others that are totally white or all white but the hump (Cauvet 1925; Mahaman 1979; Dioli 2013). This variation relates to the degree of leucism of the animal, which, with all evidence, has a genetic basis. From the available images of piebald camels (including those taken during fieldwork) and direct observations in the field, we can classify them according to the amount and distribution of white in the coat, clearly showing a characteristic progression of leucism among individuals, which starts from the snout and lower legs to progress commonly up on the head and on the belly and flanks (these seem to be the most common and shape-variegated camels), until covering all the body but the dorsal part (including the hump) (Figure 2). Camels displaying a high prevalence of white tend also to lose the contours of the solid-coloured patches, thus displaying speckled markings (Figure 3). Piebald camels with a prevalence of white on the head (i.e. most piebald camels) display blue eyes (Figure 4). Piebald camels may also have complete heterocromia of the iris, with one blue and the other brown in accordance with the coat colour surrounding the eyelids (Figure 5). In few cases, brown solid-coloured camels have blue or minnow eyes (Figures 6 and 7), and according to Sahrawi herders, this occurs in herds with piebald individuals. Toes may be white too, often in association with blue eyes.

Physiological characteristics include various degrees of deafness and visual impairments, and the notion that piebald camels with blue eyes may have audio and visual impairments is well known to African camel pastoralists (Faye et al. 2002).

Figure 2 Table of pictures of piebald camels from Western Sahara showing the different colour patterns present in piebald camels and the white colour's progression under distinct degrees of leucism (GV)

Figure 3 A camel bull of the Canary Islands showing a complex piebald pattern and speckled markings (Dioli 2013)

Figure 4 Heads of a juvenile camel (a) and of a calf (b) of Western Sahara showing leucism: pink skin and blue eye (GV, Dioli 2013)

Figure 5 Head of an immature camel of the Canary Islands with eyes of different colours (Dioli 2013)

Figure 6 Riding camel of Western Sahara with solid coat and blue eyes (D. Rossi)

Figure 7 Male riding camel of Annafi breed from Eastern Sudan with solid coat and extremely pale iris (Dioli 2013)

These impairments are the reason pastoralists regard piebald camels as reckless, stubborn, disobedient, or particularly tame, even numb. Indeed, a relation between depygmentation and similar variations in behaviour is well established in the literature for other domesticates. Sometimes, there are different levels of deafness and auditive impairments in affected animals - and apparently different predominant behaviours - in relation to the level of depygmentation: for example, Holstein cows with complete depigmented white areas on their heads are among the calmest, while those that are mostly white on the body are nervous and intractable (Grandin and Deesing 1998).[3] Similarly, Sahrawi herders describe piebald camels as having different levels of deafness, with an increased deafness according to the presence of blue eyes and white colouring of the head and toes. Conversely, 'when a boldpie camel has normal hearing, its black ears and nails are the sign of it' (Monteil 1952). Perhaps complete deafness is related to calm behaviour, whereas partial deafness or hearing distortions are related to increasing agitated and unpredictable behaviour. Complete deafness may also be at the base of Sahrawi and Tuareg herders' description of piebald camels' behaviour as 'stubborn', 'non-obedient', and 'hard in understanding orders', but also as being quiet and tame and 'following the rest of the herd' (Migeon 2006). According to Tuareg herders, 'camels with blue eyes are real nuisances. They suffer from bad eyesight at night and consequently often get lost when camel caravans move in the dark' (Curdy 2001). Some piebald bulls are attributed a 'low ability in managing the herd' due to these impairments.

But piebald camels are also appreciated for their alleged courage in crossing barren deserts (Cauvet 1925), as well as for their alleged resistance to hunger, thirst, heat, and fatigue during caravan journeys. While it is well known that camels can drink brackish and salty water, Sahrawi nomads mention that piebalds can tolerate even higher concentrations of salts in drinking water. The belief that piebald camels are more resistant to thirst and heat may be associated with the increased albedo of white and light-colour hair upon exposure to the sun.

3.2 Geography of piebald camels

Based on the available knowledge, the areas of the world and the names of the populations where contemporary breeding of piebald camels takes place are shown in Figure 8. Piebald camels are bred along a Southern Saharan fringe including Darfur (though in low numbers) and Kurdufan, Niger, Mauritania, Western Sahara, Morocco, central Algeria, and the Canary Islands. Few piebald camels are also present outside of Africa, namely in the USA and in Europe, where they have been introduced from the Canary Islands. We could not find any report, image, or direct observation of indigenous piebalds in the Middle East: there are no pied camels in Arabia

(Faye et al. 2012) nor in Somalia (where the camel was introduced in earlier historical times from southern Arabia), the rest of the Horn of Africa, or among camel populations of the Mediterranean coast. Besides the few individuals exported to Europe, the USA, and few other areas, piebald camels live and likely originated in the Saharan Africa.

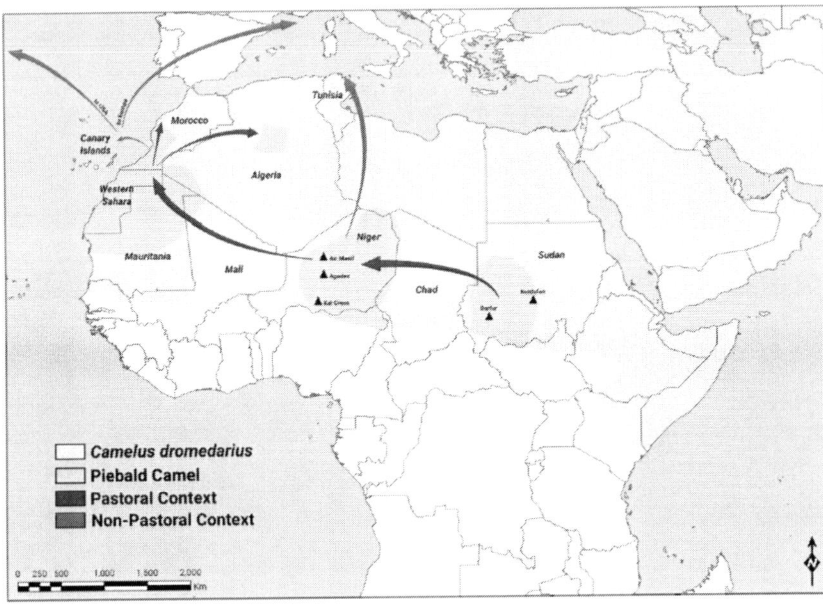

Figure 8 Areas of the world and names of the populations where contemporary breeding of piebald camels takes place (camel range is shown in softer tone)

There is little information about the presence of piebald camels in the eastern Sahara desert. Leese (1927), cited in Köhler-Rollefson (1991), reports the breeding of camels with mouse-coloured backs and necks and white bellies, faces, and legs in Kordufan and Darfur. Wilson (1978) states that 'occasional parti-coloured camels also occur' in southern Darfur.

Most of the information available in the literature about piebald camel breeding refers to the Tuareg of Niger (Figure 9). Some authors (Bernus 1969; CIRAD 2001; Pacholek et al. 2000a) report that the Kel Gress of central Niger are renown for breeding 'spotted camels', and the same is true for the Tuareg living in the region of Agadez (Chaibou and Faye 2003, Mahaman 1979; Antoine-Moussiaux et al. 2006; AA 2000). Others (Pacholek et al. 2000b; CIRAD 2001) list azarghaf (the local name for piebald camels) among the camel races/types bred in Niger and indicate that they originate in the south of the Aïr Massif. The presence of piebald camels in the region is reported also by Cauvet (1925), who describes a camel breed called 'dromadaire de l'Azbin' (Azbin is the name of the Aïr Massif in Hausa language) or 'Haoussa de l'Aïr' and reared by Tuareg tribes at the west and north of Agadez. Mahaman (1979), while classifying camel breeds of Niger, reports the existence of a breed called azarghaf, with minnow eyes and piebald coat, resistant and elected for long travels.

Figure 9 Piebald camel in Niger in 1994 (B. Faye)

Piebald camels are also known to and sometimes bred by Tuareg-neighbouring populations, mostly Hausa and Fulani of north Nigeria (e.g. in the Kano region), where the Tuareg Kel Gress move their herds during dry spells in Niger. This, for example, happened during the droughts of the 1970s: in Figure 10, a piebald camel near the Bakolori dam in Northwestern Nigeria in June 1978. These movements lie at the origin of the Nigerian pied camel breed reported in the FAO DAD-IS database.[4]

Besides the Tuareg, other pastoral populations renown for breeding piebald camels are the Hassanyia-speaking nomads of Mauritania and Western Sahara (Correra 2006; Monteil 1952), also known as Moors or, in the northwestern part, as Sahrawi. Among them, piebald breeding is an old practice that has lasted to the present day, as is the case for the Reguibat, the Oulad Delim, and the Oulad Tidrarin tribes (Mercer 1976; Caro Baroja 1955; Boyer 1962), as well as among Moorish tribes of present-day northern and coastal Mauritania (Correra 2006). Relatively small numbers of piebald camels are also bred by neighbouring (non Hassanyia-speaking) pastoralists such as the Chaamba of Central Algeria (Shinar (2004), in citing Cauneille (1968), states that they have 'dappled (patched) grey, brown, and dark copper' coloured camels) and the Marazig pastoralists of South Tunisia, reported to have camels with a white snout (Boris 1951).

Figure 10 Piebald camel near the Bakolori dam in northwestern Nigeria in June 1978 (Giorgio Volpato)

In the literature about camels in the Canary Islands, the origin, significance, and distribution of piebald camels are never addressed, although their existence among the herds of the islands is usually recognized. Camels were first introduced to the Canary Islands beginning with the fifteenth century by the Castilian conquerors; during the course of the following centuries, camels adapted to the islands' climate and were used by local peasants, particularly of Lanzarote and Fuerteventura, to power agricultural implements (e.g. ploughs, watermills), as a riding and pack animal, and for wheeled transport (Morera 1991; Schulz et al. 2010). Piebald camels were further imported to the Canary Islands from Western Sahara (possibly from individuals originally belonging to the Oulad Tidrarin, as this tribe was customarily living along the coast of Western Sahara adjacent to the Canary Islands, with which it had historical commercial contacts). Indeed, a recent study has shown a clear genetic proximity between camel populations from the Canary Islands and from western Africa (Schulz et al. 2010), and Canary Islands' piebald camels are reported as closely related to the piebald 'Western Sahara' breed (Dioli 2013), from which they likely originated. But when this happened is uncertain.

Few isolated piebald camels live in zoological gardens, ranches, and farms outside Africa. Several private ranches and farms in the USA have piebald camels, descendants of piebalds that were shipped there from some place in Morocco or the Canary Islands a few decades ago. In Europe, at least two piebald camels are present in a herd of a Dutch farm producing camel milk for the European market,[5] and some more live in a camping resort on the Mediterranean French coast, where they are used as amusement for children and for touristic parades;[6] they all come from the Canary Islands. The Oman Royal Camel Corp (Royal Court Affairs) owns two piebald camels, likely imported from western Africa, and uses them for parades (Figure 11). In Tunisia, the Farhat Hached zoological park in Rades witnessed in 2008 the birth of a 'blue-eyed dromedary of mahri breed with a rusty brown and white coat'. Media reported the news applauding at the birth of the 'very rare species', whose 'immaculate blue eyes [...] confer to the animal a singular and rare beauty', with the calf bringing 'great delight of the many children who visit the park with their parents'.[7] This 'very rare species' is said to have been introduced to the park about two years before, to be usually raised by the Tuareg living on the border between Niger and Chad and 'by Saharan tribes living between Mauritania and Morocco'.[8]

3.3 Genetics of piebald camels

The piebald character is usually transmitted by one or more recessive genes that is (are) also responsible for blue eyes, congenital deafness, and impaired vision. Involvement of more genes and mutations make the genetics of leucistic animals complex (Householder 2003), and 'the genetic pathways influencing coat coloration are still only poorly described' (Cieslak et al. 2011). Unsurprisingly, given the lack of attention to leucism in camels in the scientific literature, no investigation has up to now been conducted on the genetics of piebald camels. From the information we collected, only a tentative hypothesis can be put forward, and in order to have a deeper understanding of piebald camels' genetics, a 'herd analysis exercise' (Krätli 2008) among herds including piebald camels would be helpful.

Figure 11 Piebald camel belonging to the Oman Royal Camel Corp (Dioli 2013)

In general terms, Sahrawi herders state that a piebald bull introduced into a non-piebald herd results in a 30% to 50% of piebald calves (males in most of the cases), that a piebald calf can be produced by two solid-coloured parents if there was a piebald in the family, and that conversely two piebald parents can produce solid offspring. According to the Sahrawi, piebald coloration and associated characteristics are transmitted to calves by blue-eyed piebald fathers. If the bull is not piebald, piebald calves are born in a relation of one third or one quarter to the number of piebald mothers. Hence, piebald condition in camels seems to be sex related, with males displaying the colouring pattern more often than females. Indeed, most of the piebald camels we saw during fieldwork in Western Sahara were males. Further insights can be obtained by comparing piebald camels' genetics to that of other mammals, as it appears that there is a common genetic mechanism that determines coat colour (Wilkins et al. 2014). In Appaloosa horses, for example, the piebald expression is controlled by a single gene, which is apparently dominant, and displays the piebald pattern when two series of modifying genes are expressed: one series for the control of the expression of white (vis-à-vis the normal coat colour) and another series for the expression of spots (Householder 2003). Apparently, the expression of the white modifier and the spot genes are sex influenced, with heterozygote individuals expressing the characteristic when males, and not when females (Householder 2003), like in camels. In Dalmatian dogs, the piebald character is given by two recessive genes, with individuals displaying the phenotype when homozygous; the expression of the spotted colouring is farther related to the expression of genes for congenital deafness and blue eyes (Strain 1996), as seen in camels. The genetics of piebald camels may indeed involve not just one but more genes like in Appaloosa horses and Dalmatian dogs. Differently, in Holstein cattle, the piebald effect seems to be due to inheritance of a single piebald spotting gene, with no sex differentiation (Pape 1990), unlike in camels.

Being a condition transmitted by one (or more) recessive gene(s), leucism would appear in camel herds only at a very limited frequency in natural reproductive conditions, whereas the selection for leucism can be obtained in 'close' herds where the reproduction is selected for the appearance of recessive phenotypes. Herders favour piebald camels by selective breeding, but at the same time are aware of the risks of breed degradation and thus avoid inbreeding.[9] As a result of the interplay between piebald genetics and selective breeding subjected to these two forces, the actual prevalence of piebald camels within a herd seems to be variable but seldom reaching the majority of the herds' camels (Figure 12). This is indirectly supported by the analysis of camel herds' composition in accordance to types/breeds around Agadez in Niger done by Chaibou (2005): in out of five camel breeds, the azarghaf (piebald) is one of the main breeds along with the abzin and azawak, and it is present in 55% of the herds; however, no one herd was composed only by azarghaf camels.

Figure 12 Camel herd at a Western Sahara well with roughly 30% of piebald camels (GV)

3.4 Ethnotaxonomy

Throughout their range and among all the populations who breed them, piebald camels are known by names that recall their piebald coat or their blue eyes, less often their peculiar behavioural characteristics. Most names have an Arabic root.

Among Arabic-speaking populations (e.g. the Hassanyia-speaking Sahrawi and Moors and the Arabic-speaking Sudanese pastoralists), piebald camels are known as azraq (male), zarqa (female), or zurq (plural) (Taine-Cheikh 1989; Caro Baroja 1955). In classical and contemporary Arabic, zurq means 'blue', but also 'blue-eyed', 'blue, minnow, or glittering eyes', and 'glaucoma' (Fischer 1965). Though not in reference to piebald camels, the term has among some Maghreb populations a positive aesthetic connotation, with a female or woman called zarqa in reference to her sensual beauty (Roth 1985). In Hassanyia, the verbs stazrag and zrāg mean 'being towards piebald coloration (of camels)' and 'becoming piebald', respectively (Taine-Cheikh 2002), and azraygat their diminutive (Voisset 1989). The term zurq applied to coat colour is used also for a grey coat, especially in reference to horses and cattle (Roth 1985) but also in reference to camels, e.g. in Arabia (Abbas et al. 2000); nevertheless, to the Zaer of North Morocco, a zurq cattle is a cattle with white patches all over its body (Loubignac 1952). Interestingly, in colloquial Arabic there exists a derived term mizra:qan, used to indicate a camel delayed behind the rest of the caravan because it moves slowly (Allam 2000), which may be in relation with the tendency of some piebald camels to stray away. In Morocco and Western Sahara, the term zeroual is also used to indicate piebald camels, as well as people and camels with blue eyes.

There is a strong consistency in the terms used for piebald camels across time and space: they all originate from the Arabic zurq. The Tuareg use the zurq-derivated term azarghaf, not of Berber Tamasheq (Tuareg's language) origin. This fact takes more significance when we consider that Tamasheq has a low proportion of Arabic loanwords, and usually few of these with regard to phytonymic and zoonymic terms: items lended/adopted from one population to the other are more likely to have the same etimology, and piebald camels and their zoonym may have been historically adopted by Tamasheq-speaking Tuareg from Arabic populations.

Zurq is also the etymological origin of some Spanish words, notably 'zarco, –ca' which means light blue, azure, and 'ojizarco, –ca' which means 'blue-eyed'. Among Spanish-speaking piebald camels' breeders of the Canary Islands, names for piebald camels also refer to their coat colour or to their impairments and associated behaviour. Piebalds are known as manchado or pintado (spotted or painted) when with white patches on the body or some of its parts and as capiloto when displaying a white head and a prevalently normal-coloured body (cfr. the Spanish capirote, i.e. cattle with the head and the body of different colours; Morera 1991). Among present-day Sahrawi of Western Sahara (a former Spanish colony), besides as zurq, piebald camels are also referred to as colorín (coloured) or camellos zonzos or tontos (silly, fool camels).

3.5 Material and cultural importance

Piebald camels have different material and cultural roles among the distinct populations engaged in their breeding. These roles encompass subsistence material uses (e.g. for milk and meat production, transport, and caravans), a cultural importance in terms of aesthetics, cultural identity, and tribal/group identification, as well as the recent use of income generation out of a pastoral context.

In the Saharan areas where they are bred, piebald camels have been often used by different populations as pack camels for caravans, for transport (e.g. moving to new grazing areas), and as mounts because they are prevalently male and/or because of their resistance to thirst and heat and/or because their physiological impairments make them more tame. Among the nomads of Western Sahara and Mauritania, they were used in pre-colonial times as mounts by blacksmiths and griots as well as in caravans (Monteil 1952). They are used in caravans, particularly in areas with sandy soils (Migeon 2006), such as in Niger (Chaibou 2005; CIRAD 2001), where this breed is known 'for its resistance and courage with which it crosses the desert from the Aïr to Bilma' (Cauvet 1925). Indeed, they are used for the annual tarlamt, which are the salt caravans moving across the Ténéré desert from Agadez, Zinder, or Tahoua to the salt deposits of Fachi and Bilma in Niger (Cianchini 1999). In Morocco, they are used as mounts and to transport merchandise to towns (Driot 2009). In Nigeria, they are regarded as 'very faithful beasts of burden used essentially for personal transportation, to carry loads, draw water, and pull ploughs'.[10]

In the Canary Islands, piebald camels are bred mainly to transport tourists (particularly in the island of Lanzarote), who seem to be particularly attracted to their variegated colouring and blue eyes (Dioli 2013). Nowadays, there are about 1,000 camels in the Canary Islands (Schulz et al. 2010), and they are almost completely used in the touristic industry. Many of them are piebalds. From the Canary Islands, piebald camels have been exported during the last decades to European and American countries (Castillo and Lugo 2003). In the USA, piebald camels are much appreciated aesthetically and valued two to three times the price of a solid-coloured camel (Berry 2006).[11] The value of a piebald calf ranges between $3,000 and $6,000.[12,13] Two farms in Tennessee hire their piebald bulls for breeding, while one farm in North Carolina, specialized in breeding 'of rare and exotic animals', sells its paint camels to private people or provide them for hire for tourist 'safari' (in ranches), commercials, theatre performances, and plays.[14] Another ranch in Texas owns some fine piebalds (see pictures in the website)[15]; according to the statement in their website, they are breeding 'for bigger and better spotted camels'.

Among Sahrawi and Tuareg breeders, piebalds are appreciated for their beauty in terms of variable patterns of coloration and blue eyes. They have become a marker of cultural identity and have been incorporated in poems, literature, folk knowledge, and foundation myths. They are considered a locus of visual stimulation and a source of positive emotions. Their high variability in coat colouring patterns have become an ownership mark with similar functions to those of customary livestock brands.

The positive selection for recessive genes for productive and/or aesthetic qualities is not new to livestock breeding. Some cultures regarded leucistic (piebald) as sacred animals, and they have even been the basis of legends and folklore: for example, Egyptian tomb paintings of 1400 to 1300 BC represent spotted horses (Householder 2003); ancient nomadic populations of Egypt and Sudan were known for appreciating and venerating piebald cattle, which was often represented in the rock art of the Central Sahara (d'Alverny 1950); and a type of piebald cattle was bred by the Zulu and regarded as royal cattle and held as almost sacred (Epstein 1955). The cultural identification of specific ethnic groups with clearly defined morphological characteristics (colour, size, horns, fattiness, etc.) of their livestock is a well-known although understudied fact in the history of human-livestock relations (Hunn 2011), and the relation of some pastoralists with piebald camels is an extraordinary case study. For example, in northern Nigeria, piebald camels are used as ceremonial animals during Salla Muslim celebrations.[16] Piebald breeding can be achieved only by manipulating reproductive processes that favour the piebald character, and this genetic selection in turn becomes a cultural marker of piebald camel herders, i.e. piebald camels become means to construct and promote cultural identity, even as subjects of foundation myths. Here is a foundation myth passed down by members of the Oulad Tidrarin tribe of coastal Western Sahara (Caro Baroja 1955):

Sid Ahmed Bo Gambar, one holy ancestor of the Oulad Tidrarin, was one day praying on the beach, when suddenly an azrag [piebald] camel came out from the sea. Sid Ahmed took the camel, which was a stud, and from this camel the many azrag camels the Tidrarin owns descend

This myth attributes the origin of piebald camels to a miracle done by Sid Ahmed and explains the attachment and sense of identity that the members of the tribe have for these camels. In fact, nowadays, piebald camels are still a marker of cultural identity for the Sahrawi nomads (including the members of the former tribes of Western Sahara), who breed them particularly for their perceived beauty and the prestige derived from it.

The fact that leucism considerably varies the ratio of white to coloured hair between generations as well as between offspring from the same parents gives each individual a specific and recognizable pattern, which has often attracted pastoral populations (Rook et al. 1998). Sahrawi's and Tuareg's appreciation of piebald camels recalls the anthropology of aesthetics as discussed, for example, in the case of cattle among Nilotic populations (Coote 1994). Among the Mandari, a piebald ox is highly appreciated and, when a piebald is born, its owner is delighted and the beast is set aside for show (Buxton 1973). The high aesthetic value that the Mandari place on cattle patterning is explained by Buxton (1973) by the fact that, 'they stand out strikingly in a landscape devoid of strong color', where the individual beast provides the locus for stimulating visual experience (Coote 1994). Indeed, aesthetic explanations related to their striking and coloured appearance on an otherwise pale landscape are as appealing to camel-breeding nomads as they are for the cattle-breeding Mandari. Camels at the horizon are a common visual locus when travelling in Western Sahara. As piebald camels strike on the yellow horizon with a colour pattern variety, and as these coloured shapes vary from animal to animal, it is no wonder that a piebald camel represents a particularly appreciated visual stimulation.

Among the Tuareg, piebald camels are associated with beauty. To communicate their sense of beauty, Tuareg poets link man, camels, and pastoral resources. Piebald camels find a place in these poems. In one of them,[17] the beauty of a girl is put in analogy with that of a piebald camel (Ghabdouane and Prasse 1990). Similarly, in a recent novel describing the commitment between a man and his piebald camel written by Al-Koni (2008), himself a Tuareg raised in the desert of Libya, a Tuareg nomad is constantly lauding his piebald mahri for the unique colouring of his pelt, in spite of its undisciplined and reckless behaviour. During a passage of the novel, a local sheikh entraps the beautiful piebald to get it to impregnate the sheikh's she-camels and obtain calves from the piebald strain. In the sheikh's words, it was 'a piebald mahri as graceful as a gazelle. This line became extinct throughout the desert a hundred years ago' (Al-Koni 2008). The blue eyes and coat colour and pattern of piebald camels have historically been at the centre of different emotional and rational attachments by their herders and are also much appreciated among tourists travelling to the Canary Islands and south of France, children visiting zoological parks (e.g. in Tunisia), and owners of rare and exotic animals in the USA.

3.6 Are piebald camels a breed?

Wardeh (2004), in his classification of dromedary camels, does not refer to the existence of piebald breeds or individuals, whereas several other authors have described piebald camels as a camel breed/race (Pacholek et al. 2000b; Monteil 1952; Antoine-Moussiaux et al. 2007). From the information we collected, it seems that the piebald character may theoretically be present or introduced into different existing camel breeds, thus being a characteristic linked to different breeds (e.g. a piebald mahri) but it can also prevail over the traits of the existing non-piebald breed leading to the development of a new piebald breed. We can envision two ways through which a piebald camel breed originates: (1) A piebald camel (of a non-piebald existing breed) is born within a herd, and local herders, once it is sexually mature, favour its reproduction; (2) A piebald camel, preferably a bull, is introduced (e.g. through raid or purchase) into a herd with no piebald camels and selected for reproduction. By further selecting at the same time for the piebald character and for the characters of the original breed, the outcome in the long run will be piebald camels of the original non-piebald breed. However, protracted selection for the piebald character (rather than for the traits of the original breed) may create not a breed variant but new piebald breeds. In both cases, favouring the reproduction of piebald traits increases the pull of leucistic alleles in the herd (and then in local herds), so that in the long run, piebald camels are born with higher frequency. However, we have seen that piebald individuals are commonly a minority within herds due to the constraints of inbreeding and to the recessive character of leucism and these facts weight the balance toward considering 'piebaldness' a breed variant, maybe in the process of becoming a breed.

3.7 Origin and diffusion of piebald camels

When and from where did the breeding of piebald camels originate? Was it developed independently by different populations or spread throughout Africa from a single point and time of origin? We did not find any information about the existence of piebald Bactrian camels or of piebald dromedary camels in Asia, Australia, and the Middle East. Piebald camels are not represented in the plate showing camel coat colours in the Middle East and dated to the thirteenth century A.D. (Figure 13, from an Arab miniature by Al-Wasiti from 'Maqamat' by Al Hariri). The exclusive presence of piebald camels in Africa had been noted by Cauvet (1925), cited in Bulliet (1975), who considered piebald colouration as one among other distinctive characteristics that would prove that African and Asian dromedary camels are separate species and that the camel was domesticated in Africa independently from Arabian domestication.

However, piebald colouration is more likely to be an indication of selective breeding for the piebald character once this has appeared by chance in domestic camels (Bulliet 1975). Historical trends of camel diffusion in the Saharan Africa and present distribution of piebald camels point to Kordufan or a neighbouring area as place of origin: some time after the introduction of camels in Northern Africa about 2,500 years ago, a piebald was born within one herd and this characteristic was readily picked up by local nomads. The resulting selection for piebald camels and their spread westward, and the appreciation of these camels by different Saharan pastoral populations, are likely to be at the base of present-day distribution of piebald camels throughout Africa.

Figure 13 'Herd of camels', Arab miniature by Al-Wasiti from 'Maqamat' by Al Hariri, thirteenth century, Baghdad (Iraq), showing different camel coat colours. Note the absence of white camels, today common in many countries, and of piebald camels (courtesy BnF/National Library of France) (Dioli 2013)

When and where domestication took place is still controversial for many species, and Cieslak et al. (2011) suggest that 'at least in some cases, coat and skin color can be a valuable marker for resolving these discussions.' Although the question of camel domestication has been complicated and frustrating for many scholars, there is some agreement that the camel, after having been hunted between 6000 and 3500 BC, has become the focus of direct domestication in the third millennium BC probably in south Arabia (Clutton-Brock 1988; Bulliet 1975, Zeder 2012). Sherratt (1983) suggests that camel domestication took place in two developing zones of long-distance trade (Bactrian camel toward the east, dromedary camel towards the south-west) in relation with the growing urban area of Mesopotamia. The camel as a pack and military animal was at its greatest expansion during the Nabatean times (second century BC - first century AD), during which period camels would have been taken up by populations in the eastern Egyptian desert (e.g. the Beja).

Camels had been replacing cattle and horses throughout the southern edge of the Sahara from approximately the third century BC, and a variety of populations of camel nomads have subsequently developed from Sudan to Mauritania and Western Sahara, through Niger, Mali, and Algeria. There is some agreement among scholars that camels spread westward through Africa first from a south-Saharan route going from Sudan to Mauritania about 2,200 years ago and shortly later through a north-Saharan route, reaching western Sahara from both routes about 2,000 years ago (Blanc and Ennesser 1989; Wilson 1984; Bulliet 1975). The diffusion of camel husbandry and nomadism facilitated a growth of a trans-Saharan caravan trade in the north-south and east-west directions. Indeed, the trading route going from Sudan and the Nubian desert to Timbuktu in Mali, and passing through Darfur, Kano, lake Chad, and Gao, was active and established already in 1400 (Wolf 1982).

The camel nomads of southern Sahara developed their own distinct forms of livelihoods and culture (e.g. in uses of camels and camel products, saddle technologies, taste, and aesthetic preferences), which included a preference for piebald camels. This preference, combined with the trans-Saharan exchange networks, facilitated the diffusion of piebald camels. Indeed, ancient Egypt and Sudanese nomadic populations were known for appreciating and venerating piebald cattle (d'Alverny 1950), and a breeding preference for piebald livestock (e.g. cattle, sheep) was already established in the region before the camel's introduction. This preference may have combined with a genetic drift occurring in the relatively low number of camels supported by the Saharan environment and taking part in this westward dissemination. This genetic drift may have caused the appearance of otherwise relatively rare conditions such as leucism and the associated piebald traits. This may at the same time explain the lack of piebald camels in Arabia and Somalia, where camels were present in huge numbers since the dawn of camel domestication and where leucism would have been quickly damped out by the much broader gene pool. All this suggests that the pioneer piebald camel breeders were relatively few, their livestock herds small in size, and their ability to control breeding well developed (Bulliet, pers. comm.).

From western Sudan, piebald camels would have spread (through raids, exchanges, and purchases) westward and their breeding being taken up by the southern Tuareg first and by Moorish populations of Western Sahara and Mauritania then (Figure 8). For example, according to Mahaman (1979), '[azarghaf camels] can be held by other ethnic groups who purchase them from the Tuaregs to use them essentially as pack animals.' This supports the hypothesis that piebald camels of Western Sahara find their origin from purchases at the markets of Niger, as expressed by older living members of the Reguibat tribe: the first piebald camel - they say - was a bull brought from the east about four centuries ago (between the sixteenth and the seventeenth centuries).

The paths of camel diffusion support the idea that similar paths have been used by piebald camels in their diffusion and the contemporary distribution of piebald camels supports the idea that they spread through a south-Saharan route. Through this route, piebald camels were adopted and bred by populations who had already shown an appreciation for other piebald livestock. During the following centuries, piebald camels spread through the Sahara from east to west of Africa, and further into the Canary Islands, Europe, and the USA. Overall, the history of piebald camels have much in common with the history of other piebald breeds around the world, as, for example, the one of splashed white Icelandic horses.18 Splashed white blue-eyed horses of Iceland are the outcome of a genetic drift that isolated the population coupled with a recent upsurge in an aesthetical preference for these traits among Icelandic horse breeders.

3.8 Trends in piebald camels today

Based on the total number of camels known to occur in piebald breeding areas, on piebald genetics, and on direct observations, we roughly estimate that there may be less than 5,000 piebald camels in the world. Most are in Niger and Western Sahara, bred by the Tuareg and the Sahrawi, respectively. In different areas, piebald camels are subjected to distinct trends affecting their numbers and material and cultural importance.

At least three different major contemporary trends can be identified as involving piebald camels: (1) their decline due to the abandonment of nomadic livelihoods (due to droughts, wars, destocking, etc.) and due to camel commodification and market pressures against piebald camels; (2) their disengagement from a pastoral economy and their use as source of income in the touristic and entertainment industries due to their aesthetical qualities; and (3) a resurgence of piebald camel breeding as a marker of political and cultural identities.

The first trend seems to be taking place in Sudan due to expansion of crop farms, droughts, and wars (Musa et al. 2006), as well as in Niger due to market pressures. In both areas, the demise of caravans in favour of motorized transport during the last century has surely contributed to a progressive reduction of piebald camel husbandry, the more so considering their widespread use in caravans. In Niger, there seems to be an ongoing process of further disengagement by many Tuareg herders. According to Pacholek et al. (2000a), piebald camels in the south of the Aïr are threatened by 'absorption' by the other two non-piebald local camel breeds, the azawak and the manga. This is attributed to the fact that 'the monetarization of pastoral societies pushes herders to turn to the breeding of types with the highest market value', and as piebald camels are less valued (e.g. for milk and meat production) on local markets, herders introduce non-piebald bulls into their herds (Pacholek et al. 2000b).

The second trend is evident in the Canary Islands and the USA, where piebald camels are actively bred and used in the touristic and recreational industries. Ranches in the USA have started raising piebald camels during the last two decades: they are appreciated aesthetically; they have higher monetary value than solid-coloured camels and are actively sold or hired to a variety of actors (Berry 2006), thus further spreading in new areas of the world, albeit out of a pastoral context.

The Sahrawi exemplify the third process: during the twentieth century, with colonialism, droughts, and then the Morocco-Polisario war over Western Sahara (1975 to 1991), piebald camels (and nomads) confronted a drastic reduction in numbers. During the war, camels were killed and herds bombed and others were abandoned in the desert by their fleeing owners and either died of hunger or thirst or were used by soldiers for meat provision. In the refugee camps where the Sahrawi ended up living, camels were absent and the social entities that used them as objects of cultural representation (i.e. tribes, fraction) destroyed overnight. But starting with the 1990s, Sahrawi refugees regained access to camels and camel husbandry, to grazing territories, and to traditional camel-related knowledge, and re-established relations with local ecologies (Volpato and Howard 2014; Volpato and Puri 2014; Volpato et al. 2013). Sahrawi refugees and nomads recovered the breeding of piebald camels for the same material and cultural reasons but with a new cultural referent: the Sahrawi people. Piebald camels became a marker of Sahrawi identity and a statement of cultural and political freedom, besides being appreciated for their aesthetical qualities. Piebald camels are a fairly common encounter nowadays in Western Sahara, particularly in its inland part under Sahrawi-Polisario control (Figure 14).

Figure 14 A group of piebald calves in Western Sahara (A. Broglia)

Although camel breeds have been historically important in the ability of pastoral populations to adapt and better exploit the desert environment, with increasing commoditization of camels and camel products, there is the risk that some locally adapted camel breeds may be diminishing in population and slowly disappearing (Blanc and Ennesser 1989). With the possibility that negative trends may prevail in different areas of Africa, it is urgent to study piebald camels and cultures around them. In reference to Nigerien piebald camels, studies report that it is undergoing a 'progressive disappearing of a type [of camel] with marked characteristics...and maybe unique...This danger deserves to be taken in consideration in regard to the safeguard of...biodiversity' (Pacholek et al. 2000b), as well as cultural diversity.

4. CONCLUSIONS

This study has addressed the history, geography, and anthropology of piebald camels. Based on fieldwork among the Sahrawi of Western Sahara, direct observations across north and east Africa and the Middle East, and a literature review, the case illustrates how piebald camels are an integral part of the story of the camel and of camel pastoralists in Africa and how piebald breeds are an interesting case to understand the dynamic relations of pastoralists with their livestock. Piebald camels probably originated in the eastern Sahara hundreds of years ago from a mutation that was picked up and selected for by local pastoralists who already had a preference for piebald livestock. In the ensuing centuries, piebald camels spread westward throughout Africa while being adopted by distinct pastoral tribes and groups across present-day Sudan, Niger, Mali, Mauritania, Western Sahara, and Morocco. They were bred for their alleged docility, for resistance to heat and thirst, and for their aesthetical value and were used in caravans and as mounts, as well as for subsistence purposes (e.g. milk and meat production) and for cultural reasons (e.g. as marker of cultural identity). During the last decades, piebald camels have been exported to the Canary Islands, Europe, and the USA, where they are used out of a pastoral context in the touristic and entertainment industries due to their perceived beauty and tameness.

The findings of this study are significant for understanding trends related to breeds and breeders in contexts of historical and contemporary change in pastoral systems. The findings also contribute to providing insights into the more general process of human selection for piebald colorations in domestic animals. This process can have different drivers and motives in different socio-ecological contexts and time. Piebald camels are under threat due to wars and droughts that kill herds and undermine pastoral livelihoods, e.g. in Darfur. They are threatened by increasing commodification of camels and camel products and consequent herd homogenization towards breeds with high marketing value, e.g. in Niger. Piebald camels and their husbandry have seen a resurgence during the last two decades as a symbol of political and cultural identity, in Western Sahara. Further studies are needed to fully elucidate the biology and history of piebald camels and their roles and dynamics in the present and in historical time. Fieldwork in other geographical areas (other than Western Sahara) where piebald camels are bred would be of great help in this sense. Also, a genetic study may help to identify different strains and to reconstruct piebald camels' temporal and geographical evolution and diffusion. A research into old written sources, mainly of Arabic origin, could help clarify the timing when piebald camels came to the Saharan Africa. Further studies would also be interesting to understand the place of piebald camels as an aesthetic locus among desert pastoralists. Such studies could make an interesting contribution to the field of anthropology of aesthetics.

We have pointed out that, in distinct breeding areas, piebald camels are subjected to negative trends, which increase the possibility of losing important genetic traits as well as the associated biocultural diversity (e.g. genetic, knowledge, and management). If these negative trends prevail, we risk losing the results of centuries-old engagement of humans in piebalds' breeding. The full story of piebald camels may be lost before having the chance to be told.

ENDNOTES

[1]A mahri or mehari is a Sahelian riding camel

[2]It is important to clarify the distinction between piebald animals whose white areas are depigmented (e.g., Dalmatian dogs, Overo and Tobiano horses), and those that have a white coat color but are fully pigmented (e.g., Arab horse, Brahman cattle) (Granding and Deesing 1998); only the former are leucistic animals.

[3]Holstein cows with blue eyes are nowadays selected against and excluded from inclusion in the European Union register of accepted breeds.

[4]http://www.fao.org/dad-is/

[5]www.kamelenmelk.nl

[6]http://www.marina-plage.com/v2/

[7]http://www.letemps.com.tn/article.php?ID_art=16296

[8]http://allafrica.com/stories/200805121059.html

[9]Inbreeding this kind of animals might cause neurological abnormalities, as it has been shown for dogs and cats (Grandin and Deesing 1998).

[10]http://www.fao.org/dad-is/

[11]www.camellosafari.com

[12]http://www.camelphotos.com/cgi-bin/ads.cgi?ct=baby_dromedary&md=second&id=286&min=&max

[13]http://www.camelphotos.com/cgi-bin/ads.cgi?ct=baby_dromedary&md=second&id=243&min=&max

[14]http://www.ferncroftfarms.com/Home_Page.html

[15]http://www.hiviewfarms.com/images/exotics2.htm#Camels

[16]http://www.fao.org/dad-is/

[17]Ezzara, tu es plus belle que les jeunes chamelles au milieu desquelles se dresse un étalon brun Celui aux taches noires sur la croupe et la tête, ce chameau pie aux (aisselles) blanches…'

[18]http://icelandmonitor.mbl.is/news/nature_and_travel/2016/06/12/the_incredible_blue_eyed_horses_of_iceland/

ACKNOWLEDGEMENTS

We are very thankful to many Sahrawi refugees and nomads for their patience, kindness, and hospitality during the interviews. We also thank the Italian NGO Africa'70 and the Sahrawi Veterinary Services of the SADR (Sahrawi Arab Democratic Republic) for logistical support and cooperation. We thank Alessandro Broglia, Davide Rossi, Saleh Mohamed Saleh, and Sidahmed Fadel for helping during fieldwork and Anna Littlejohn for editing the manuscript. Our thanks go also to Prof. Richard Bulliet for his insights on piebald camels and his comments on the manuscript.

FUNDING

Funds were granted to GV by the Ceres Research School of the Wageningen University, The Netherlands, as part of his PhD research on recovery and adaptation of subsistence practices and ethnobiological knowledge among the Sahrawi refugees. ADN was supported by the BBSRC DTA grant (1094618 and BBS/E/I/00001642) and by the BBSRC Institute Strategic Programme Grant on Livestock Viral Diseases to The Pirbright Institute, UK.

ETHICS APPROVAL AND CONSENT TO PARTICIPATE

In every case, prior informed consent was obtained verbally before the interview was conducted, according to the ethical guidelines adopted by the American Anthropological Association (1998) and by the International Society of Ethnobiology (2006).

REFERENCES

1. AA. 2000. Proceedings of the international workshop on the camel calf. Revue d Elevage et de Medecine Veterinaire des Pays Tropicaux 53(2): 1–138.
2. AAA. 2009. Code of Ethics of the American Anthropological Association. http://s3.amazonaws.com/rdcms-aaa/files/production/public/FileDownloads/pdfs/issues/policy-advocacy/upload/AAA-Ethics-Code-2009.pdf. Accessed 18 Jan 2017.
3. Abbas, B., A.A. Al Qarawi, and A. Al Hawas. 2000. Survey on camel husbandry in Qassim region, Saudi Arabia: Herding strategies, productivity and mortality. Revue d Elevage et de Medecine Veterinaire des Pays Tropicaux 53(3): 293–298.
4. Al-Koni, I. 2008. Gold dust. Translated by E. Colla. London: Arabia Books.

5. Allam, J. 2000. A sociolinguistic study on the use of color terminology in Egyptian colloquial and classical Arabic. In Diversity in Language. Contrastive Studies in English and Arabic Theoretical and Applied Linguistics, ed. Z. Ibrahim, N. Kassabgy, and S. Aydelott, 77–92. Cairo and New York: The American University in Cairo Press.
6. Antoine-Moussiaux, N., B. Faye, and G. Vias. 2006. Connaissances ethnovétérinaires des pathologies camélines dominantes chez les Touaregs de la région d'Agadez (Niger). http://camelides.cirad.fr/fr/science/pathotouareg1.html:1–21.
7. Antoine-Moussiaux, N., B. Faye, and G. Vias. 2007. Tuareg ethnoveterinary treatments of camel diseases in Agadez area (Niger). Tropical Animal Health and Production 39: 83–89.
8. Bellone, R.R. 2010. Pleiotropic effects of pigmentation genes in horses. Animal Genetics 41: 100–110.
9. Bernus, E. 1969. Maladies humaines et animales chez les Touaregs sahéliens. Journal de la Société des Africanistes 39(1): 111–138.
10. Berry, R. 2006. Spotted camels. http://camelphotos.com/spotted.html. Accessed 27 Jan 2017.
11. Blanc, C.P., and Y. Ennesser. 1989. Approche zoogéograpique de la différenciation infraspécifique chez le dromadaire Camelus dromedarius Linné, 1766 (Mammalia: Camelidae). Revue d Elevage et de Medecine Veterinaire des Pays Tropicaux 42(4): 573–587.
12. Boris, G. 1951. Le chameau chez les Marazig. Notes lexicographiques, 57–72. I.B.L.A.
13. Boyer, J.R. 1962. Contribution à l'étude de l'élevage camelin au Sahara Occidental. Le Regueibat et son chameau. Diplome d'Etat, Faculté de Médecine de Paris, Ecole Nationale Veterinaire D'Alfort.
14. Brooks, S.A., and E. Bailey. 2005. Exon skipping in the KIT gene causes a Sabino spotting pattern in horses. Mammalian Genome 16: 893–902.
15. Brooks, N., I. Chiapello, S. Di Lernia, N. Drake, M. Legrand, C. Moulin, and J. Prospero. 2005. The climate-environment-society nexus in the Sahara from prehistoric times to the present day. The Journal of North African Studies 10(3–4): 253–292.
16. Brooks, S.A., T.L. Lear, D.L. Adelson, and E. Bailey. 2007. A chromosome inversion near the KIT gene and the Tobiano spotting pattern in horses. Cytogenetic and Genome Research 119: 225–230.
17. Bulliet, R. 1975. The camel and the wheel. Cambridge: Harvard University Press.
18. Bulliet, R. 2005. Hunters, herders, and hamburgers. The past and future of human-animal relationships. New York: Columbia University Press.
19. Buxton, J. 1973. Religion and healing in Mandari. Oxford: Clarendon.
20. Caro Baroja, J. 1955. Estudios Saharianos. Madrid: Consejo Superior de Investigaciones Cientificas.
21. Castillo, I. del, and B. de Lugo. 2003. De la llegada de los camellos a las Canarias. http://www.camellosafari.com/camellos/camellos.html. Accessed 27 Jan 2017.
22. Cauneille, A. 1968. Les Chaanba (leur nomadisme), évolution de la tribu durant l'administration française. Paris: Éditions du Centre national de la recherche scientifique.
23. Cauvet, G. 1925. Le Chameau (anatomie, physiologie, races extérieur, alimentation, maladies, rale économique). Paris: Librairie J. B. Bailliere et Fils.
24. Chaibou, M. 2005. Productivité Zootechnique du Desert: Le Cas du Bassin Laitier d'Agadez au Niger. Docteur es Sciences, Biologie des Systèmes intégrés, Agronomie, Environnement, Université de Montpellier II.
25. Chaibou, M., and B. Faye. 2003. Production latiére des chamelles Abzin élevées par les Touaregs nomades du Niger. Niamey: Lait du chamelle pour l'Afrique. Atelier sur la filiére caméline en Afrique.
26. Cianchini, D. 1999. L'ultima carovana del sale.. Africa.
27. Cieslak, M., M. Reissmann, M. Hofreiter, and A. Ludwig. 2011. The colours of domestication. Biological Reviews 86(4): 885–899.
28. CIRAD. 2001. La répartition géographique des races des grands camélidés dans le monde. Centre de coopération internationale en recherche agronomique pour le développement, accessed 17 Oct.

29. Clutton-Brock, J. 1988. The walking larder: Patterns of domestication, pastoralism, and predation.
30. Cooper, M.P., N. Fretwell, S.J. Bailey, and L.A. Lyons. 2005. White spotting in the domestic cat (Felis catus) maps neat KIT on feline chromosome B1. Animal Genetics 37: 163–165.
31. Coote, J. 1994. 'Marvels of everyday vision': The anthropology of aesthetics and the cattle-keeping Nilotes. In Anthropology, Art, and Aesthetics, ed. J. Coote and A. Shelton, 245–273. Oxford: Oxford University Press.
32. Correra, A. 2006. Dynamique de l'utilisation des ressources fourrageres par les dromadaires des pasteurs nomades du Parc National du Banc d'Arguin (Mauritanie). PhD, Ecologie et gestion de la biodiversité. Museum National d'Histoire Naturelle de Paris.
33. Curdy, A. 2001. Never trust a camel with blue eyes. Field Exchange 13.
34. d'Alverny, F. 1950. Vestiges d'art rupestre au Tibesti oriental. Journal de la Société des Africanistes 20(2): 239–272.
35. Dioli, M. 2013. Pictorial guide to traditional management, husbandry and diseases of the one-humped camel.. Photographic CD-ROM.
36. Dioli, M. 2016. Towards a rational camel breed judging: A proposed standard of a camel (Camelus dromedarius) milk breed. Journal of Camel Practice and Research 23(1): 1–12.
37. Driot, C. 2009. Etude epidemiologique et histopathologique de la gale sarcoptique et de la teigne chez le dromadaire dans le sud marocain. Université Paul Sabatier de Toulouse, Diplome d'Etat Docteur Veterinaire.
38. Driscoll, C.A., D.W. Macdonald, and S.J. O'Brien. 2009. From wild animals to domestic pets, an evolutionary view of domestication. PNAS 106: 9971–9978.
39. Epstein, H. 1955. Domestication features in animals as functions of human society. Agricultural History 29(4): 137–146.
40. Farah, Z., and A. Fischer. 2004. Milk and meat from the camel. Handbook of products and processing. Zurich: vdf Hochschulverlag AG.
41. Faye, B., M.A. Kriska, M. Launois, G. Laveissiére, and T.M. Luong. 2002. Le dromadaire pédagogique. Montpellier: CIRAD.
42. Faye, B., S. Grech, and T. Korchani. 2004. Le dromadaire, entre féralisation et intensification. Anthropozoologica 39(2): 7–14.
43. Faye, B., H.R. Abdallah, F.S. Almathen, B. Harzallah, and S.E. Al-Mutairi. 2012. Camel biodiversity. Camel phenotypes in Saudi Arabia. FAO.
44. Fischer, W. 1965. Farb- und Formbezeichnungen in der Sprache der altarabischen Dichtung. Wiesbaden: Harrassowitz.
45. Francis, R.C. 2015. Domesticated. Evolution in a man-made world. New York: W.W. Norton & Company.
46. Gauly, M., J. Vaughan, S.K. Hogreve, and G. Erhrdt. 2005. Brainstem auditory evoked potential assessment of auditory function and congenital deafness in llamas (Lama glama) and alpacas (L pacos). Journal of Veterinary International Medicine 19: 756–760.
47. Gauthier-Pilters, H., and A.I. Dagg. 1981. The camel. Its evolution, ecology, behavior and relationship to man. Chicago: The University of Chicago Press.
48. Ghabdouane, M., and K. Prasse. 1990. Poeme touaregs de l'Ayr. Copenhagen: CNI Publications.
49. Giuffra, E., G. Evans, A. Tornsten, R. Wales, A. Day, H. Looft, G. Plastow, and L. Andersson. 1999. The belt mutation in pigs is an allele at the dominant white (I/KIT) locus. Mammalian Genome 10: 1132–1136.
50. Grandin, T., and M.J. Deesing. 1998. Genetics and animal welfare. In Genetics and the behaviour of domestic animals, ed. T. Grandin, 319–341. San Diego: Academic.
51. Householder, D.D. 2003. The genetics of equine coat color. Texas: A&M University, Department of Animal Science.
52. Hunn, E. 2011. Ethnozoology. In Ethnobiology, ed. E.N. Anderson, D. Pearsall, E. Hunn, and N. Turner, 83–96. Hoboken: Wiley-Blackwell.
53. Hurn, S. 2012. Humans and other animals. London: PlutoPress.
54. International Society of Ethnobiology, 2006. International Society of Ethnobiology Code of Ethics (with 2008 additions). http://www.ethnobiology.net/what-we-do/core-programs/ise-ethics-program/code-of-ethics/code-in-english/.

55. Julivert, M. 2003. El Sáhara. Tierras, pueblos y culturas. Valéncia: Universitat de Valéncia.
56. Kakar, A.R., K. de Verdier, and M. Younas. 2011. Rapid change of strategy is necessary for development of dromedary camel pastoralism in the Cholistan desert of Pakistan. Pastoralism Research Policy and Practice 1: 3.
57. Köhler-Rollefson, I.U. 1991. Camelus dromedarius. Mammalian Species 375: 1–8.
58. Krätli, S. 2008. Cattle breeding, complexity and mobility in a structurally unpredictable environment: The Wodaabe herders of Niger. Nomadic Peoples 12(1): 11–41.
59. Larson, G., and D.Q. Fuller. 2014. The evolution of animal domestication. Annual Review of Ecology, Evolution, and Systematics 45: 115–136.
60. Leese, A.S. 1927. A treatise on the one humped camel in health and disease. Stamford: Haynes.
61. Loubignac, V. 1952. Textes arabes des Zaer (Transcription, traduction, notes et lexique). Paris: Librarie Orientaliste et Américaine.
62. Ludwig, A., M. Pruvost, M. Reissmann, N. Benecke, G.A. Brockmann, P. Castanos, M. Cieslak, S. Lippold, L. Llorente, A.-S. Malaspinas, M. Slatkin, and M. Hofreiter. 2009. Coat color variation at the beginning of horse domestication. Science 324: 485.
63. Mahaman, O. 1979. Contribution a l'étude du dromadaire et de sa pathologie infectieuse. Diplome d'etat, Faculté de Médecine et de Pharmacie, University of Dakar.
64. Mercer, J. 1976. Spanish Sahara. London: George Allen & Unwin Ltd.
65. Migeon, C. 2006. Ma vie de chameau. Magazine Roadbook, 54–62
66. Monteil, V. 1952. Essai sur le chameau au Sahrara Occidental, Etudes Mauritaniennes, vol. 2. Saint-Louis du Senegal: Centre IFAN - Mauritanie.
67. Morera, M. 1991. La tradición del camello en Canarias. Anuario de Estudios Atlánticos 37: 167–204.
68. Musa, H.H., E. Shuiep, and I.E.M. El-Zubeir. 2006. Camel husbandry among pastoralists in Darfur, Western Sudan. Nomadic Peoples 10(2): 101–105.
69. Pacholek, X., R. Lancelot, M. Lesnoff, and S. Messad. 2000a. Performances de croissance des chamelons élevés dans la zone pastorale nigérienne. Revue d Elevage et de Medecine Veterinaire des Pays Tropicaux 53(2): 189–197.
70. Pacholek, X., G. Vias, B. Faye, and O. Faugére. 2000b. Elevage camelin au Niger. Référentiel zootechnique et sanitaire. Niamey: Project de Renforcement Institutionnel et Technique de la Filiére Cameline.
71. Pape, H. 1990. The inheritance of the piebald spotting pattern and its variation in Holstein-Friesian cattle and in Landseer-Newfoundland dogs. Genetica 80: 115–128.
72. Price, E.O. 2002. Animal domestication and behavior. Wallingford: CABI Publishing.
73. Rook, A., D.S. Wilkinson, F.J.B. Ebling, R.H. Champion, and J.L. Burton, eds. 1998. Textbook of dermatology. 4th ed: Blackwell Scientific Publications.
74. Roth, A. 1985. Introduction à l'étude des systèmes de désignation de la couleur dans les parlers arabes du Maghreb in Dictions du prisme. Les couleurs: Désignations et valeurs. Littérature orale arabo-berbère 16–17: 21–65.
75. Sandmeyer, L.S., C.B. Breaux, S. Archer, and B.H. Grahn. 2007. Clinical and electroretinographic characteristics of congenital stationary night blindness in the Appaloosa and the association with the leopard complex. Veterinary Ophtalmology 10: 368–375.
76. Schulz, U., I. Tupac-Yupanqui, A. Martínez, S. Méndez, J.V. Delgado, M. Gómez, S. Dunner, and J. Cañón. 2010. The Canarian camel: A traditional dromedary population. Diversity 2: 561–571.
77. Seitz, J.J., S.M. Schmutz, T.D. Thue, and F.C. Buchanan. 1999. A missense mutation in the bovine MGF gene is associated with the roan phenotype in Belgian Blue and Shorthorn cattle. Mammalian Genome 10: 710–712.
78. Sherratt, A. 1983. The secondary exploitation of animals in the Old World. World Archeology 15(1): 90–104.
79. Shinar, P. 2004. Modern Islam in the Maghrib. Jerulasem: The Max Schloessinger Memorial Foundation, The Hebrew University of Jerulasem.
80. Soler, N., C. Serra, J. Escola, and J. Unge. 1999. Sahara Occidental: Pasado y Presente de un Pueblo. Girona: Universidad de Girona.

81. Strain, G.M. 1996. Aetiology, prevalence and diagnosis of deafness in dogs and cats. British Veterinary Journal 152: 17–36.
82. Stritzel, S., A. Wohlke, and O. Distel. 2009. A role of the microphthalmia-associated transcription factor in congenital sensorineural deafness and eye pigmentation in Dalmatian dogs. Journal of Animal Breeding and Genetics 126: 59–62.
83. Taine-Cheikh, C. 1989. Le ḥassāniyya: Autopsie d'un dialecte vivant. Matériaux arabes et sudarabiques (GELLAS) Nouvelle série 2: 59–93.
84. Taine-Cheikh, C. 2002. Les valeurs du préfixe s- en Hassaniyya et les conditions de sa grammaticalisation, 2002nd ed. Spain: AIDA 5th Conference Proceedings.
85. Trut, L., I. Oskina, and A. Kharlamova. 2009. Animal evolution during domestication: The domesticated fox as a model. BioEssays 31: 349–360.
86. Voisset, G. 1989. Enquête sur la littérature mauritanienne: Formes et perspectives. Revue du monde musulman et de la Méditerranée 54: 188–199.
87. Volpato, G., and P. Howard. 2014. The material and cultural recovery of camels and camel husbandry among Sahrawi refugees of Western Sahara. Pastoralism: Research, Policy and Practice 4: 7.
88. Volpato, G., and R. Puri. 2014. Dormancy and revitalization: The fate of ethnobotanical knowledge of camel forage among Sahrawi nomads and refugees of Western Sahara. Ethnobotany Research & Applications 12: 183–210.
89. Volpato, G., D. Rossi, and D. Dentoni. 2013. A reward for patience and suffering - ethnomycology and commodification of desert truffles among Sahrawi refugees and nomads of Western Sahara. Economic Botany 67(2): 147–160. doi:10.1007/s12231-013-9234-7.
90. Wardeh, M.F. 2004. Classification of the dromedary camels. Journal of Camel Science 1: 1–7.
91. Webb, A.A., and C.L. Cullen. 2010. Coat color and coat color pattern-related neurologic and neuro-ophthalmic diseases. Canadian Veterinary.Journal 51: 653–657.
92. Wilkins, A.S., R.W. Wrangham, and W.T. Fitch. 2014. The "domestication syndrome" in mammals: A unified explanation based on neural crest cell behavior and genetics. Genetics 197: 795–808.
93. Wilson, R.T. 1978. Studies on the livestock of Southern Darfur, Sudan. Tropical Animal Health and Production 10: 19–25.
94. Wilson, R.T. 1984. The camel. London: Longman.
95. Wolf, E.R. 1982. Europe and the people without history. Los Angeles: University of California Press.
96. Woolf, C.M. 1991. Common white facial markings in bay and chestnut Arabian horses and their hybrids. Journal of Hered. 82: 167–169.
97. Zeder Melinda A. 2012. Pathways to animal domestication. Biodiversity in agriculture: Domestication, evolution, and sustainability:227.

CHAPTER 2

Genetic assessment of captive red panda (*Ailurus fulgens*) population

Arun Kumar[1],Upashna Rai[2],Bhupen Roka[2],Alankar K. Jha[2] and P. Anuradha Reddy[1*]

[1]*CSIR-Centre for Cellular and Molecular Biology, Uppal Road, Hyderabad 500 07, India.*

[2]*Padmaja Naidu Himalayan Zoological Park, Darjeeling 734101, India.*

ABSTRACT

Red panda *(Ailurus fulgens)* is threatened across its range by detrimental human activities and rapid habitat changes necessitating captive breeding programs in various zoos globally to save this flagship species from extinction. One of the ultimate aims of ex situ conservation is reintroduction of endangered animals into their natural habitats while maintaining 90 % of the founder genetic diversity. Advances in molecular genetics and microsatellite genotyping techniques make it possible to accurately estimate genetic diversity of captive animals of unknown ancestry. Here we assess genetic diversity of the red panda population in Padmaja Naidu Himalayan Zoological Park, Darjeeling, which plays a pivotal role in ex situ conservation of red panda in India. We generated microsatellite genotypes of fifteen red pandas with a set of fourteen loci. This population is genetically diverse with 68 % observed heterozygosity (H_O) and mean inbreeding (F_{IS}) coefficient of 0.05. However population viability analysis reveals that this population has a very low survival probability (<2 %) and will rapidly loose its genetic diversity to 37 % mainly due to small population size and skewed male-biased sex ratio. Regular supplementation with a pair of adult individuals every five years will increase survival probability and genetic diversity to 99 and 61 % respectively and will also support future harvesting of individuals for reintroduction into the wild and exchange with other zoos.

Keywords: Red panda, Captive breeding, Genetics, Population viability

1. BACKGROUND

Red panda (*Ailurus fulgens*), also known as the lesser panda, is one of earth's living fossils and its ancestors can be traced back to more than ten million years ago across Eurasia (Mayr 1986). Presently red panda populations are sporadically distributed in bamboo forests of Himalayan and Heng-Duan mountains in Nepal, India, Bhutan, Myanmar and Southwestern China (Su et al. 2001; Li et al. 2005). These populations continue to drastically decline across their habitats due to hunting, poaching, habitat loss and fragmentation (Wei et al. 1999; Choudhury 2001; Jha 2011). Red panda is classified as vulnerable by IUCN and is listed in Appendix I of CITES (Convention on International Trade in Endangered Species of Wild Flora and Fauna) since its wild population is estimated to be less than 10,000 mature individuals (Wang et al. 2008). Although endemic to eastern Himalayas and a flagship species for conservation of this fragile ecosystem, little is known about the genetic diversity of red panda in the wild, as it is shy, scansorial and dwells in altitudinal ranges of 1500–4800 m making it a very difficult animal to study (Choudhury 2001). Recent research on demography, phylogeography and genetic diversity of red panda (Su et al. 2001; Li et al. 2005; Hu et al. 2011) opens more opportunities to study current trends in wild red panda populations.

Several ex situ breeding programs have been initiated worldwide to protect this iconic species as its future survival relies on implementation of active conservation measures. Ex situ management and captive breeding for species conservation have grown enormously in recent years, and are the ultimate alternatives to in situ conservation, for preservation and recovery of endangered species aiming towards their reintroduction in the wild (Ballou 1992; Snyder et al. 1996; Ramirez et al. 2006) along with other aspects like education, research and fund raising. Founding and managing populations in captivity improve chances of long term survival of the species and can be used to supplement wild populations (Ramirez et al. 2006; Araki et al. 2007). Though widely accepted as a conservation tool, captive breeding/management of small populations comprises loss of genetic diversity, difficulty in achieving breeding, disease epidemics, adaptation to selection pressures in captivity, inbreeding and genetic drift (O'Grady et al. 2006; Zeoli et al. 2008; Ouborg et al. 2010). Genetic drift and inbreeding are known to increase homozygosity and accumulation of deleterious alleles thus decreasing both long and short term viabilities of populations by reducing the individual's fitness and adaptive potential to changing environments (Ouborg et al. 2010). Genetic consequences of captive breeding can nevertheless be managed in pedigreed populations by preventing breeding between related individuals and by introducing new animals to conserve genetic variation and gene pool of the representative (founder) animals. Recently however molecular genetic management measures have been employed to estimate relatedness and variations for many captive populations with individuals of unknown ancestry, (Russello and Amato 2004; Walling et al. 2011).

Advances in molecular genetics and genomics and consequent use of small informative fragments of DNA, like microsatellites, can help infer levels of genetic variation, inbreeding and relatedness within a population. Microsatellites are hypervariable loci, and are popular markers for population and conservation genetic studies at the intraspecific level due to their high mutation rates (Zachos et al. 2009). Since these markers are versatile, cost effective and the results are reproducible, genotyping of individual animals by simple PCR amplification is a sensible method to estimate genetic diversity, population size and structure, migration rates, kinship and parentage of various endangered species (Hoffman and Amos 2005; Guichoux 2011). The status and efficiency of demographic and genetic management measures for a captive population can then be evaluated and graded by Population Viability Analysis or PVA (Akçakaya and Sjögren-Gulve 2000; Brook et al. 2000; Holmes et al. 2007; Greenwald 2010). PVA estimates the extinction risk and effective population size of endangered species, by modeling effects of demographic and genetic stochasticities, and influences of management measures on the endangered population (Yang et al. 2007). It is a sensitive population model, and is often applied with care for those species where the availability of demographic data is minimal (Greenwald 2010).

India has very small captive populations of approximately twenty-five red panda housed at the Padmaja Naidu Himalayan Zoological Park (PNHZP), Darjeeling, and Himalayan Zoological Park (HZP), Gangtok. These animals are periodically exchanged with international zoos as part of the Species Survival Plan to maintain genetic diversity among captive bred individuals. The two Indian zoo populations of red panda have the distinctive priority of being located within the range states of its wild population distribution; and importantly being part of international captive breeding program, have the potential to provide a link between captive and wild populations (Glatston 2011). Therefore these comparatively small captive populations play a very important role in conservation of red panda (Glatston 2011). In this study we describe the genetic diversity of captive red pandas in PNHZP, Darjeeling, with a panel of fourteen microsatellite loci and project the viability of this population to maintain founder genetic diversity in the captive-bred individuals for a period of 100 years.

2. METHODS

2.1 Sampling and DNA isolation

Padmaja Naidu Himalayan Zoological Park (PNHZP), Darjeeling plays a pivotal role in red panda captive breeding in India, and had fifteen individuals, including three founders (two males and one female) and twelve captive-born at the time of our study (2012–2013). Fresh faecal samples of all fifteen red pandas (five females and ten males) were collected in sterile bottles with 90 % ethanol. Out of these animals, blood samples from four males and three females were also collected in EDTA-coated vacutainer tubes. Samples were acquired as per the Padmaja Naidu Himalayan Zoological Park and Central Zoo Authority, Govt. of India, ethical guidelines and procedures. No animal was injured as a result of this study. All samples were transported to the laboratory and stored at −20 °C until DNA isolation. Genomic DNA from blood was isolated by standard phenol–chloroform method (Sambrook et al. 1989). DNA was isolated from the external epithelial layer of fecal samples with QIAamp DNA stool kit (Qiagen, USA) as per the manufacturer's instructions.

2.2 Microsatellite amplification

The following loci were used for microsatellite genotyping—AF2, AF4, AF6, AF7, AF21, AF23, AF24 (Liang et al. 2007) RP4, RP5, RP6, RP8, RP9, RP11, RP12 (Wu et al. 2008) with an average reported polymorphism of five to seven alleles per locus. All forward primers were labeled at the 5′ end with either HEX or FAM fluorescent dye (Bioserve Inc, India). Briefly, a 15 μl singleplex reaction contained 200 mM of each dNTP, 10X Ex-Taq Buffer (Takara, Japan), 10× BSA, and 0.75U Ex-Taq Polymerase (Takara, Japan), 1.5 mM each of forward and reverse primers, and ~10 ng of template DNA. PCR amplifications were carried out in an epgradient S Mastercycler (Eppendorf, Germany) with initial denaturation at 94 °C for 10 min and 40 cycles of 94 °C for 30 s, annealing at Tm of each primer (Table 1) for 20 s, 72 °C for 20 s, followed by a final extension for 10 min at 72 °C.

Table 1 Statistics on captive red panda genotypes obtained from fourteen microsatellite markers

Locus	Repeat	Range (bp)	Tm	References	k	N	H_O	H_E	PIC	HW	F(IS)
AF2	(CTAT)11	120–144	57 °C	Liang et al. (2007)	5	13	0.538	0.655	0.581	0.2158	0.184
AF4	(GATA)15	231–255	56 °C	Liang et al. (2007)	6	14	0.857	0.817	0.757	0.1232	−0.051
AF6	(GATA)16	205–253	54 °C	Liang et al. (2007)	5	15	0.667	0.726	0.645	0.8253	0.085
AF7	(GATA)11	213–243	60 °C	Liang et al. (2007)	5	14	0.786	0.743	0.674	0.8539	−0.059
AF23	(TATC)13	133–161	62 °C	Liang et al. (2007)	6	14	0.786	0.741	0.671	0.9241	−0.063
AF24	(TATC)10	225–243	62 °C	Liang et al. (2007)	4	11	0.727	0.675	0.586	0.7636	−0.081
RP5	(CA)23	110–130	53 °C	Wu et al. (2008)	4	15	0.667	0.671	0.586	0.7103	0.007
RP6	(CA)13	155–167	53 °C	Wu et al. (2008)	3	14	0.286	0.455	0.386	0.0445	0.381
RP11	(CA)12	105–127	54 °C	Wu et al. (2008)	6	12	0.917	0.772	0.697	0.8248	−0.198
RP12	(CA)16	131–147	52 °C	Wu et al. (2008)	3	15	0.467	0.618	0.517	0.0791	0.252
AF21	(GATA)11	176–208	58 °C	Liang et al. (2007)	7	10	0.9	0.821	0.748	0.399	−0.102
RP4	(GT)19	198–212	54 °C	Wu et al. (2008)	5	14	0.714	0.796	0.729	0.0727	0.107
RP8	(CA)28	163–173	56 °C	Wu et al. (2008)	4	11	0.455	0.662	0.587	0.0291	0.324
RP9	(CA)20	125–143	55.5 °C	Wu et al. (2008)	6	9	0.778	0.85	0.774	0.8824	0.089
							0.6818	0.7144	0.6384		0.048

Tm annealing temperature, *K* number of alleles per locus, *N* number of Individuals typed at each locus, H_O observed heterozygosity, H_E expected heterozygosity, *PIC* polymorphism information content, *HW* Hardy–Weinberg equilibrium

All reactions were performed for a minimum of three repeats following a multi-tube approach (Taberlet et al. 1996) and were set up in a hood cleaned with bleach and alcohol, and irradiated with UV light to eliminate PCR contaminants. A negative control was included in all PCR reactions. Post amplification, PCR products were visualized by agarose gel electrophoresis, were electrophoresed with allelic size standard LIZ 500 (Applied Biosystems, USA) in an ABI Prism 3730 genetic analyzer (Applied Biosystems, USA), and alleles were scored with GeneMapper v3.2 (Applied Biosystems, USA).

2.3 Allele frequency analysis

Allele frequency and identity analyses were implemented as described by Reddy et al. (2012) to match genotypes of individuals from blood and faecal DNA with Identity test in CERVUS 3.0 (Kalinowski et al. 2007). Probability of identifying an unrelated individual (P_{ID}) and sibling (P_{SIB}) from a pair of genotypes was determined from allele frequencies, and minimum number of loci required to differentiate siblings was restricted to nine loci. Mismatched genotypes were re-examined manually for scoring errors, and in some cases the respective loci were genotyped again. All loci were scored manually for genotyping errors, amplification success and certainty of PCR repetitions in Microsoft EXCEL spreadsheets.

2.4 Genetic diversity analysis

Consensus genotypes were used to determine overall expected (H_E) and observed (H_O) heterozygosities using allele frequency function in CERVUS 3.0. Polymorphism Information Content (PIC) of heterozygous loci was also obtained from allele frequencies. Alleles deviating from Hardy–Weinberg equilibrium (HWE) were tested with Genepop v4 (Rousset 2008) with default Markov chain parameters. Linkage disequilibrium (LD) estimates between loci and inbreeding coefficient (F_{IS}) in the population were determined using FSTAT (Goudet 1995).

2.5 Viability analysis

Population viability analyses of the captive red panda population were simulated with Vortex v10 (Lacy 1993, 2000). All simulations were projected for a period of 100 years with 1000 burn-ins.

2.5.1 Baseline scenario

Basic demographic parameters were used to project the baseline scenario in viability analysis. Red panda is a polygynous mammal and generally both sexes mature at 18–24 months and reproduce till 12 years of age. These animals brood once a year and have 1–4 offspring per brood with a 1:1 sex ratio. Broods with 1 or 2 cubs are more frequent, and litters with 4 offspring are very rare (Glatston 2006). The captive population described in this study is skewed and consists of five females and ten males in reproductively active age. Default inbreeding value of 3.14 lethals per diploid individual was considered as per the mean values estimated from 40 different captive species of mammals (O'Grady et al. 2006). Red pandas in captivity exhibit high mortality rates till 2 years of age. All simulations were projected with default scenarios along with genetic management without supplementing to/harvesting from the existing population. A file on allele frequencies calculated from microsatellite data was added under genetic management in Vortex *v*10. Carrying capacity of PNHZP was predicted to be a maximum of 50 red pandas for future increase of population.

2.5.2 Supplementation scenario

In order to maintain 90 % of the genetic diversity present in founder/wild population for a period of 100 years, captive populations need supplementation on a timely basis either from the wild or from other zoos. Although an ideal scenario requires supplementation with two females and one male once every year or 2 years, the practical reality is that only one animal on an average can be acquired every 2 or 3 years due to logistics problems and difficulties in obtaining permits. Therefore simulations were projected by supplementing one adult male and one adult female once in 5 years until 100th year.

2.5.3 Supplementation and harvesting scenario

PNHZP has an important role to play in captive breeding of red panda especially since it is located in a range state of its wild population distribution. This adds a new dimension and responsibility on PNHZP to provide wild-caught animals to international zoos to infuse wild blood into their stocks and also to release captive-bred animals to replenish dwindling wild populations. Simulations were projected by harvesting one adult male and one adult female once in 5 years till 100th year while supplementing individuals as mentioned above.

2.5.4 Scenarios with catastrophes

Although protected and cared for in captivity, there can be several local or widespread catastrophes impeding an animal's growth. So far the red panda population at PNHZP has not been affected by any catastrophic event resulting in major loss of population. However we cannot rule out future possibilities especially since Darjeeling has high incidences of landslides, heavy rainfall, etc., and there is always a chance for outbreaks of viral diseases like canine distemper. Effects of catastrophes were included in baseline scenario as well as supplementation and harvesting scenarios, and these were projected to analyze the probability of population extinction at each incidence of catastrophic events. Although landslides and heavy rainfall have not majorly affected reproductive and survival efficiencies of PNHZP red panda population so far, we projected a reduction in reproductive efficiency by 3 % and survival efficiency by 1 % during these events. Disease outbreak can be a serious phenomenon that can wipe out the entire population. However the frequency or probability of occurrence of an outbreak is less and may happen once in 10 years in contrast to environmental catastrophes mentioned above which can occur every year.

3. RESULTS

3.1 Microsatellite statistics

We genotyped all fifteen red pandas at PNHZP at a minimum of eleven microsatellite loci, and got an average dropout rate of 9.8 % at fourteen loci and 89 % amplification success rate (Table 2). We needed an average of four PCR repetitions to obtain genotypes with 99 % certainty at each locus (data not shown). Probability of identifying an individual (P_{ID}) from a pair of unrelated individuals is 0.0001 with a minimum of six least variable loci and 9.7×10^{-13} with all fourteen loci, while probability of identifying siblings (P_{SIB}) from pair of animals is 0.001 with nine least variable loci and 1.06×10^{-5} with all fourteen loci. Identity analysis in CERVUS 3.0 correctly matched genotypes obtained from faecal and blood DNA samples of each animal. Genotypes obtained with a minimum of eleven loci were sufficient to generate genetic diversity parameters effectively.

3.2 Genetic diversity analysis

All fourteen microsatellites were shown to be polymorphic by Allele frequency analysis in CERVUS 3.0 with mean polymorphism information content (PIC) of 0.64, and 3–7 alleles per locus (Table 1). Expected (H_E) and Observed (H_O) heterozygosities were 0.714 and 0.681 respectively (Table 1). Genepop v4 (Rousset 2008) analysis with default Markov chain parameters showed that all the loci were in Hardy–Weinberg equilibrium ($p < 0.001$) (Table 1). Inbreeding and linkage analysis with FSTAT (Goudet

1995) indicated that none of the loci are linked to each other and the present PNHZP red panda population has a mean inbreeding value of 0.048 (Table 1).

Table 2 Locus-wise PCR success and allelic dropout rates, and number of PCR repeats required to obtain genotypes with 99 % certainty

Locus	PCR success (%)	Dropout rate (%)	Certainty of repeats
AF2	92.68	9.84	4
AF4	88.51	5.97	3
AF6	94.95	6.45	3
AF7	87.78	11.48	4
AF21	60.26	4.55	3
AF23	88.51	8.20	3
AF24	84.44	18.42	4
RP4	96.15	1.79	3
RP5	97.22	0.00	2
RP6	82.35	9.38	4
RP8	94.44	10.34	4
RP9	89.33	15.22	4
RP11	95.24	16.22	4
RP12	92.22	19.44	4
Mean	88.86	9.81	

3.3 Population viability analysis

According to baseline scenario simulations, the captive red panda population at PNHZP is a slow growing population. Deterministic growth rate (assuming no stochastic fluctuations, no inbreeding depression, no supplementation and no harvest) is 10.3 % ($r = 0.103$, $\lambda = 1.109$), with a mean generation time of 4.27 years for both females and males, while retaining only 37 % gene diversity after 100 years. Survival probability of this population without any additional intervention is less than 2 %. Gene diversity can however be maintained at 61 % (nearly 90 % of the initial diversity of 68 %) by supplementing this population with one reproductively active female and male once every 5 years till 100th year.

Supplementation also helps in increasing the population size to about 45 individuals in 25 years (Fig. 1). Harvesting individuals by translocation them to other populations or releasing them into wild along with supplementing new individuals reduces the risk of mating related individuals. Removing one female and one male from the population every 5 years in conjunction with supplementation retains the gene diversity at 61 % for 100 years. However the final number of animals is less than 40 and never reaches the full carrying capacity of PNHZP. The present red panda population is very vulnerable to environmental stochasticities and even a slight reduction in reproductive efficiency by 3 % and survival efficiency by 1 % in the baseline scenario reduces the rate of population growth to 8.6 % ($r = 0.086$ and $\lambda = 1.091$). When effects of these environmental factors are added to a population with supplementation and harvesting, the growth rate slows down to 7.0 % and the total number of animals never go beyond 25 individuals (Fig. 1).

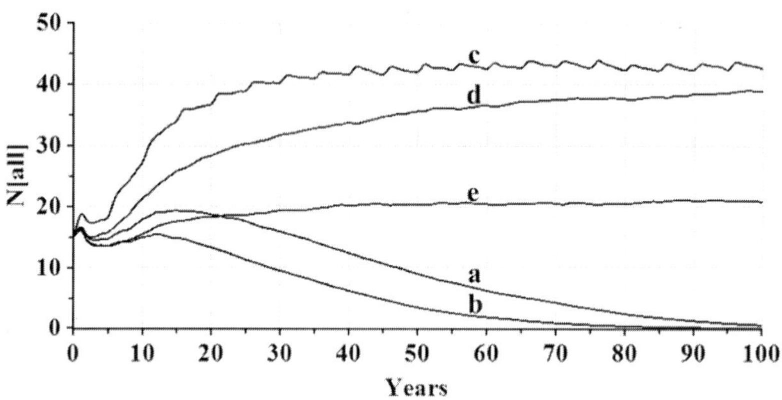

Figure. 1 PVA graph indicating projected number of individuals (N) corresponding to simulated scenario for a period of 100 years. Scenarios are **A** baseline; **B** baseline with catastrophes; **C** supplementation; **D** supplementation and harvesting; **E** supplementation and harvesting with catastrophes

4. DISCUSSION

This study attempts to understand genetic representation and population viability of captive red pandas in PNHZP, Darjeeling. We generated genetic profile of fifteen red pandas by genotyping DNA isolated from both blood and faecal samples at fourteen microsatellite loci. Although PCR reactions were performed in singleplex, these microsatellite primers show overall

PCR success (88.9 %) and dropout (9.81 %) rates comparable with multiplex protocols described earlier (Morin et al. 2001; Piggott 2004; Arandjelovic et al. 2009). Observed heterozygosity (HO) level of 0.68 (Table 1) indicates that the captive red panda population at PNHZP has a genetic variation in the range of wild red panda populations in Singhalila National Park (HO = 0.63) and Neora Valley National Park (HO = 0.75) screened with the same set of microsatellite markers (data not shown).

PNHZP is pivotal in ex situ conservation of red panda in India as it is located within the country's red panda distribution range, and is also the main Indian zoo participating in red panda Species Survival Plan. PNHZP also holds the record of successfully reintroducing captive-bred red panda in Singalila National Park in 2004 (Jha 2011). Presently the PNHZP red panda population is small and therefore extremely susceptible to demographic stochasticities. This population is also highly skewed and male-biased which predilects it to further instabilities and extinction risk. Viability analysis indicates that this is a slow growing population ($r = 0.103$, $\lambda = 1.109$), with a mean generation time of 4.27 years for both females and males, and will retain only 37 % of its gene diversity after 100 years. Extinction probability of this population is nearly 98 % and active measures are needed for its growth and to maintain its present genetic diversity. Although supplementation simulations show that addition of three individuals (two females and one male) will help in improving and maintaining genetic diversity close to the original value (data not shown), and also in rapidly attaining population growth, this is not practically possible. Obtaining animals either from the wild or from other zoos entails several logistic problems like permits, finances, disease surveillance, accidents during transit, etc. and ultimately the animal's ability to adapt to and reproduce in its new environment. A more conservative estimate which is also practiced presently is addition of a pair of animals, every 5 years. This will help in retaining gene diversity at 61 % and in increasing the present population to about 40 individuals within 25 years (Fig. 1).

As PNHZP is located within a range state it becomes relatively easy for it to build up genetically diverse stock population from rescued wild red pandas which can later be used for reintroduction into the wild or can be exchanged with other zoos/subpopulations (Leus 2011). While precarious now due to skewed sex ratio and small population size, harvesting from PNHZP population should be feasible after increasing the number of breeding females. Even in its present condition and with regular supplementation, removing one female and one male from the population every 5 years will retain gene diversity at 61 % for 100 years. However, unless the supplementation rate is increased, the population size will always remain small and never quite reach the full carrying capacity of PNHZP (Fig. 1). This situation can rapidly deteriorate in cases of local environmental catastrophes like landslides, and in disease outbreaks.

The Central Zoo Authority of India proposed a Conservation Breeding Initiative in 2005 wherein zoos would actively participate in captive breeding of highly threatened species with less than 2500 individuals left in the wild, especially species with localised distributions (Sharma et al. 2008). At least 250 properly bred and physically, genetically and behaviourally healthy individuals of each target species should be managed in captivity globally, of which at least 100 individuals should be managed within India, to act as insurance in case of loss of the species in wild (Leus 2011). The proposed target of 100 red pandas managed in Indian zoos will be possible with involvement and active participation of Himalayan Zoological Park (HZP), Gangtok and Bharat Ratna Pt. Govind Ballabh Pant High Altitude Zoo, Nainital, and by regularly adding new founders from the wild.

5. CONCLUSIONS

Padmaja Naidu Himalayan Zoological Park (PNHZP), Darjeeling has the largest captive population of red panda in India and is principally responsible for conservation breeding of this species in India. Although supported by two other satellite zoos in Gangtok and Nainital, their respectively populations are presently too small to influence the breeding program. Being located in the species' range states, Darjeeling and Gangtok's captive red panda populations have tremendous potential to grow, maintain the genetic diversity of the international captive population and restock dwindling wild populations.

ACKNOWLEDGEMENTS

We gratefully acknowledge Dr S. Shivaji for taking a keen interest and supporting this work in its initial stages.

ETHICAL APPROVAL

Blood samples used in this study were acquired from the Padmaja Naidu Himalayan Zoological Park, Darjeeling, India following the Padmaja Naidu Himalayan Zoological Park and Central Zoo Authority, Govt. of India, ethical guidelines and procedures. No animal was injured as a result of this study.

REFERENCES

1. Akçakaya HR, Sjögren Gulve P (2000) Population viability analysis in conservation planning: an overview. Ecol Bull 48:9–21
2. Araki H, Cooper B, Blouin MS (2007) Genetic effects of captive breeding cause a rapid, cumulative fitness decline in the wild. Science 318:100–103
3. Arandjelovic M, Guschanski K, Schubert G, Harris TR, Thalmann O, Siedel H, Vigilant L (2009) Two-step multiplex polymerase chain reaction improves the speed and accuracy of genotyping using DNA from noninvasive and museum samples. Mol Ecol Res 9:28–36
4. Ballou J (1992) Genetic and demographic considerations in endangered species captive breeding and reintroduction programs. In: McCullough D, Barrett R (eds) Wildlife populations. Springer, Netherlands, pp 262–275
5. Brook BW, O'Grady JJ, Chapman AP, Burgman MA, Akçakaya HR, Frankham R (2000) Predictive accuracy of population viability analysis in conservation biology. Nature 404:385–387
6. Choudhury A (2001) An overview of the status and conservation of the red panda Ailurus fulgens in India, with reference to its global status. Oryx 35:250–259
7. Glatston AR (2006) International studbook for the red or lesser panda (Ailurus fulgens). In: Glatston AR (ed) International studbook for the red or lesser panda (Ailurus fulgens). Royal Rotterdam Zoological and Botanical Gardens, Rotterdam, p 282
8. Glatston AR (2011) Chapter 17—red pandas in zoos today; the history of the current captive population. In: Glatston AR (ed) Red panda. William Andrew Publishing, Oxford, pp 303–321
9. Goudet J (1995) FSTAT (version 1.2): a computer program to calculate F-statistics. J Hered 86:485–486
10. Greenwald KR (2010) Genetic data in population viability analysis: case studies with ambystomatid salamanders. Anim Conserv 13:115–122
11. Guichoux E (2011) Current trends in microsatellite genotyping. Mol Ecol Resour 11:591–611
12. Hoffman JI, Amos W (2005) Microsatellite genotyping errors: detection approaches, common sources and consequences for paternal exclusion. Mol Ecol 14:599–612
13. Holmes EE, Sabo JL, Viscido SV, Fagan WF (2007) A statistical approach to quasi-extinction forecasting. Ecol Lett 10:1182–1198
14. Hu Y, Guo YU, Qi D, Zhan X, Wu HUA, Bruford MW, Wei F (2011) Genetic structuring and recent demographic history of red pandas (Ailurus fulgens) inferred from microsatellite and mitochondrial DNA. Mol Ecol 20:2662–2675
15. Jha AK (2011) Chapter 25—release and reintroduction of captive-bred red pandas into Singalila National Park, Darjeeling, India. In: Glatston AR (ed) Red panda. William Andrew Publishing, Oxford, pp 435–446
16. Kalinowski ST, Taper ML, Marshall TC (2007) Revising how the computer program cervus accommodates genotyping error increases success in paternity assignment. Mol Ecol 16:1099–1106
17. Lacy R (1993) VORTEX: a computer simulation model for population viability analysis. Wildl Res 20:45–65
18. Lacy RC (2000) Structure of the VORTEX simulation model for population viability analysis. Ecol Bull 48:191–203
19. Leus K (2011) Chapter 19—the global captive population of the red panda—possibilities for the future. In: Glatston AR (ed) Red panda. William Andrew Publishing, Oxford, pp 335–356
20. Li M, Wei F, Goossens B, Feng Z, Hidetoshi BT, Bruford MW, Funk SM (2005) Mitochondrial phylogeography and subspecific variation in the red panda (Aiulurus fulgens): implications for conservation. Mol Phylogenet Evol 36:78–89

21. Liang XU, Zhang ZH, Zhang L, Zhang WP, Shen FJ, Yang Z, Hou R, He W, Wei K, Yue BS (2007) Isolation and characterization of 16 tetranucleotide microsatellite loci in the red panda (Ailurus fulgens). Mol Ecol Notes 7:1012–1014
22. Mayr E (1986) Uncertainty in science: is the giant panda a bear or a raccoon? Nature 323:769–771
23. Morin PA, Chambers KE, Boesch C, Vigilant L (2001) Quantitative polymerase chain reaction analysis of DNA from noninvasive samples for accurate microsatellite genotyping of wild chimpanzees (Pan troglodytes verus). Mol Ecol 10:1835–1844
24. O'Grady JJ, Brook BW, Reed DH, Ballou JD, Tonkyn DW, Frankham R (2006) Realistic levels of inbreeding depression strongly affect extinction risk in wild populations. Biol Conserv 133:42–51
25. Ouborg NJ, Pertoldi C, Loeschcke V, Bijlsma RK, Hedrick PW (2010) Conservation genetics in transition to conservation genomics. Trends Genet 26:177–187
26. Piggott MP (2004) Effect of sample age and season of collection on the reliability of microsatellite genotyping of faecal DNA. Wildl Res 31:485–493
27. Ramirez O, Altet L, Enseñat C, Vilà C, Sanchez A, Ruiz A (2006) Genetic assessment of the Iberian wolf Canis lupus signatus captive breeding program. Conserv Genet 7:861–878
28. Reddy PA, Gour DS, Bhavanishankar M, Jaggi K, Hussain SM, Harika K, Shivaji S (2012) Genetic evidence of tiger population structure and migration within an isolated and fragmented landscape in Northwest India. PLoS One 7:e29827
29. Rousset F (2008) GENEPOP'007: a complete re-implementation of the genepop software for Windows and Linux. Mol Ecol Resour 8:103–106
30. Russello MA, Amato G (2004) Ex situ population management in the absence of pedigree information. Mol Ecol 13:2829–2840
31. Sambrook E, Fritsch F, Maniatis T (1989) Molecular cloning. Cold Spring Harbor Laboratory Press, Cold Spring Harbor
32. Sharma BR, Akhthar N, Gupta BK (eds) (2008) Proceedings of international conference on India's conservation breeding initiative. Central Zoo Authority, New Delhi
33. Snyder NFR, Derrickson SR, Beissinger SR, Wiley JW, Smith TB, Toone WD, Miller B (1996) Limitations of captive breeding in endangered species recovery. Conserv Biol 10:338–348
34. Su B, Fu Y, Wang Y, Jin L, Chakraborty R (2001) Genetic diversity and population history of the red panda (Ailurus fulgens) as inferred from mitochondrial DNA sequence variations. Mol Biol Evol 18:1070–1076
35. Taberlet P, Griffin S, Goossens B, Questiau S, Manceau V, Escaravage N, Waits LP, Bouvet J (1996) Reliable genotyping of samples with very low DNA quantities using PCR. Nucleic Acids Res 24(16):3189–3194
36. Walling CA, Nussey DH, Morris A, Clutton-Brock TH, Kruuk LE, Pemberton JM (2011) Inbreeding depression in red deer calves. BMC Evol Biol 11:318
37. Wang X, Choudhury A, Yonzon P, Wozencraft C, Zaw T (2008) Ailurus fulgens. The IUCN red list of threatened species. Version 2014.3. www.iucnredlist.org. Downloaded on 18 March 2015
38. Wei F, Feng Z, Wang Z, Hu J (1999) Current distribution, status and conservation of wild red pandas Ailurus fulgens in China. Biol Conserv 89:285–291
39. Wu H, Wu H, Zhan XJ, Guo Y, Zhang Z, Zhu L, Yan L, Li M, Wei FW (2008) Isolation and characterization of 12 novel microsatellite loci for the red panda (Ailurus fulgens). Conserv Genet 10:523–525
40. Yang Z, Hu J, Liu N (2007) The influence of dispersal on the metapopulation viability of Giant Panda (Aliuropoda melanoleuca) in the Minshan Mountains. Acta Zool Acad Sci Hung 53:169–184
41. Zachos FE, Hajji GM, Hmwe SS, Hartl GB, Lorenzini R, Mattioli S (2009) Population viability analysis and genetic diversity of the endangered red deer Cervus elaphus population from Mesola, Italy. Wildl Biol 15:175–186
42. Zeoli LF, Sayler RD, Wielgus R (2008) Population viability analysis for captive breeding and reintroduction of the endangered Columbia basin pygmy rabbit. Anim Conserv 11:504–512

CHAPTER 3

Genetic considerations for mollusk production in aquaculture: current state of knowledge

Marcela P. Astorga

Institutotode Acuicultura, Universidad Australde Chile, Puerto Montt, Chile

In 2012, world mollusk production in aquaculture reached a volume of 15,171,000 tons, representing 23% of total aquaculture production and positioning mollusks as the second most important category of aquaculture products (fishes are the first). Clams and oysters are the mollusk species with the highest production levels, followed in descending order by mussels, scallops, and abalones. In view of the increasing importance attached to genetic information on aquaculture, which can help with good maintenance and thus the sustainability of production, the present work offers a review of the state of knowledge on genetic and genomic information about mollusks produced in aquaculture. The analysis was applied to mollusks which are of importance for aquaculture, with emphasis on the 5 species with the highest production levels. According to FAO, these are: Japanese clam *Ruditapes philippinarum*; Pacific oyster *Crassostrea gigas*; Chilean mussel *Mytilus chilensis*; Blood clam *Anadara granosa* and Chinese clam *Sinonovacula constricta*. To date, the genomes of 5 species of mollusks have been sequenced, only one of which, *Crassostrea gigas*, coincides with the species with the greatest production in aquaculture. Another important species whose genome has been sequenced is *Mytilus galloprovincialis*, which is the second most important mussel in aquaculture production, after *M. chilensis*. Few genetic improvement programs have been reported in comparison with the number reported in fish species. The most commonly investigated species are oysters, with at least 5 genetic improvement programs reported, followed by abalones with 2 programs and mussels with one. The results of this work will establish the current situation with respect to the genetics of mollusks which are of importance for aquaculture production, in order to assist future decisions to ensure the sustainability of these resources.

Keywords: mollusk, genomic, aquaculture, genetic improvement, genetic, bivalves

1. AQUACULTURE

The aquaculture industry consists of 3 main groups with different production volumes: fish, with a volume of 44.15 million tons in 2012, representing 66.3% of production; mollusks with 15.17 million tons equivalent to 22.8%; crustaceans, 6.44 million tons, 9.7%. The production of other types was 0.86 million tons, or 1.3% (FAO, 2014). Based on these values, mollusks form the second largest group; however the levels of biotechnological research and the technologies applied are very low compared to the high level of development in cultivated fish species.

Biotechnology and genetic research have developed strongly in the last decade, and a large database now exists for fish species of importance for aquaculture (NCBI, Genbank, EMBL). These new research lines have permitted great progress and increased efficiency. Research of this kind has led to the development of many applications for fish production, with an increase in the number of techniques for genetic improvement, and a wider range of genetic variability of the resources for application in studies of traceability, pedigree, degrees of relatedness, and the search for molecular markers for Marked Assisted Selection. These developments are well reviewed in Hulata (2001), Liu and Cordes (2004), and Martínez (2007); genetic improvement programs are reviewed in Gjedrem (2012) and Gjedrem et al. (2012). In mollusk production, however, the development of such knowledge has been minimal (for Latin American mollusks see Astorga, 2008).

The present work therefore reviews the current state of scientific knowledge on the genetics of mollusk species which are important for aquaculture. The amount of existing information on the genomes of these resources will be assessed—this is needed to establish a baseline for knowledge-based decision-making.

Table 1. Aquaculture production and economic value of the 5 mollusk species with production greater than 250,000 tons (FAO, 2013), with review of genetic and genome information.

Species	Common name	Production '000 Ton.	Total value MUSS	WebScience publications	PubMed publications	Sequences in GenBank	EST in NCBI	Protein in NCBI
Ruditapes philippinarum	Japanese clam	3681	3438	25	273	30,442	6,469	1046
Crassostrea gigas	Pacific oyster	637	1323	200	1821	11,775	206,647	29,004
Mytilus chilensis	Chilean mussel	288	1148	15	35	119	7	27
Anadara granosa	Blood clam	393	468	6	48	641	2,278	158
Sinonovacula constricta	Chinese clam	744	670	5	23	16,383	5,296	204

The genetic information is based on the number of genetics-related publications in Web of Science databases, NCBI, PubMed, databases in GenBank, EST (Expressed Sequences Tag) regions and coded proteins.

Such knowledge would also enable us to define brood-stock selection programs, generate the bases for genetic improvement programs and finally assess the state of a given resource in order to ensure its long term conservation and sustainability.

2. MOLLUSK PRODUCTION

World mollusk production in 2012 was 15.17 million tons. Total aquaculture production of mollusks is represented by 5 main groups: clams, oysters, mussels, scallops, and abalones. Clam production during 2011 was 4.929 million tons; this was followed by oysters with 4.519 million tons; mussels with 1.802 million tons; scallops with 1.52 million tons; and finally abalones with aquaculture production of 395,000 tons (FAO, 2013).

There are 5 principal mollusk species produced in aquaculture (i.e., with production of over 250,000 tons), all bivalves, which correspond to only 5 species of mollusks reported in the FAO statistics (2013). Three of these are clams: *Ruditapes philippinarum* or Japanese clam; *Anadara granosa (=Tegillarca granosa)*, Blood clam or Blood cockle from Malaysia; and *Sinonovacula constricta* or Chinese clam. The other two are the oyster *Crassostrea gigas* and the Chilean mussel *Mytilus chilensis*. The detailed production for each of these is shown in Table 1, including the economic value generated by the production of these 5 resources with the highest volume in shellfish aquaculture. In this review the state of knowledge on the genetics of each of these 5 species will be evaluated.

3. GENETIC RESOURCES

Genetic resources is the name given to all the genetic or genome information for a group of representative individuals of a species which constitutes a natural resource (CBD Nagoya Protocol, 2010). Knowledge of the genetic resources of species has become increasingly important with the discovery of the importance of genetic indicators for species conservation, obtaining sufficient raw material to generate successful improvement programs, the sustainability of natural banks and the appropriate management of genetic resources (Kenchington et al., 2003).

Correct use of genetic resources allows the species to be developed sustainably, maintaining it for future generations; this information is also the basis for new genetic improvement programs in species in early or advanced stages of aquaculture (Kenchington et al., 2003; Guo, 2009; Lind et al., 2012). To start mollusk production in a hatchery, Gaffney (2006) proposes that high levels of genetic diversity must be ensured among the brood-stock. Proper levels of variability must then be maintained over time, by monitoring genetic variability to avoid excessive inbreeding (Gaffney, 2006). These factors—which require the application of biotechnologies or the design of genetic improvement plans—have been identified as important for starting aquaculture production.

Genetic improvement programs require reproductive management throughout the whole cycle of the species. Mollusk production is still carried out in extensive aquaculture facilities, where the initial production stages depend completely on the natural environment and on seed obtained from natural banks for the fattening stage, which makes it difficult to design breeding programs. On the other hand, the costs of obtaining seed from the natural environment are drastically lower than those of seed production in a hatchery, threatening the profitability of mollusk production companies and their ability to invest in genetic improvement programs. These programs may become very profitable, but only if the sale value of the product is very high, as occurs with the oyster, or if the genetic gain is directly related with a reduction in production costs.

4. GENETIC INFORMATION ON THE MAIN AQUACULTURE SPECIES

(Sort by economic value of production according to Table 1).

4.1 Ruditapes Philippinarum

Genetic information on the Malaysian clam *Ruditapes philippinarum* (Figure 1A), also known as *Venerupis philippinarum*, was searched using the ISI Web of Science and the National Center of Biotechnology Information (NCBI) search engines and the PubMed database. This allowed us to establish the low level of existing genetic information, despite the fact that it is one of the 5 mollusk species with the highest production in the world and is located in the first place of economic value production. Only around 30 papers discussing the genetics of this species were found (see database at Annex 1), although we consider that the number is not always representative of the information available. The research focused mainly on the genetic characterization of populations and the evaluation of genetic differentiation and population structure (42%), followed by searches for molecular markers (17%), and evaluation of hybrids with congeneric species (13%). A few papers describe research related to the genetic response associated with the immune system, gene expression, and transcriptome (total 21%), and finally 2 works address heritability and selection measurement (8%). However, although the main focus of research is on molecular genetics and molecular biology, a large number of works was found associated with study of gene expression as it relates to the immune response or to exposure to contaminants; there were at least 19 papers from 2012 to date, showing a strong focus on the transcriptome of this species. On searching for genome information in the National Center for Biotechnology Information (NCBI), data was found on 30,442 DNA, and RNA sequences, 6469 EST (Expressed Sequence Tag) sequences, and 1046 protein sequences (Table 1). Little genetic information is available on this species, despite the fact that it has been cultivated in hatcheries since the mid 1970s (Zhang and Yan, 2006). The information is insufficient to generate a basis for developing genetic improvement programs. The heritability of shell length has been calculated in the natural and cultured population at the larval and juvenile stages (Yan et al., 2014); the values found were 0.22 for larvae and 0.39 for juveniles in natural populations, and 0.17 and 0.87 for cultured populations. These results suggest that selection should be highly effective in this species, and that selecting either a natural or a cultured population to obtain faster growth should be successful. Zhao et al. (2012) established that divergent selection is effective and promotes shell length changes in the Manila clam, finding relatively moderate heritability values of 0.165 and 0.260, but quite high genetic gain values, with an increase in growth of 3.55% over the control line. Zhao et al. (2012) found that a base population with high genetic variability should be established as the basis for a divergent selection program. This would allow stable production of high quality seed to ensure the sustainability of aquaculture of this resource.

4.2 Crassostrea Gigas

Unlike the species described above, a large quantity of genetic research has been done in the oyster *Crassostrea gigas* (Figure 1B), associated with the search for molecular markers and the genetic characterization of populations. In the area with potential for application in aquaculture, works were found associated with selection response, heritability measurement, and searches for QTL and chromosome manipulation, highlighting the work of Guo et al. (2012), Ge et al. (2014) and Zhong et al. (2014). There were also many works associated with gene expression in relation to environmental stressors and transcriptome analysis (citations from 2010 to date in Annex 2). This agrees with the large amount of information found in the NCBI database on EST (206,647) as compared to the other species with high aquaculture production (Table 1). In general databases of scientific publications, at least 450 citations from the last 10 years were found relating to the genetics of this species. It is therefore one of the bivalve mollusks with the most highly developed genetic and biotechnological research. Furthermore, the nuclear and mitochondrial genomes of this species have been sequenced, which is the case with only 5 species of mollusk and 2 species of bivalve mollusk. There is a specialized database on this oyster, called Gigasbase, run by the IFREMER group in France. At least five genetic improvement programs have been developed for this species (Rye et al., 2010) although Gjedrem et al. (2012) mention only 3 selective breeding programs.

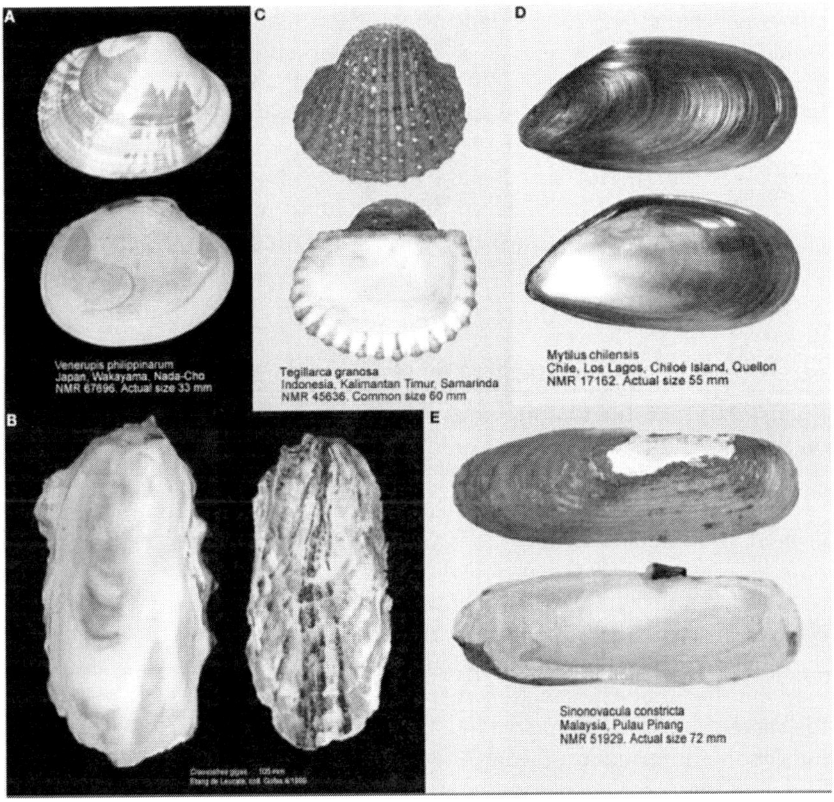

Figure 1. Photograph of the 5 mollusk species with the greatest production in aquaculture. (A) Ruditapes philippinarum or Venerupis philippinarum. (B) Crassotrea gigas. (C) Mytilus chilensis. (D) Tegillarca granosa or Anadara granosa. (E) Sinonovacula constricta. All images extracted from WoRMS (www.marinespeciers.org) with use authorized. Images (A,C–E) by Joop Trausel and Frans Slieker, Author: Natural History Museum Rotterdam. Image (B) by Serge Gofras, Muséum national d'histoire naturelle (MNHN).

There is a large volume of information for the implementation of new genetic improvement programs in the species. For example, we found a work assessing the heritability of shell pigmentation (Evans et al., 2009) which reported high values for narrow sense heritability (h^2 = 0.59). This suggests that total left-shell pigmentation in *C. gigas* is strongly influenced by additive genetic variation and therefore amenable to selection. Another characteristic which has been extensively studied in oysters is the genetic response to the high summer mortality observed in hatcheries (Dégremont et al., 2010). High realized heritabilities (h^2r) values have been estimated for high survival groups (above 0.88), higher than the values for low survival groups (0.55–0.68), and the authors conclude that these values would permit good results in a selective breeding program to improve survival (Dégremont et al., 2010). We also found estimates of growth heritability, with high values reported by Li et al. (2011) with 0.40; 0.33; and 0.15 and Wang et al. (2012)—who also report high heritability values for growth rate (0.457; 0.312; and 0.332), which has allowed a solid basis to be established for the further development of genetic improvement programs for this species, ensuring greater genetic gain in the short term.

4.3 Mytilus Chilensis

There is less scientific information published on the mussel *Mytilus chilensis* (Figure 1C) than the two previous species; in August 2014, only 15 papers on the genetics of this species could be found in the Web of Science (Annex 3). Aquaculture production is higher for this species than for other mussels, yet the amount of scientific information published is smaller—for example, around 100 scientific publications were found in the same database for the mussel *Mytilus galloprovincialis*. The largest proportion (50%) of the research published on *M. chilensis* is associated with the genetic characterization and population structure of this species distributed along the Chilean coast. Works can be found on the identification of molecular markers, heritability, selection response and gene expression, all in much the same proportions and small quantities. This is corroborated by a search for genome information in the NCBI databases, where only 119 data were found on sequenced genes, and only 7 EST were identified (Table 1). Of the 5 top species in aquaculture production, *M. chilensis* presents the least amount of genome information and its genome has not been sequenced. However, it should be noted that the genome of the Mytilid species *Mytilus galloprovincialis*, has been sequenced; this species, cultivated principally around the coast of Spain, presents the next highest aquaculture production among mussels.

There is one record of a genetic improvement program for mussels (Rye et al., 2010; Gjedrem et al., 2012); however, it is not being applied in production because cultivation is based on seed obtained from the natural environment and not from hatcheries. Nevertheless, works exist which estimate heritability and response to selection in this species. High heritability values have been found for a variety of characteristics such as growth in shell height and live weight (0.2–0.9) (Toro et al., 2004a). In larvae these parameters present a heritability range between 0.38 and 0.84 (Toro et al., 2004b). Finally, Alcapan et al. (2007) assessed the effects of environment and ageing on the heritability of body size in *M. chilensis*; they observed great variability in heritability values, conditioned by the site and the age of the individuals. These data would allow good results to be obtained in improvement programs, however the high production costs of seed obtained from hatcheries as opposed to the natural environment remains a limiting factor for this species. This makes the implementation of these improvement programs uncompetitive for producers in the short term. Nevertheless, the possibility cannot be ruled out that it may become necessary in future to make seed of high genetic quality available to this production sector.

4.4 Anadara Granosa

In the case of the Blood cockle *Anadara granosa* or *Tegillarca granosa* (Figure 1D), at least 4 publications were found on genetic characterization of populations, and two more relating to the search for molecular markers (Annex 4). There is therefore very little genetic information published for this species, despite its high production value. However, when we searched for genome information, we found more research into EST regions in this species (2278) (Table 1) than in *Mytilus chilensis* (only 7 EST). No records were found of current genetic improvement programs in this species, nor any information on baseline data which would allow the development potential of such programs to be assessed. On the other hand, the natural population structure has been studied (Ni et al., 2012; Wang et al., 2013) and this is a pre-requirement for selecting brood-stock and forming families for an improvement program. Wang et al. (2013) reported high genetic variability values, meaning that high diversity exists for brood-stock selection. However, they found genetic divergence in various sites on the coast of China presenting lower genetic variability, possibly caused by the admixture of artificially produced seed. This type of information suggests that genetic improvement programs need to include cross-breeding in their design in order to prevent genetic variability from being reduced, perhaps by using a large number of breeding individuals or a group which is representative of the genetic variability of the population. Ni et al. (2012) observe high genetic structuring in populations, indicating a high risk of erosion and possible local extinction. Populations need to be protected to prevent loss of genetic variability in the species.

4.5 Sinonovacula Constricta

Of the 5 top species in aquaculture production, the one with the least scientific information published on genetics is the Chinese clam *Sinonovacula constricta* (Figure 1E). In the Web of Science databases to August 2014, only three publications were found on searches for molecular markers plus two on the genetic evaluation of populations (Annex 5) (Table 1). However, there is a large number of sequenced genes in the NCBI databases, many more than for some other species analyzed. In this respect it ranks third among the top 5 species (Table 1), with the highest number of genes sequenced among the main 5 species produced in aquaculture (16,383 genes). There are no records of genetic improvement programs in this species, nor any publications of estimators for the implementation of such programs. Very little is known about its population structure. Niu et al. (2012) observed high genetic variability in 10 populations on the Chinese coast; however, they found genetic differentiation between sites, and propose the possible presence of cryptic species. This aspect should be studied since it is important to identify the cultivated species correctly and establish different management methods if more than one species exists.

5. CONCLUDING REMARKS

On the basis of this review it is apparent that there is no obvious relationship between the economic importance of a species, in terms of the volume produced in aquaculture, and the amount of scientific information available about genetics/genomic resources, when restricted to the five most produced species. In this case our analysis was carried out for research into genetics, which is the basis for genetic improvement programs in aquaculture species. The scientific information found is summarized, with the databases used, in Table 1. From this we see that the most studied species, using all the genetic information indicators, is *Crassotrea gigas*. There is great interest in this species and it has been introduced into many parts of the world for cultivation. Among mussels, the small amount of genetic information available on *Mytilus chilensis* is striking. At the same time the congeneric species *Mytilus galloprovincialis* is the most studied mussel; furthermore the latter is one of only two mussel species whose genome has been sequenced, and there are 41,294 records of DNA sequences, 19,756 of EST and 2,424 coded proteins. The species with the largest number of DNA sequences is the oyster *Crassostrea angulata*, with 81,015 DNA sequences, even more than *C. gigas*. On the other hand, *C. gigas* is the bivalve mollusk with the most EST sequences identified.

The quantity of genetic information available on bivalve mollusks varies widely, with great differences between species, even though their production volumes in aquaculture are similar. There is also still little information with which to generate solid bases for genetic improvement programs such as are observed in fish or other cultivated species. The only exception is *C. gigas*, for which adequate base information exists and genetic improvement programs are currently being implemented. Sufficient information exists on the mussel *M. chilensis* to provide a basis for more improvement programs, but the low use of reproduction in controlled environments is a limiting factor in this species. For the other three species there is virtually no basic genetic information on which predicting their response to selection programs. Further research is required to generate a sufficient basis for selective reproduction, and also to clarify information on natural populations, identifying possible cryptic species or strong differentiation between populations. The lack of knowledge about species with high production in aquaculture could lead to population bottlenecks, since if the brood-stock management plan is unsuitable, or seed extraction from local populations is not handled correctly, the populations which sustain these resources could be drastically reduced. This could result in a severe diminution of genetic diversity, to the point where it might be difficult for the species to recover, meaning that in future these resources which are so important for aquaculture would not be sustainable.

REFERENCES

1. Alcapan, A. C., Nespolo, R. F., and Toro, J. E. (2007). Heritability of body size in the Chilean blue mussel *(Mytilus chilensis* Hupe 1854): effects of environment and ageing. *Aquacult. Res.* 38, 313–320. doi: 10.1111/j.1365-2109.2007.01678.x
2. Astorga, M. P. (2008). "Estado actual del uso de marcadores moleculares en moluscos bivalvos de importancia para la acuicultura," in *Estado Actual Del Cultivo y Manejo de Moluscos Bivalvos y su Proyección Futura: Factores Que Afectan Su Sustentabilidad en America Latina*, eds A. Lovatelli, A. Farias, and I. Uriarte Actas de Pesca de la FAO. N 12 (Roma: FAO), 277–287.
3. CBD Nagoya Protocol. (2010). *Access to Genetic Resources and the Fair and Equitable Sharing of Benefits Arising from their Utilization.* Nagoya: CBD. Available online at: www.cbd.int/decision/cop/?id=12267
4. Dégremont, L., Bédier, E., and Boudry, P. (2010). Summer mortality of hatchery-produced Pacific oyster spat (*Crassostrea gigas*). II. Response to selection for survival and its influence on growth and yield. *Aquaculture* 299, 21–29. doi: 10.1016/j.aquaculture.2009.11.017
5. Evans, S., Camara, M. D., and Langdon, C. J. (2009). Heritability of shell pigmentation in the Pacific oyster, *Crassostrea gigas*. *Aquaculture* 286, 211–216. doi: 10.1016/j.aquaculture.2008.09.022
6. FAO. (2013). *FAO Yearbook. Fishery and Aquaculture Statistics 2011.* Rome: Food and Agriculture Organization of the United Nations, 76.
7. FAO. (2014). *The State of World Fisheries and Aquaculture 2014. Opportunities and challenges.* Rome: Food and Agriculture Organization of the United Nations, 223.

8. Gaffney, P. M. (2006). The role of genetics in shellfish restoration. *Aquat. Living Resour.* 19, 277–282. doi: 10.1051/alr:2006028
9. Ge, J., Li, Q., Yu, H., and Kong, L. (2014). Identification and mapping of a SCAR marker linked to a locus involved in shell pigmentation of the Pacific oyster (*Crassostrea gigas*). *Aquaculture* 434, 249–253 doi: 10.1016/j.aquaculture.2014.08.027
10. Gjedrem, T. (2012). Genetic improvement for the development of efficient global aquaculture: a personal opinion review. *Aquaculture* 344, 12–22. doi: 10.1016/j.aquaculture.2012.03.003
11. Gjedrem, T., Robinson, N., and Rye, M. (2012). The importance of selective breeding in aquaculture to meet future demands for animal protein: a review. *Aquaculture* 350, 117–129. doi: 10.1016/j.aquaculture.2012.04.008
12. Guo, X. (2009). Use and exchange of genetic resources in molluskan aquaculture. *Rev. Aquacult.* 1, 251–259. doi: 10.1111/j.1753-5131.2009.01014.x
13. Guo, X., Li, Q., Wang, Q. Z., and Kong, L. F. (2012). Genetic mapping and QTL analysis of growth-related traits in the Pacific oyster. *Mar. Biotechnol.* 14, 218–226. doi: 10.1007/s10126-011-9405-4
14. Hulata, G. (2001). Genetic manipulations in aquaculture: a review of stock improvement by classical and modern technologies. *Genetica* 111, 155–173. doi: 10.1023/A:1013776931796
15. Kenchington, E., Heino, M., and Nielsen, E. E. (2003). Managing marine genetic diversity: time for action? *ICES J. Mar. Sci.* 60, 1172–1176. doi: 10.1016/S1054-3139(03)00136-X
16. Li, Q., Wang, Q., Liu, S., and Kong, L. (2011). Selection response and realized heritability for growth in three stocks of the Pacific oyster *Crassostrea gigas*. *Fish. Sci.* 77, 643–648. doi: 10.1007/s12562-011-0369-0
17. Lind, C. E., Ponzoni, R. W., Nguyen, N. H., and Khaw, H. L. (2012). Selective breeding in fish and conservation of genetic resources for aquaculture. *Reprod. Dom. Anim.* 47, 255–263. doi: 10.1111/j.1439-0531.2012.02084.x
18. Liu, Z. J., and Cordes, J. F. (2004). DNA marker technologies and their applications in aquaculture genetics. *Aquaculture* 238, 1–37. doi: 10.1016/j.aquaculture.2004.05.027
19. Martínez, V. (2007). "Marker-assisted selection in fish and shellfish breeding schemes," in *Marker-Assisted Selection–Current Status and Future Perspectives in Crops, Livestock, Forestry and Fish*, ed E. P. Guimarães (Rome: FAO), 329–362.
20. Ni, G., Li, Q., Kong, L., and Zheng, X. (2012). Phylogeography of the bivalve *Tegillarca granosa* in coastal China: implications for management and conservation. *Mar. Ecol. Prog. Ser.* 452, 119–130. doi: 10.3354/meps09624
21. Niu, D.-H., Feng, B.-B., Liu, D.-B., Zhong, Y.-M., Shen, H.-D., and Li, J.-L. (2012). Significant genetic differentiation among ten populations of the razor clam *Sinonovacula constricta* along the coast of china revealed by a microsatellite analysis. *Zool. Stud.* 51, 406–414.
22. Rye, M., Gjerde, B., and Gjedrem, T. (2010). "Genetic improvement programs for aquaculture species in developed countries," in *Proceedings of the 9th World Congress on Genetics Applied to Livestock Production*. Available online at: http://www.kongressband.de/wcgalp2010/assets/pdf/0963.pdf. (Accessed January 12, 2012).
23. Toro, J. E., Alcapán, A. C., Vergara, A. M., and Ojeda, J. A. (2004b). Heritability estimates of larval and spat height in the *Chilean blue* mussel (*Mytilus chilensis* Hupe 1854) produced under controlled laboratory conditions. *Aquacult. Res.* 35, 56–61. doi: 10.1111/j.1365-2109.2004.00985.x
24. Toro, J. E., Alcapan, A. C., Ojeda, J. A., and Vergara, A. M. (2004a). Selection response for growth rate (shell height and live weight) in the chilean blue mussel (*Mytilus chilensis* hupe 1854). *J. Shellfish Res.* 23, 753–757.
25. Wang, Q. Z., Li, Q., Kong, L. F., and Yu, R. H. (2012). Response to selection for fast growth in the second generation of Pacific Oyster (*Crassostrea gigas*). *J. Ocean Univ. China* 11, 413–418. doi: 10.1007/s11802-012-1909-7
26. Wang, Y. J., Zeng, Q. G., and Xu, L. N. (2013). Population structure of the blood clam (*Tegillarca granosa*) in China based on microsatellite markers. *Genet. Mol. Res.* 12, 892–900. doi: 10.4238/2013.April.2.6

27. Yan, X., Huo, Z., Yang, F., and Zhang, G. (2014). Heritability of larval and juvenile growth for two stocks of *Manila clam Ruditapes philippinarum*. *Aquacult. Res.* 45, 484–490. doi: 10.1111/j.1365-2109.2012.03250.x
28. Zhang, G., and Yan, X. (2006). A new three-phase culture method for *Manila clam, Ruditapes philippinarum*, farming in northern China. *Aquaculture* 258, 452–461. doi: 10.1016/j.aquaculture.2006.04.046
29. Zhao, L., Yan, X., Huo, Z., and Yang, F. (2012). Divergent selection for shell length in the *Manila Clam, Ruditapes philippinarum*. *J. World Aquacult. Soc.* 43, 878–884. doi: 10.1111/j.1749-7345.2012.00612.x
30. Zhong, X., Li, Q., Guo, X., Yu, H., and Kong, L. (2014). QTL mapping for glycogen content and shell pigmentation in the Pacific oyster *Crassostrea gigas* using microsatellites and SNPs. *Aquacult. Int.* 22, 1877–1889. doi: 10.1007/s10499-014-9789-z

CHAPTER 4

Genomic analysis for managing small and endangered populations: a case study in Tyrol Grey cattle

Gábor Mészáros[1*], Solomon A. Boison[1], Ana M. Pérez O'Brien[1], Maja Ferenčaković[2], Ino Curik[2], Marcos V. Barbosa Da Silva[3], Yuri T. Utsunomiya[4], Jose F. Garcia[4] and Johann Sölkner[1]

[1]*Division of Livestock Sciences, University of Natural Resources and Life Sciences, Vienna, Austria*

[2]*Department of Animal Science, University of Zagreb, Zagreb, Croatia*

[3]*Empresa Brasileira de Pesquisa Agropecuária, Juiz de Fora, Brazil*

[4]*UNESP-Universidade Estadual Paulista, Jaboticabal, Brazil*

Analysis of genomic data is increasingly becoming part of the livestock industry. Therefore, the routine collection of genomic information would be an invaluable resource for effective management of breeding programs in small, endangered populations. The objective of the paper was to demonstrate how genomic data could be used to analyse (1) linkage disequlibrium (LD), LD decay and the effective population size (Ne_{LD}); (2) Inbreeding level and effective population size (Ne_{ROH}) based on runs of homozygosity (ROH); (3) Prediction of genomic breeding values (GEBV) using small within-breed and genomic information from other breeds. The Tyrol Grey population was used as an example, with the goal to highlight the potential of genomic analyses for small breeds. In addition to our own results we discuss additional use of genomics to assess relatedness, admixture proportions, and inheritance of harmful variants. The example data set consisted of 218 Tyrol Grey bull genotypes, which were all available AI bulls in the population. After standard quality control restrictions 34,581 SNPs remained for the analysis. A separate quality control was applied to determine ROH levels based on Illumina GenCall and Illumina GenTrain scores, resulting into 211 bulls and 33,604 SNPs. LD was computed as the squared correlation coefficient between SNPs within a 10 mega base pair (Mb) region. ROHs were derived based on regions covering at least 4, 8, and 16 Mb, suggesting that animals had common ancestors approximately 12, 6, and 3 generations ago, respectively. The corresponding mean inbreeding coefficients (F_{ROH}) were 4.0% for 4 Mb, 2.9% for 8 Mb and 1.6% for 16 Mb runs. With an average generation interval of 5.66 years, estimated Ne_{ROH} was 125 ($Ne_{ROH>16\ Mb}$), 186 ($Ne_{ROH>8\ Mb}$) and 370 ($Ne_{ROH>4\ Mb}$) indicating strict avoidance of close inbreeding in the population. The LD was used as an alternative method to infer the population history and the Ne. The results show a continuous decrease in Ne_{LD}, to 780, 120, and 80 for 100, 10, and 5 generations ago, respectively. Genomic selection was developed for and is working well in large breeds. The same methodology was applied in Tyrol Grey cattle, using different reference populations. Contrary to the expectations, the accuracy of GEBVs with very small within breed reference populations were very high, between 0.13–0.91 and 0.12–0.63, when estimated breeding values and deregressed breeding values were used as pseudo-phenotypes, respectively. Subsequent analyses confirmed the high accuracies being a consequence of low reliabilities of pseudo-phenotypes in the validation set, thus being heavily influenced by parent averages. Multi-breed and across breed reference sets gave inconsistent and lower accuracies. Genomic information may have a crucial role in management of small breeds, even if its primary usage differs from that of large breeds. It allows to assess relatedness between individuals, trends in inbreeding and to take decisions accordingly. These decisions would be based on the real genome architecture, rather than conventional pedigree information, which can be missing or incomplete. We strongly suggest the routine genotyping of all individuals that belong to a small breed in order to facilitate the effective management of endangered livestock populations.

Keywords: breed management, endangered breeds, SNP chip, linkage disequilibrium, runs of homozygosity, genomic selection

1. INTRODUCTION

In the last decade, technological advancement has allowed for the genotyping of large numbers of single nucleotide polymorphisms (SNP) in the genome. The increase in SNP density was accompanied with decrease in price for the commercial SNP-chips, standard sets of SNPs selected, and sold by genotyping companies in large numbers, dominating animal, and plant breeding research in many countries.

Traditionally microsatellite markers were used for genotyping animals in population genetics studies. A popular set of microsatellites endorsed by the Food and Agriculture Organization (FAO) is widely used to evaluate genetic diversity in farm animals, especially endangered breeds (Baumung et al., 2004; Groeneveld et al., 2010). As a technological follow up, a set of SNPs could be used for a similar purpose. An advantage of the SNP markers is their occurrence on standard genotyping panels. Pooling of genotypes and comparison of different populations is feasible, contrary to the microsatellites, where (partially) different panels could be genotyped each time. The disadvantage of the current SNP panels from breed diversity perspective is their development in direction of commercial application in the most common species and breeds, with little research undertaken to prepare assays to replace the ISAG-FAO microsatellite panels.

The application of the SNP markers in animal breeding however, goes beyond population genetics. The early adopters were the large breeding organizations managing breeds with many animals and large financial capital. After the general success of the genomic selection approach (Meuwissen et al., 2001) the utilization of the genomic information has increased considerably. Today genomic breeding values (GEBV) are routinely used for making selection decisions, which has resulted in reducing the generation interval and increasing genetic gain compared to classical progeny testing systems in dairy cattle populations (Hutchison et al., 2014).

Genomic selection was an incentive to genotype nearly every young bull in many large cattle populations. This incentive is missing in smaller breeds because a large population size is generally perceived as a requirement to estimate reliable GEBV. Although there were numerous studies using SNP data in many small breeds, these are rather isolated efforts to demonstrate an interesting phenomenon or describe other interesting general aspect of a particular breed. Even though there are a relatively low number of animals to be genotyped in small populations, there is a general lack of routine genotyping in small breeds.

The objective of the paper is to demonstrate how genomic data could be used to ascertain population structure in small and endangered breeds, evaluate GEBV, and assess the range of potential applications from the breed management perspective. To tackle this goal, we have used the Tyrol Grey breed as an example to demonstrate some of the potential uses of genomic data in small and endangered populations. Some of the potential uses of genomic data are not applicable to the Tyrol Grey breed, but they are still extremely useful in the breed management context. We discuss such uses in the last part of the paper in order to give a comprehensive overview about the potential of genomics in small and endangered breeds.

2. MATERIAL AND METHODS

2.1 Data and Quality Control

The Tyrol Grey cattle is an endangered cattle breed with population size of consisting of 3785 breeding animals as of 2013 (ÖNGENE, www.oengene.at, 2014). We were able to genotype all available sires due to its small population size. From the available 218 Tyrol Grey AI bulls, we have genotyped 99 animals with the Illumina® BovineSNP50 BeadChip (50 K) and 119 animals with the Illumina® BovineHD BeadChip (HD) with about 770 K SNPs.

Only the 49,394 SNPs appearing on both 50 K and HD chips were retained, as the 50 K chip is a standard genotyping platform used for routine genotyping in taurine cattle. SNP markers with unknown positions and those on sex chromosome were excluded.

Two separate quality checks of the data were undertaken. The first quality check was done and the data used for estimating linkage disequilibrium (LD), effective population size (Ne) and also in genomic prediction approaches. The second quality check followed the approach of Ferenčaković et al. (2013). This was a more stringent quality check to help reduce error that might occur when estimating genomic inbreeding from runs of homozygosity (ROH).

The first quality check of the available SNP data was undertaken with the following criteria. SNPs with call rate less than 90% and Hardy-Weinberg equilibrium Fishers's exact *P*-value below 10^{-6} were removed using PLINK v1.07 (Purcell et al., 2007). SNP markers with minor allele frequency (MAF) < 0.05 and those mapped to the same physical positions were also deleted. Samples with call rate lower than 90% were also discarded. After quality control, 34,581 SNPs remained. The second quality check has been applied in the calculation of ROH as we did not exclude SNPs with low MAF, with high LD or those deviating from HWE. Genotyping errors were reduced by discarding SNPs with Illumina GenCall score ≤ 0.7, SNPs with Illumina GenTrain score ≤ 0.4 and animals with more than 5% missing genotypes. The same quality control settings has been used in Ferenčaković et al. (2013). The analyses were based on 211 bulls each genotyped for the same 33,604 SNPs with average distance of 73.655 kb between adjacent SNPs (from 23 to 1,955,291 bp), all placed on 29 autosomes. ROH segments were identified as a part of the genome in which 15 or more consecutive homozygous SNPs at a density of one SNP on every 100 kb are not more than 1 Mb apart. ROH calculations were done using the algorithm implemented in SNP and Variation Suite (v7.6.8 Win 64; Golden Helix, Bozeman, MT, USA www.goldenhelix.com).

2.2 Linkage Disequilibrium

The squared correlation (r^2) was used to measure the LD. The r^2 values were calculated using PLINK as pairwise comparisons of markers on the same chromosome, separated by less than 10 Mb. The decay of LD was analyzed using bins of 100 kb for the maximum distance between SNP pairs. Marker bins below 100 kb for a 50 K SNP panel generally generate very small numbers of pairwise LD values. PLINK calculates as r^2 between two SNPs:

$$r_{LD}^2 = \left(\frac{\sum_{i=1}^{n} (g_{ij} - \bar{g}_j)(g_{im} - \bar{g}_m)}{\sqrt{\sum_{i=1}^{n} (g_{ij} - \bar{g}_j)^2} * \sqrt{\sum_{i=1}^{n} (g_{im} - \bar{g}_m)^2}} \right)^2$$

Where n = number of individuals with non-missing genotype; g is the genotype allele count of 0,1, and 2 for AA, AB, and BB, respectively for individual i of SNP j and SNP m.

The calculations of effective population size (Ne_{LD}) were based on McEvoy et al. (2011). The Ne was based on the LD values:

$$E\left(r_{LD}^2\right) \approx \frac{1}{\alpha + 4Ne_{LD}c}$$

Where $E(r_{LD}{}^2)$ is the expected squared correlation of allele frequencies at a pair of loci, α is 2 when the impact of mutation is considered and 1 otherwise. Variable c is the genetic distance between loci in Morgans. The Ne was calculated as:

$$Ne_{LD} \approx \frac{1}{4c} * \left(\frac{1}{r_{LD}^2} - \alpha \right)$$

Assuming that the population has been constant in size, the approximation of Ne_{LD} is true for t generations ago, where t = 1.2c (Hayes et al., 2003). It has been noted that LD patterns from shorter inter marker distances were informative about the Ne in the more distant past, while markers separated by longer distances are informative about the recent Ne. The relationships describing the historical development of Ne_{LD} should be considered only as an approximation, as LD patterns might be affected by variety of factors (de Roos et al., 2008).

2.3 Runs of Homozygosity

The inbreeding coefficients (F_{ROH}) were calculated from the formula proposed by McQuillan et al. (2008); $F_{ROH} = L_{ROH}/L_{AUTOSOME}$, where L_{ROH} is the total length of all ROH in the genome of an individual while $L_{AUTOSOME}$ refers to the specified length of the autosomal genome covered by SNPs on the chip (here 2,499,624,571 bp). We calculated three coefficients; $F_{ROH>4\ Mb}$, $F_{ROH>8\ Mb}$, and $F_{ROH>16\ Mb}$ defined by the minimum ROH lengths being higher than 4, 8, or 16 Mb, respectively. Under the assumption that 1 cM = 1 Mb, calculated inbreeding coefficients are expected to correspond to the reference ancestral population being remote approximately 12 ($F_{ROH>4\ Mb}$), 6 ($F_{ROH>8\ Mb}$), and 3 ($F_{ROH>16\ Mb}$) generations. For more detailed explanation see Howrigan et al. (2011) and Curik et al. (2014).

The calculation of the effective population size (Ne_{ROH}) was based on the equation $Ne_{ROH} = 1/2\Delta F$, where ΔF was calculated as regression coefficient b, representing change of $F_{ROH>4\ Mb}$ per 1 year (regression of $F_{ROH>4\ Mb}$ on the birth year), multiplied by the generation interval 5.66, previously calculated from the Tyrol Grey pedigree by using the software ENDOG, version 4.6 (Gutiérrez and Goyache, 2005), together with pedigree inbreeding coefficient (F_{PED}).

Autozygosity islands were defined as regions where SNPs had extreme ROH frequency (outliers according to boxplot distribution, see Figure 6).

Computation of descriptive statistics (PROC Means), bootstrap confidence intervals (SAS Macro), regression analysis (PROC REG), and graphical illustrations (PROC Boxplot, SAS Macro) were performed by the SAS software v 9.3.

2.4 Genomic Selection in Small Breeds

Single and multi-breed scenarios were considered to derive within and across breed GEBV. For the multi-breed scenarios the German-Austrian genotype pool of 6730 Fleckvieh and 1415 Brown Swiss bulls was used to extend the training set, as Fleckvieh is the major breed in Austria and Brown Swiss has common history with the Tyrol Grey. Breeding values (EBV), deregressed breeding values (dEBV) and their corresponding reliabilities for 10 major production and functional traits in Austria were provided by *Zuchtdata EDV- Dienstleistungen GmbH*, Austria. The deregression procedure of the EBVs removed the contribution of relatives other than daughters to the breeding value, based on the methodology of Garrick et al. (2009).

GEBV was estimated by fitting a polygenic effect assuming that every marker has a constant variance (GBLUP) (Meuwissen et al., 2001) i.e., assuming that each marker explains an equal proportion of the total genetic variance (σ^2_g). The GBLUP model was:

$$Y = \mathbf{1}_n g + e$$

$$Y = \text{EBV or dEBV}$$

μ = overall mean;

Z = design matrix allocating records to breeding values;

G = vector of random additive genetic effect using the genomic relationship matrix (G)

coming from $N(0, G\sigma^2_g)$;

e = vector of random residual errors $N(0, R\sigma^2_e)$, where R was diagonal matrix with weight calculated as $r^2/(1 - r^2)$

To study the predictability of the above model, three strategies were used to group the animals into reference and validation sets, with main focus on the genomic evaluation of the Tyrol Grey breed.

1) Single breed scenario: Only Tyrol Grey bulls were used. The validation sets consisted from young bulls born after 2003, with the older Tyrol Grey bulls in the reference set.
2) Multi-breed scenario: The validation sets consisted from young Tyrol Grey bulls born after 2003, the reference sets included the rest of the Tyrol Grey bulls, as well as the Brown Swiss or the Fleckvieh or both Brown Swiss and Fleckvieh.
3) Across breed scenario: All Tyrol Grey bulls were put into the validation set. The reference set consisted of the Fleckvieh and/or Brown Swiss.

To be able to discuss the results obtained from the multi-breed and across breeds' scenarios, Eigen vectors, and values are computed on an estimated genomic relationship matrix with the three breeds. Principal component analysis plots are provided in Figure 1.

Prediction accuracy was measured as the correlations between the resulting GEBVs and pseudo-phenotype EBV. Bootstrapping procedure (sampling with replacement) was used to calculate the standard error of the correlation between the GEBV and EBV. The estimated GEBV were bootstrapped 10,000 times (this value appeared to give stable results) and the bootstrap GEBVs are correlated to the EBVs. The standard errors were calculated from the 10,000 estimated accuracies. This procedure gives us a fair estimate of the degree of dispersion of the estimated correlations. Although other cross validation procedure like random splitting procedures could have been employed, we chose to use forward prediction which is more relevant in breeding. In addition, limited number of individuals in the validation set also supports the idea of bootstrapping to calculate standard errors of the correlation estimates.

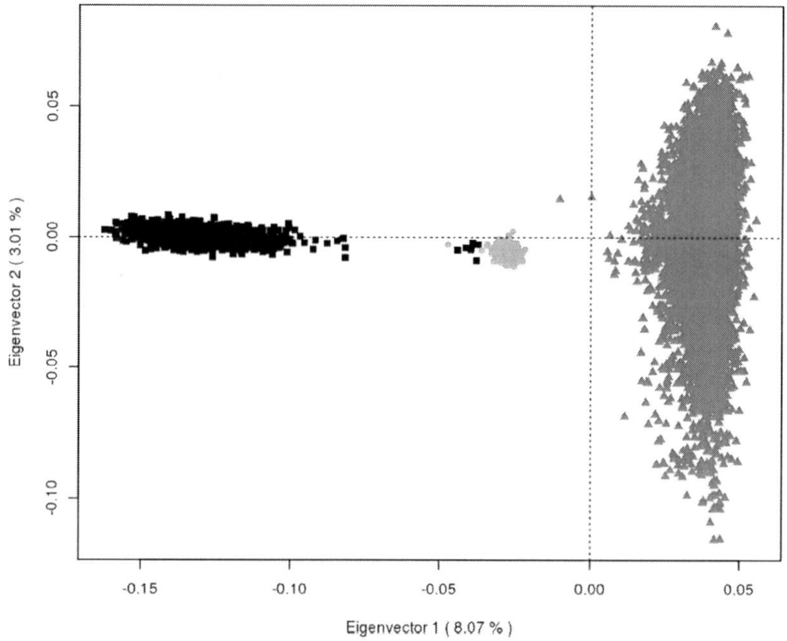

Figure 1. Principal component analysis of the tyrol grey (in green), brown swiss (in black), and fleckvieh (in brown) breeds; the amount of explained variance by the first two eigenvectors is shown in brackets.

3. RESULTS

3.1 Linkage Disequilibrium and Effective Population Size

LD was computed as squared correlations (r_{LD}^2) for all SNP pairs within chromosomes. The LD was high for markers close to each other, but decayed quickly with growing inter marker distance (Figure 2). The r_{LD}^2 was around 0.7 for very short inter-marker distances below 10 kb, but was 0.1 for marker distances at ~150 kb. After this it followed by a moderate decay until 10 Mb (r^2_{LD} = 0.03). The average inter-marker distance in our study was ~75 kb, with average r_{LD}^2 of 0.192 ± 0.254 for adjacent markers.

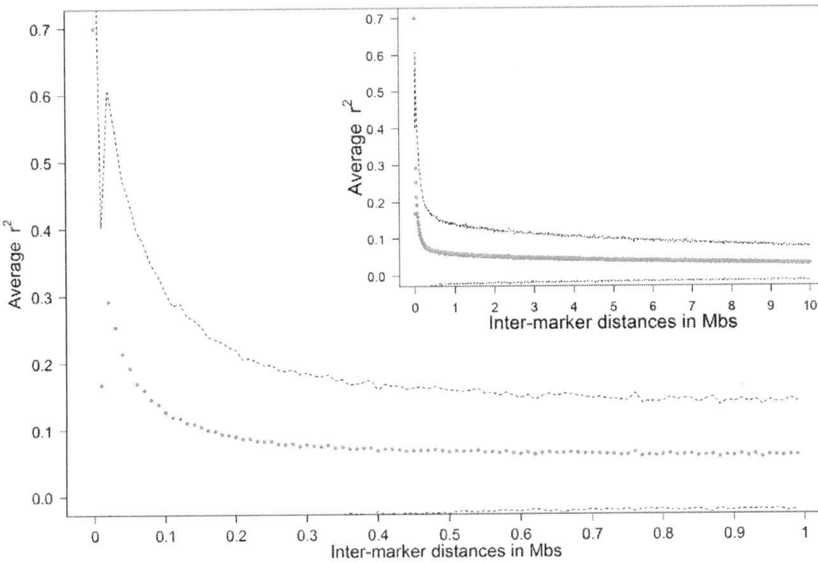

Figure 2. Average ld decay in the tyrol grey population, dashed lines show the standard deviation boundary.

The LD was used to estimate historical Ne (Figure 3). The method of calculation allows varying genetic distance and mutation occurrence, leading to slightly different results. In our case we calculated historical Ne based on the most likely scenario, i.e., considering mutations to occur and genetic distance per unit of physical distance (cM/Mb) of 1.25, according to Arias et al. (2009). The Ne_{LD} was around 200 about 20 generations in the past and declined to about 80 in the following 15 generations. The standard deviations of estimates show the uncertainty caused by slightly differing results from multiple LD windows pointing to the same generation.

3.2 Runs of Homozygosity

There was a considerable difference among animals in number of ROH segments and the length of the genome covered by these ROH segments

(Figure 4). For example when there was only a single ROH segment, this could be 8 to 60 Mb long. The cumulative length of the ROH segments of 60 Mb could be due to a single ROH segment, or the sum of 10 smaller ROH segments (Figure 4). Similar distributions were observed for other animals, with higher differences between total lengths of homozygous regions as the number of ROH increased. The age of inbreeding is defined as the time to the common ancestor and is quantified with the length of the ROH segments. Thus, the minimal ROH length of 4 Mb implies a common ancestor dating 12 generations in the past. Similarly the minimal ROH length of 8 and 16 Mb implies a common ancestor dating 6 and 3 generations ago, respectively.

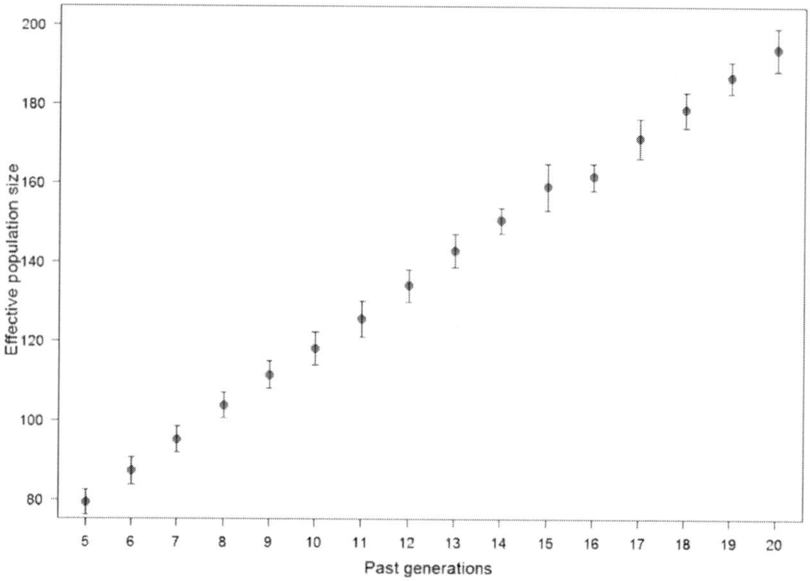

Figure 3. Means and standard deviations for the historical ne_{ld} in tyrol grey cattle, accounting for mutation and ratio of genetic per physical distance of 1.25.

The summary statistics for the three ROH ($F_{ROH>4\ Mb}$, $F_{ROH>8\ Mb}$, and $F_{ROH>16\ Mb}$) and one pedigree (F_{PED}) inbreeding coefficients are presented in Table 1 and in Figure 5. Considering pedigree depth of the Tyrol Grey population, the values obtained are in agreement with the assumption that $F_{ROH>4\ Mb}$, $F_{ROH>8\ Mb}$, and $F_{ROH>16\ Mb}$ correspond to the reference ancestral population where the common ancestors are approximately considered to be 12, 6, and 3 generations remote as well as with values obtained in other populations (Ferenčaković et al., 2013). Animals with extreme pedigree inbreeding, for example after the threshold where $F_{PED} > 0.05$, were precisely identified by $F_{ROH>8\ Mb}$. The two peaks in Figure 5 are caused by the fact that $F_{ROH>16\ Mb}$ values cannot be smaller than 0.006, i.e., 16 Mb divided by genome covered by SNPs on the chip.

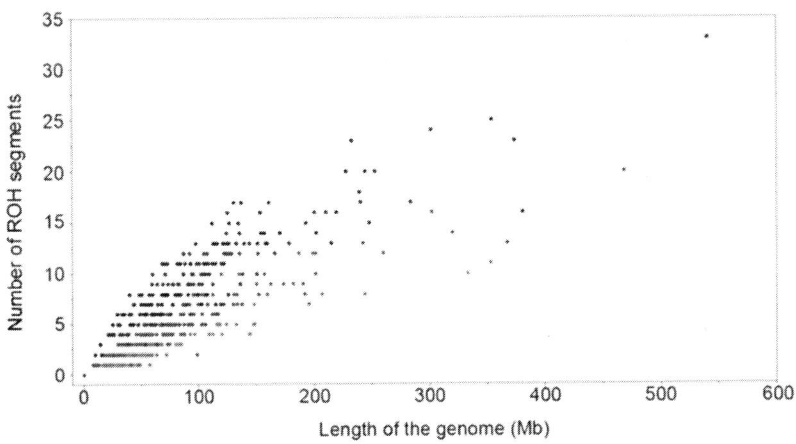

Figure 4. Number of roh segments and the length of the genome covered by roh segments (minimum roh length set to 4 mb in black, 8 mb in blue and 16 mb in red).

Table 1. Levels of inbreeding (f) with lower and upper 95% confidence intervals (l95ci, u95ci), change of inbreeding per generation (δf) and inbreeding effective population size [ne, with ne = 1/(2δf)].

Statistic	ROH > 4 Mb	ROH > 8 Mb	ROH > 16 Mb	Pedigree
F	0.040	0.029	0.016	0.024
L95CI	0.036	0.025	0.014	0.021
U95CI	0.044	0.032	0.019	0.027
ΔF	0.004	0.003	0.001	0.005
Ne	125	186	370	102

The relationship between the number of ROH segments and the length of the genome covered by ROH is shown in Figure 4. A considerable difference among animals has been found in number of ROH segments and the length of the genome covered. Animals with the same total ROH inbreeding ($F_{ROH>4\ Mb}$) might have a different number of ROH segments but with different lengths, which is a consequence of the different distances from the common ancestors.

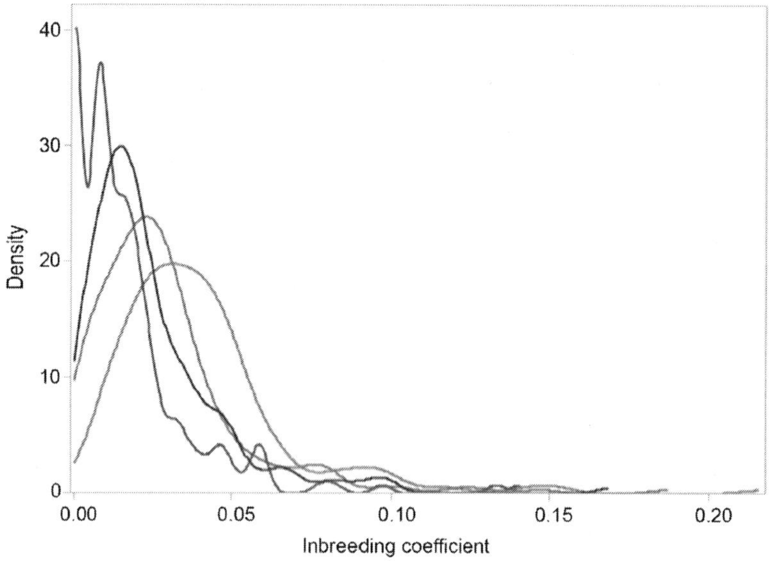

Figure 5. Distributions of three roh ($f_{roh>4\ mb}$ in magenta; $f_{roh>8\ mb}$ in brown and $f_{roh>16\ mb}$ in green) and one pedigree (f_{ped} in black) inbreeding coefficients for the tyrol grey cattle.

Ne_{ROH} derived from change of inbreeding levels per generation (ΔF) is lowest when estimated from pedigree information and increases with restriction to longer ROH segments (see Table 1). The very high Ne_{ROH} ($Ne_{ROH>16\ Mb}$ = 370) indicates strict avoidance of close inbreeding (like half sib, parent-offspring or first cousin mating) by the Tyrol Grey cattle breeders.

We have identified three regions with outlying ROH frequencies (4 Mb threshold) on chromosomes 5, 6, and 8 (Figure 6). Regions with increased ROH frequencies, the highest genomic autozygosity, are most likely consequences of selection as shown by Kim et al. (2013) and in computer simulations by Curik et al. (2002). The first region with the highest signal on BTA 8 starts at position 32,450,361 (BTB-00258020) and ends at position 46,041,080 (SNP BTA-28204-no-rs). Second region positioned on BTA 6 starts at position 36,277,967 (BTA-97637-no-rs) and ends at position 41,123,393 (SNP BTB-00406718). Finally, the region on BTA 5 starts at position 34,101,843 (BTB-01495784) and ends at position 42,918,584 (BTA-73464-no-rs). There are 147, 23, and 38 genes with known or unknown function within the signal regions on BTA 8, 6, and 5, respectively.

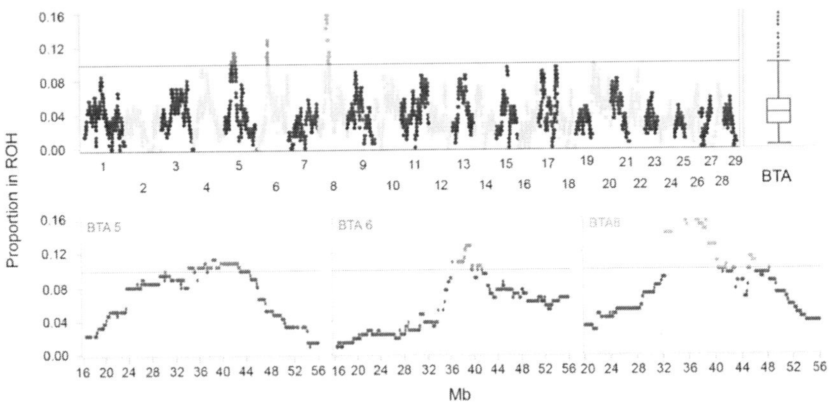

Figure 6. Autozygosity islands, regions with extreme roh frequency in tyrol grey cattle (minimum roh length set to 4 mb).

3.3 Genomic Selection

Genomic breeding values for Tyrol Grey bulls using major production and functional traits were computed. For the production traits the breeding values (EBV) and deregressed breeding values (dEBV) for milk yield, fat yield and content, protein yield and content were considered. For functional traits EBVs and dEBVs for longevity, persistency, maternal fertility, somatic cell count, and milking speed were included.

Single breed evaluations were used with only old bulls and young bulls born before 2003 as reference population. The validation animals consisted of bulls born after 2003. The number of animals in the validation set differed based on the trait, but in general they were between 36 and 49 when EBVs were used, and between 29 and 42 when dEBVs were used as response variable. The results are shown in Figures 7A,B. In general the average accuracies ranged from 0.13 to 0.91. Accuracy for the production traits in kg (fat kg, protein kg) was lower than using their direct counterpart measured in percent (fat% and protein%). This has largely been attributed to higher heritabilities for fat and protein content, compared to fat and protein production. For almost all functional traits however, the correlations were much higher compared to any of the previously reported results in the literature. In fact on average they were even higher than that of the production traits. The follow up bootstrapping generated large standard errors for all traits.

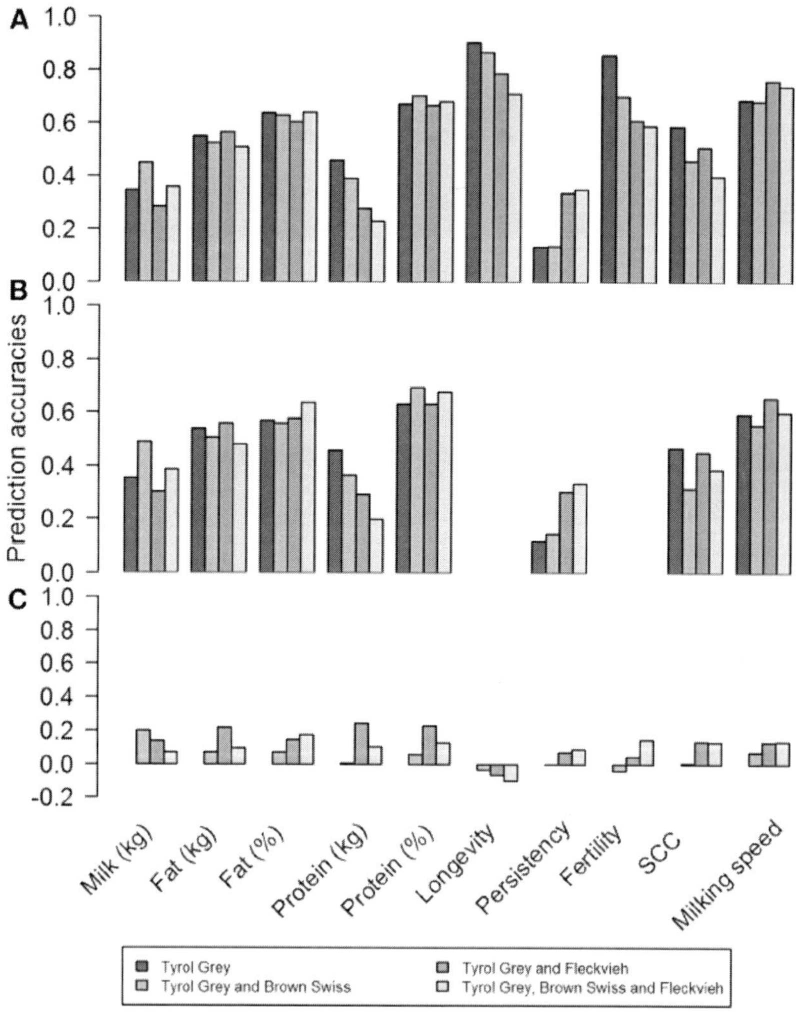

Figure 7. Correlations between estimated breeding values and genomic breeding values based on (a) ebv for single and multi-breed reference sets; (b) debv for single and multi-breed reference sets; (c) ebv for across breed reference sets.

In order to improve accuracies for production traits, a multi-breed approach was undertaken by adding genotypes to reference population from other breeds. Just like in the single breed scenario, validation individuals consisted of bulls born after 2003. In theory the increase in the size of reference population should increase the prediction accuracy. However, the gains and losses in accuracies varied considerably, depending on the trait, when either Brown Swiss, Fleckvieh or both populations) were added to the Tyrol Grey reference. In general, for all functional traits, adding other breeds resulted in lower accuracies, except for persistency.

When EBVs were used as response variable (Figure 7A) the impact of adding Fleckvieh into the reference set was favorable for persistency. For all other traits the results did not improve or were even worse with mixed reference sets. The benefits, if any, were not consistent across traits, showing a different pattern for each trait. The longevity and maternal fertility traits could not be evaluated due to lack of bulls with deregressed breeding values with reliability over 0.3 (Figure 7B).

An additional scenario to predict GEBV of Tyrol Grey bulls from another breed (Fleckvieh and Brown Swiss bulls) was studied. Unlike the multi-breed approach, an across breed scenario meant that, the reference population to estimate marker effect were either Fleckvieh or Brown Swiss bulls, while the validation set was the entire population of Tyrol Grey bulls. The correlations between the EBVs and estimated $GEBV_{EBV}$ were very low (Figure 7C). In general, the correlations were somewhat higher with Fleckvieh bulls in the reference set, but still remained below 0.25 in all cases. For longevity, predicting GEBVs from both estimates of marker effects from Brown Swiss and Fleckvieh resulted in negative accuracies. Moreover, the accuracies obtained in with this scenario were lower than that of the single breed or multi-breed approach.

Bootstrapping was used to assess the degree of confidence in the GEBV accuracies. It showed very wide confidence intervals for estimated GEBV for almost all traits (Table 2). Contrary to expectations, the confidence intervals for both longevity and fertility remained very high.

In addition to estimating the correlation between GEBVs and EBVs, we also computed the correlation between GEBVs and parent averages (Table 3). High correlations signify that the estimated GEBVs only predict the part of EBVs estimated as parent averages [0.5 (*EBVsire* + *EBVdam*)]. The correlations between the GEBVs and the parent averages for EBVs and dEBVs were very moderate to high. These correlations are the highest for longevity and fertility, indicating that the high GEBV accuracies were driven by parent averages. In other words, there is no advantage of GEBVs over parent averages for longevity and fertility, and relatively little advantage for other traits in the Tyrol Grey population.

Table 2. Mean accuracies of gebvs computed from ebvs and debvs from the single breed scenario and their 95% confidence intervals computed based on 10,000 bootstrap samples.

Traits	r(EBV,GEBV)			r(dEBV,GEBV)		
	Mean	Lower CI	Upper CI	Mean	Lower CI	Upper CI
Milk yield	0.345	0.027	0.675	0.354	0.050	0.668
Fat yield	0.549	0.349	0.755	0.539	0.335	0.755
Fat content	0.637	0.403	0.902	0.569	0.311	0.864
Protein yield	0.460	0.205	0.727	0.458	0.193	0.741
Protein content	0.673	0.493	0.869	0.632	0.445	0.830
Longevity	0.906	0.859	0.950	-	-	-
Persistency	0.133	−0.101	0.364	0.117	-0.139	0.373
Fertility	0.858	0.777	0.944	-	-	-
SCC	0.589	0.383	0.804	0.468	0.188	0.747
Milking speed	0.688	0.521	0.856	0.594	0.343	0.849

Table 3. Average reliabilities of validation animals and correlations between parent averages based on ebv and gebv/pa based on debv and gebv.

Traits	Average EBV reliability	r(PA-EBV, GEBV-EBV)	r(PA-dEBV, GEBV-dEBV)
Milk yield	0.74	0.67	0.64
Fat yield	0.74	0.77	0.73
Fat content	0.74	0.79	0.71
Protein yield	0.74	0.73	0.69
Protein content	0.74	0.82	0.76
Longevity	0.26	0.81	NA
Persistency	0.70	0.39	0.35
Fertility	0.30	0.92	NA
SCC	0.58	0.75	0.62
Milking speed	0.57	0.88	0.86

4. DISCUSSION

4.1 Genomic Analysis in Tyrol Grey Cattle

Traditionally the research interests in small and endangered populations are in genetic diversity parameters and breed conservation efforts. The justification is to describe breeding resources which might be important for coping with future needs and for facilitating the sustainable use of marginal areas (Toro et al., 2009). Microsatellites were a popular tool to describe genetic diversity (Baumung et al., 2004). In addition to microsatellites, SNP markers have been used to describe genetic diversity via parameters like allelic richness, heterozygosity/homozygosity levels (Makina et al., 2014), or LD and the associated Ne_{LD} (Hill, 1981; Hayes et al., 2003; Tenesa et al., 2007; Medugorac et al., 2009; Flury et al., 2010).

LD, measured as the correlation between alleles, is a fundamental concept in molecular genetics, while a large number of genomic methodologies are highly dependent on it (McKay et al., 2007; Pérez O'Brien et al., 2014). A typical LD pattern was observed in our study, with high LD for markers close to each other, quickly decaying with increasing inter-marker distance. Similar patterns were observed also in other studies (de Roos et al., 2008; Flury et al., 2010; Qanbari et al., 2010). In addition to the genome wide scale it is possible to utilize the LD information on the gene level. An example of this approach was the description of the entire genetic variability of a meat tenderness gene with only 16 polymorphic SNPs and 18 haplotypes in three French cattle breeds (Marty et al., 2010).

LD can be used to calculate Ne_{LD} (Hill, 1981), even when the pedigree information is missing or it is incomplete. As the Ne size is sometimes used as a criterion to determine the endangerment status of a breed and thus always of interest. The Ne_{LD} relies on assumed impact of mutation and recombination distance (McEvoy et al., 2011), thus neglecting mutation rate or approximating the recombination distance to 1 Mb ≈ 1 cM leads to different outcomes (Corbin et al., 2012).

We note here that, calculation of Ne based on genomic data is deemed controversial. The Ne_{LD} was nearly the same in two Finnish pig populations when compared to pedigree data (Uimari and Tapio, 2011), much lower in a Swiss cattle breed (Flury et al., 2010) and strongly biased upwards in a Spanish pony population (Goyache et al., 2011). Simulation studies showed a downward bias for Ne_{LD} (Sved et al., 2013). As there are several theoretical conflicts in the estimation procedure, extreme caution is advised when calculating Ne_{LD} (Goyache et al., 2011; Corbin et al., 2012).

Several other methods were developed to overcome some of the limitations of Ne_{LD}. The most popular approaches are chromosome segment homozygosity (Hayes et al., 2003) and calculation of Ne based on inbreeding rate per generation calculated from ROH (MacLeod et al., 2013, 2009; Curik et al., 2014). Here we have presented the estimation of several Ne_{ROH} depending on the three ROH length thresholds. The method is direct extension of the estimation of Ne based on pedigree inbreeding coefficient and has not been evaluated empirically. While the values obtained for $Ne_{ROH>4\ Mb}$ are close to, although somewhat higher, Ne_{LD} and Ne_{PED} broader empirical evaluation of the method is required for its comprehensive understanding. We would like to point out that Ne_{ROH} and Ne_{LD} are two conceptually different estimates that can supplement and/or substitute Ne_{PED} estimates and provide useful information for the conservation management of a population in question. The Ne based on genomic information could be directly applied, for example in determining the risk status of breeds, as the current method used by FAO relies solely on number of male and female animals.

In the genomic selection era, SNP information is predominantly used for breeding value estimation. The popularity of genomic selection (Meuwissen et al., 2001; Hayes et al., 2009) resulted to the routine genotyping of young bulls in several large breeds (e.g., Holstein, Fleckvieh). These genotypes accumulate to an ever growing reference population which is subsequently re-used to estimate SNP effects to improve the accuracy of GEBV (Van Raden et al., 2011a). These large reference populations allow the genomic selection to be so successful (Misztal, 2011). Given the small population size of the Tyrol Grey cattle, and many other small breeds, the size of the reference population will not be high enough to meet standards of large breeds, especially in reference population size. In order to increase the reference population size other breeds are sometimes added to the breed of interest (multi-breed) or used entirely alone (across breed) to calculate marker effects in genomic evaluations.

Estimated breeding values and deregressed breeding values for a range of traits were used to assess the feasibility of genomic selection is the Tyrol Grey breed. Surprisingly high accuracy of GEBV was obtained in the single breed evaluations (Figure 7A) and especially for longevity and fertility. Based on our criteria of discarding records with dEBV reliability below 0.30, GEBV were not estimated for longevity and fertility when using dEBVs as pseudo-phenotypes. The reason for the high accuracy of prediction for $GEBV_{EBV}$ was that, reliabilities of EBVs for young bulls were low. With low reliabilities, EBVs were similar to parent averages. This was affirmed by the high correlation between EBVs and parent averages (Table 3). Similarly large reliabilities were reported by Morota et al. (2014) when GEBV were correlated to low reliable EBVs.

In small breeding populations, the opportunities to obtain a sufficiently large number of daughters to generate highly reliable EBVs in a progeny testing scheme are limited. The problem is compounded especially for lowly heritable traits. For example, with trait heritability of 0.34 (milk yield), about 100 daughter records are need achieve reliabilities of 0.90. With heritability of 0.12 for longevity and 0.02 for fertility in our data set, about 300 and 1840 daughter records would be need. As shown in this study, predictive ability of a forward prediction scheme using young bulls as validation set was unusually high from PA driven EBVs. Lower reliabilities has been reported for the same traits with similar heritability in other large population breeds such as Fleckvieh (Ertl et al., 2014) or Holstein (Olson et al., 2012). The results from the study affirms the idea that, validation animals should have reliable EBVs if predictive ability is computed based on the correlation between GEBV and EBV. An alternative to this problem would be to use a single-step GBLUP approach (Legarra et al., 2014). Reliabilities would be computed based on the inverse of the diagonal element of the MME (Henderson, 1975). Reliabilities computed using the single step GBLUP approach could be compared to reliabilities of parent averages. Potential benefit of use of genomic information could be directly measured.

Multi-breed reference populations for genomic prediction are highly dependent on the LD and structure and genetic distance between breeds. The accuracy of genomic prediction could be substantially improved when the breeds are genetically very close or when animals of the same breed from multiple countries are pooled (Lund et al., 2014). Also using Bayesian variable selection instead of the BLUP approach could be beneficial in case of more distantly related breeds (Erbe et al., 2012; Bolormaa et al., 2013; Zhou et al., 2014). Lower accuracies in multi-breed genomic evaluation can be attributed to extent and differences in LD between markers and QTL (Goddard and Hayes, 2009), phase and allele substitution effects of QTLs (Spelman et al., 2002; Thaller et al., 2003).

In order to demonstrate the across breed genomic evaluation in Tyrol Grey cattle the German-Austrian Fleckvieh and Brown Swiss genotype pools were used in the evaluation. Using the large reference population composed of these two breeds to estimate GEBV was not successful. As shown in Figure 7C, the accuracies were very low for all traits. The accuracies have improved when part of the Tyrol Grey bulls were included into the reference set. This multi-breed reference however, did not have a clear advantage over the accuracies from the small breed reference set, similarly to Karoui et al. (2012).

A large population size is generally perceived as a requirement to estimate reliable GEBV, as we highlighted in the introduction of this paper. When the population is below of a critical mass the EBVs will be driven by parent averages, therefore genomic selection techniques will bring little new information into genetic evaluation of small breeds, as demonstrated in our paper. On the other hand, if reliable pedigrees are not available in a certain breed, i.e., no conventional breeding value estimation can be done, the breeding values estimated based on genomic data are a secure way to improve the breed.

4.2 Additional Uses of Genomic Data for Management of Small and Endangered Breeds

Even if genomic selection methods produce uncertain results in small breeds, there are a number of other reasons why a routine genotyping of the population would be beneficial. The identification of relatedness and inbreeding levels in the population has one of the biggest advantages from the practical perspective. The genomic relationship matrix can uncover family structures and infer relatedness within the population (Supplement Figure 1), even if the pedigree information is missing or it is incomplete (Calus et al., 2011). The correctness of the existing pedigrees can be verified comparing genomic information, e.g., by checking for Mendelian inconsistencies to identify incorrect parent-offspring relationships.

Similarly to the genomic relatedness it is possible to calculate the inbreeding coefficient. Compared to non-genomic approaches, here the knowledge of the pedigree is not needed, and so equally good results can be produced for animals, whose pedigree is dubious, incomplete, or entirely missing. Runs of homozygosity extend the analysis of relatedness between two individuals by identifying long homozygous segments, supposedly coming from the same ancestor somewhere in the past, inferring the individual's inbreeding coefficient. Based on the length of the segments it is possible to identify the number of generations to the common ancestor. In this study, the mean level of inbreeding estimated from three different ROH lengths ($Ne_{ROH>4\ Mb}$, $Ne_{ROH>8\ Mb}$, and $Ne_{ROH>16\ Mb}$) ranged from 0.016 to 0.029 and were comparable to other studies. In addition, there were only four individuals with outlying inbreeding coefficients higher than 0.125, indicating that potential risks could have been even more reduced with genomic information available. The utilization of genomic information to control inbreeding as well as to reduce early embryonic loss or appearance of congenital genetic defects due to recessive haplotypes in homozygous state (see more detailed discussion below) seems promising.

Crossbreeding is a very common strategy to increase the productivity of a breed or to introduce a desirable quality from another breed. The levels of crossbreeding are traditionally computed based on pedigree information. The pedigree approach assumes that the genetic composition of individuals with the same type of ancestry information is equal. This assumption does not hold however, as recombinations alter the composition of ancestral chromosomes, resulting into different admixture levels (Bryc et al., 2010). The Girolando cattle for example were bred to achieve a 5/8 of Holstein and 3/8 of Gir cross. Based on the pedigree information the expected Holstein admixture level is 62.5%. The real admixture levels based on SNP data can vary as much as 49–85% (Orazietti et al., 2014, unpublished). The adaptability of breeds can be also increased by producing optimal composites for a specific region. For example, introducing the alleles that are responsible for the trypanotolerance in Baoule cattle into the genomes of the trypano susceptible zebu populations in Burkina Faso would be a great advantage (Soudré et al., 2013). In other small populations the crossbreeding with large commercial populations could be a concern due to the loss of purebred stock. In all cases SNP chip data provides reliable estimates of the admixture levels which facilitates the selection of the desirable genotypes for breeding purposes (Frkonja et al., 2012). Furthermore, the genomic information could be used to purge the foreign genome from a small population (Amador et al., 2014).

A less frequent, but a much more critical utilization of genomic data is detection of lethal or sub-lethal genotypes. The obvious case is when a disorder is found in a population and an attempt is made to discover its source and genetic background by *ad hoc* genotyping of affected individuals. A very good example for this *ad hoc* approach was the disorder similar to bovine progressive degenerative myeloncephalopathy (weaver syndrome) in the Tyrol Grey population. As the purebred population is small, the disorder would have had a devastating effect. The region with the causative mutation was identified combining homozygosity mapping (Charlier et al., 2008) and other genome wide association techniques in 14 affected and 27 control animals. More detailed analysis allowed pinpointing the causal mutation in the mitofusin (MFS2) gene. Routine genotyping of breeding animals identifies any carriers and will purge the population from this mutation within a short period (Drögemüller et al., 2011).

The previous case demonstrated an efficient an identification of causal variants for a known disorder. If the disorder itself or its symptoms were less obvious however, the detection of affected animals may be much more difficult. To detect these cases it is possible to screen the whole population genotype data. Alleles with relatively high heterozygote frequency in the population, but without the occurrence of both homozygotes indicate lethal variants. Eleven candidate haplotypes were detected using this technique in the North American Holstein, Jersey, and Brown Swiss population, some of them with confirmed phenotypic effects (Van Raden et al., 2011b). Similar technique was used to identify homozygote deficient haplotypes with potentially negative effects on fertility traits in Nordic Holstein (Sahana et al., 2013) and Jersey (Sonstegard et al., 2013). In most cases the frequency of carrier animals with harmful genomic regions in heterozygous state is relatively low, but it can also be surprisingly widespread. In Finnish Red cattle a region associated with embryonic death had a frequency of 32% in the population, due to its positive effect on milk yield (Kadri et al., 2014). In general, the genotype screening allows the detection of new disorders or to confirm the causative sites of known defects. These disorders and defects can be then avoided in subsequent generations by planned mating of carriers and non-carriers, or even eradication of certain disorders from the breed by restricted usage of carrier genotypes.

5. CONCLUSIONS

In a very short time, high-throughput molecular information has become a standard tool in animal breeding. Routine genotyping of the entire male population in small breeds is often not in place, although it would be feasible due to the small population size and extreme reduction in genotyping price. Our results suggest that genomic selection is not readily applicable in small breeds even with very large reference populations in a multi breed setting. There are numerous other utilizations of the genomic information however, that make routine genotyping not only beneficial but outright desirable for the management of small breeds. Apart from various genetic diversity measures, the identification of regions identical by descent instead of approximations according to the pedigree will help to better understand relatedness and inbreeding in the population. Furthermore, the pool of genotypes for the entire breed enables to continuously scan the population and allow a swift reaction in identifying carriers of lethal or potentially harmful haplotypes. The new information can be used to eliminate undesirable alleles through the mating process. Similarly, the breed proportions due to admixture could be estimated with the goal to fix a desirable ratio or to preserve the purity of the breed.

While our paper describes an example from the Tyrol Grey population, we would like to stress that the recommendations are valid for all small and endangered breeds. The genotyping of SNP markers is a mature and well understood technology, with uses that can complement, improve or even replace approaches for breed management. Therefore, we suggest the continuous collection of genotypes and their use in breed monitoring and improvement.

ACKNOWLEDGMENTS

In memory of Otto Hausegger (1964–2014), managing director of the Tiroler Grauviehzuchtverband. The authors would like to thank the Tiroler Grauviehzuchtverband for the cooperation in the genotyping of the Tyrol Grey population. We are also grateful to Förderverein Biotechnologieforschung, Rinderbesammungsgenossenschaft Memmingen, Gesellschaft zur Förderung der Fleckviezucht in Niederbayern, Nutzvieh GmbH Miesbach, Rinderunion Baden-Württemberg eG, Zentrale Arbeitsgemeinschaft Österreichischer Rinderzüchter, Arbeitsgemeinschaft Süddeutscher Rinderzucht- und Besamungsorganisationen for providing the Fleckvieh and Brown Swiss genotype data and ZuchtData EDV-Dienstleistungen GmbH for providing the phenotype information. The Vienna Scientific Cluster is acknowledged for providing computing resources for part of the analyses. The authors are grateful for the financial support of the Austrian Ministry for Transport, Innovation and Technology (BMVIT) and the Austrian Science Fund (FWF) via the project TRP46-B19.

REFERENCES

1. Amador, C., Hayes, B. J., and Daetwyler, H. D. (2014). Genomic selection for recovery of original genetic background from hybrids of endangered and common breeds. *Evol. Appl.* 7, 227–237. doi: 10.1111/eva.12113
2. Arias, J. A., Keehan, M., Fisher, P., Coppieters, W., and Spelman, R. (2009). A high density linkage map of the bovine genome. *BMC Genet.* 10:18. doi: 10.1186/1471-2156-10-18
3. Baumung, R., Simianer, H., and Hoffmann, I. (2004). Genetic diversity studies in farm animals – a survey. *J. Anim. Breed. Genet.* 121, 361–373. doi: 10.1111/j.1439-0388.2004.00479.x
4. Bolormaa, S., Pryce, J. E., Kemper, K., Savin, K., Hayes, B. J., Barendse, W., et al. (2013). Accuracy of prediction of genomic breeding values for residual feed intake and carcass and meat quality traits in Bos taurus, Bos indicus, and composite beef cattle. *J. Anim. Sci.* 91, 3088–3104. doi: 10.2527/jas.2012-5827
5. Bryc, K., Velez, C., Karafet, T., Moreno-Estrada, A., Reynolds, A., Auton, A., et al. (2010). Genome-wide patterns of population structure and admixture among Hispanic/Latino populations. *Proc. Natl. Acad. Sci. U.S.A.* 107, 8954–8961. doi: 10.1073/pnas.0914618107
6. Calus, M. P. L., Mulder, H. A., and Bastiaansen, J. W. M. (2011). Identification of Mendelian inconsistencies between SNP and pedigree information of sibs. *Genet. Sel. Evol.* 43:34. doi: 10.1186/1297-9686-43-34
7. Charlier, C., Coppieters, W., Rollin, F., Desmecht, D., Agerholm, J. S., Cambisano, N., et al. (2008). Highly effective SNP-based association mapping and management of recessive defects in livestock. *Nat. Genet.* 40, 449–454. doi: 10.1038/ng.96
8. Corbin, L. J., Liu, A. Y. H., Bishop, S. C., and Woolliams, J. A. (2012). Estimation of historical effective population size using linkage disequilibria with marker data. *J. Anim. Breed. Genet.* 129, 257–270. doi: 10.1111/j.1439-0388.2012.01003.x
9. Curik, I., Ferenčaković, M., and Sölkner, J. (2014). Inbreeding and runs of homozygosity: a possible solution to an old problem. *Livest. Sci. Genomics Appl. Livest. Prod.* 166, 26–34. doi: 10.1016/j.livsci.2014.05.034

10. Curik, I., Sölkner, J., and Stipic, N. (2002). Effects of models with finite loci, selection, dominance, epistasis and linkage on inbreeding coefficients based on pedigree and genotypic information. *J. Anim. Breed. Genet.* 119, 101–115. doi: 10.1046/j.1439-0388.2002.00329.x
11. Drögemüller, C., Reichart, U., Seuberlich, T., Oevermann, A., Baumgartner, M., Kühni Boghenbor, K., et al. (2011). An unusual splice defect in the Mitofusin 2 gene (MFN2) is associated with degenerative axonopathy in tyrolean grey cattle. *PLoS ONE* 6:e18931. doi: 10.1371/journal.pone.0018931
12. de Roos, A. P. W., Hayes, B. J., Spelman, R. J., and Goddard, M. E. (2008). Linkage disequilibrium and persistence of phase in Holstein–Friesian, Jersey and Angus Cattle. *Genetics* 179, 1503–1512. doi: 10.1534/genetics.107.084301
13. Erbe, M., Hayes, B. J., Matukumalli, L. K., Goswami, S., Bowman, P. J., Reich, C. M., et al. (2012). Improving accuracy of genomic predictions within and between dairy cattle breeds with imputed high-density single nucleotide polymorphism panels. *J. Dairy Sci.* 95, 4114–4129. doi: 10.3168/jds.2011-5019
14. Ertl, J., Edel, C., Emmerling, R., Pausch, H., Fries, R., and Götz, K.-U. (2014). On the limited increase in validation reliability using high-density genotypes in genomic best linear unbiased prediction: observations from Fleckvieh cattle. *J. Dairy Sci.* 97, 487–496. doi: 10.3168/jds.2013-6855
15. Ferenčaković, M., Sölkner, J., and Curik, I. (2013). Estimating autozygosity from high-throughput information: effects of SNP density and genotyping errors. *Genet. Sel. Evol.* 45:42. doi: 10.1186/1297-9686-45-42
16. Flury, C., Tapio, M., Sonstegard, T., Drögemüller, C., Leeb, T., Simianer, H., et al. (2010). Effective population size of an indigenous Swiss cattle breed estimated from linkage disequilibrium. *J. Anim. Breed. Genet.* 127, 339–347. doi: 10.1111/j.1439-0388.2010.00862.x
17. Frkonja, A., Gredler, B., Schnyder, U., Curik, I., and Sölkner, J. (2012). Prediction of breed composition in an admixed cattle population. *Anim. Genet.* 43, 696–703. doi: 10.1111/j.1365-2052.2012.02345.x
18. Garrick, D. J., Taylor, J. F., and Fernando, R. L. (2009). Deregressing estimated breeding values and weighting information for genomic regression analyses. *Genet. Sel. Evol.* 41:55. doi: 10.1186/1297-9686-41-55
19. Goddard, M. E., and Hayes, B. J. (2009). Mapping genes for complex traits in domestic animals and their use in breeding programmes. *Nat. Rev. Genet.* 10, 381–391. doi: 10.1038/nrg2575
20. Goyache, F., Álvarez, I., Fernández, I., Pérez-Pardal, L., Royo, L. J., and Lorenzo, L. (2011). Usefulness of molecular-based methods for estimating effective population size in livestock assessed using data from the endangered black-coated Asturcón pony. *J. Anim. Sci.* 89, 1251–1259. doi: 10.2527/jas.2010-3620
21. Groeneveld, L. F., Lenstra, J. A., Eding, H., Toro, M. A., Scherf, B., Pilling, D., et al. (2010). Genetic diversity in farm animals – a review. *Anim. Genet.* 41, 6–31. doi: 10.1111/j.1365-2052.2010.02038.x
22. Gutiérrez, J. P., and Goyache, F. (2005). A note on ENDOG: a computer program for analysing pedigree information. *J. Anim. Breed. Genet.* 122, 172–176. doi: 10.1111/j.1439-0388.2005.00512.x
23. Hayes, B. J., Bowman, P. J., Chamberlain, A. J., and Goddard, M. E. (2009). Invited review: genomic selection in dairy cattle: progress and challenges. *J. Dairy Sci.* 92, 433–443. doi: 10.3168/jds.2008-1646
24. Hayes, B. J., Visscher, P. M., McPartlan, H. C., and Goddard, M. E. (2003). Novel multilocus measure of linkage disequilibrium to estimate past effective population size. *Genome Res.* 13, 635–643. doi: 10.1101/gr.387103
25. Henderson, C. R. (1975). Best linear unbiased estimation and prediction under a selection model. *Biometrics* 31, 423–447. doi: 10.2307/2529430
26. Hill, W. G. (1981). Estimation of effective population size from data on linkage disequilibrium. *Genet. Res.* 38, 209–216. doi: 10.1017/S0016672300020553

27. Howrigan, D. P., Simonson, M. A., and Keller, M. C. (2011). Detecting autozygosity through runs of homozygosity: a comparison of three autozygosity detection algorithms. *BMC Genomics* 12:460. doi: 10.1186/1471-2164-12-460
28. Hutchison, J. L., Cole, J. B., and Bickhart, D. M. (2014). Short communication: use of young bulls in the United States. *J. Dairy Sci.* 97, 3213–3220. doi: 10.3168/jds.2013-7525
29. Kadri, N. K., Sahana, G., Charlier, C., Iso-Touru, T., Guldbrandtsen, B., Karim, L., et al. (2014). A 660-Kb deletion with antagonistic effects on fertility and milk production segregates at high frequency in Nordic Red cattle: additional evidence for the common occurrence of balancing selection in livestock. *PLoS Genet.* 10:e1004049. doi: 10.1371/journal.pgen.1004049
30. Karoui, S., Carabaño, M. J., Díaz, C., and Legarra, A. (2012). Joint genomic evaluation of French dairy cattle breeds using multiple-trait models. *Genet. Sel. Evol.* 44:39. doi: 10.1186/1297-9686-44-39
31. Kim, E.-S., Cole, J. B., Huson, H., Wiggans, G. R., Van Tassell, C. P., Crooker, B. A., et al. (2013). Effect of artificial selection on runs of homozygosity in U.S. Holstein Cattle. *PLoS ONE* 8:e80813. doi: 10.1371/journal.pone.0080813
32. Legarra, A., Christensen, O. F., Aguilar, I., and Misztal, I. (2014). Single Step, a general approach for genomic selection. *Livest. Sci. Genomics Appl. Livest. Prod.* 166, 54–65. doi: 10.1016/j.livsci.2014.04.029
33. Lund, M. S., Su, G., Janss, L., Guldbrandtsen, B., and Brøndum, R. F. (2014). Genomic evaluation of cattle in a multi-breed context. *Livest. Sci. Genomics Appl. Livest. Prod.* 166, 101–110. doi: 10.1016/j.livsci.2014.05.008
34. MacLeod, I. M., Larkin, D. M., Lewin, H. A., Hayes, B. J., and Goddard, M. E. (2013). Inferring demography from runs of homozygosity in whole-genome sequence, with correction for sequence errors. *Mol. Biol. Evol.* 30, 2209–2223. doi: 10.1093/molbev/mst125
35. MacLeod, I. M., Meuwissen, T. H. E., Hayes, B. J., and Goddard, M. E. (2009). A novel predictor of multilocus haplotype homozygosity: comparison with existing predictors. *Genet. Res.* 91, 413–426. doi: 10.1017/S0016672309990358
36. Makina, S. O., Muchadeyi, F. C., van Marle-Koster, E., MacNeil, M. D., and Maiwashe, A. (2014). Genetic diversity and population structure among six cattle breeds in South Africa using a whole genome SNP panel. *Front Genet.* 5:333. doi: 10.3389/fgene.2014.00333
37. Marty, A., Amigues, Y., Servin, B., Renand, G., Levéziel, H., and Rocha, D. (2010). Genetic variability and linkage disequilibrium patterns in the bovine DNAJA1 gene. *Mol. Biotechnol.* 44, 190–197. doi: 10.1007/s12033-009-9228-y
38. McEvoy, B. P., Powell, J. E., Goddard, M. E., and Visscher, P. M. (2011). Human population dispersal "Out of Africa" estimated from linkage disequilibrium and allele frequencies of SNPs. *Genome Res.* 21, 821–829. doi: 10.1101/gr.119636.110
39. McKay, S. D., Schnabel, R. D., Murdoch, B. M., Matukumalli, L. K., Aerts, J., Coppieters, W., et al. (2007). Whole genome linkage disequilibrium maps in cattle. *BMC Genet.* 8:74. doi: 10.1186/1471-2156-8-74
40. McQuillan, R., Leutenegger, A.-L., Abdel-Rahman, R., Franklin, C. S., Pericic, M., Barac-Lauc, L., et al. (2008). Runs of homozygosity in European populations. *Am. J. Hum. Genet.* 83, 359–372. doi: 10.1016/j.ajhg.2008.08.007
41. Medugorac, I., Medugorac, A., Russ, I., Veit-Kensch, C. E., Taberlet, P., Luntz, B., et al. (2009). Genetic diversity of European cattle breeds highlights the conservation value of traditional unselected breeds with high effective population size. *Mol. Ecol.* 18, 3394–3410. doi: 10.1111/j.1365-294X.2009.04286.x
42. Meuwissen, T. H. E., Hayes, B. J., and Goddard, M. E. (2001). Prediction of total genetic value using genome-wide dense marker maps. *Genetics* 157, 1819–1829.
43. Misztal, I. (2011). FAQ for genomic selection. *J. Anim. Breed. Genet.* 128, 245–246. doi: 10.1111/j.1439-0388.2011.00944.x
44. Morota, G., Boddhireddy, P., Vukasinovic, N., Gianola, D., and De Nise, S. (2014). Kernel-based variance component estimation and whole-genome prediction of pre-corrected phenotypes and progeny tests for dairy cow health traits. *Front Genet.* 5:56. doi: 10.3389/fgene.2014.00056

45. Olson, K. M., Van Raden, P. M., and Tooker, M. E. (2012). Multibreed genomic evaluations using purebred Holsteins, Jerseys, and Brown Swiss. *J. Dairy Sci.* 95, 5378–5383. doi: 10.3168/jds.2011-5006
46. Pérez O'Brien, A. M., Mészáros, G., Utsunomiya, Y. T., Sonstegard, T. S., Garcia, J. F., Van Tassell, C. P., et al. (2014). Linkage disequilibrium levels in Bos indicus and Bos taurus cattle using medium and high density SNP chip data and different minor allele frequency distributions. *Livest. Sci. Genomics Appl. Livest. Prod.* 166, 121–132. doi: 10.1016/j.livsci.2014.05.007
47. Purcell, S., Neale, B., Todd-Brown, K., Thomas, L., Ferreira, M. A. R., Bender, D., et al. (2007). PLINK: a tool set for whole-genome association and population-based linkage analyses. *Am. J. Hum. Genet.* 81, 559–575. doi: 10.1086/519795
48. Qanbari, S., Pimentel, E. C. G., Tetens, J., Thaller, G., Lichtner, P., Sharifi, A. R., et al. (2010). The pattern of linkage disequilibrium in German Holstein cattle. *Anim. Genet.* 41, 346–356. doi: 10.1111/j.1365-2052.2009.02011.x
49. Sahana, G., Nielsen, U. S., Aamand, G. P., Lund, M. S., and Guldbrandtsen, B. (2013). Novel harmful recessive haplotypes identified for fertility traits in nordic holstein cattle. *PLoS ONE* 8:e82909. doi: 10.1371/journal.pone.0082909
50. Sonstegard, T. S., Cole, J. B., Van Raden, P. M., Van Tassell, C. P., Null, D. J., Schroeder, S. G., et al. (2013). Identification of a nonsense mutation in CWC15 associated with decreased reproductive efficiency in Jersey cattle. *PLoS ONE* 8:e54872. doi: 10.1371/ journal. pone.0054872
51. Soudré, A., Ouédraogo-Koné, S., Wurzinger, M., Müller, S., Hanotte, O., Ouédraogo, A. G., et al. (2013). Trypanosomosis: a priority disease in tsetse-challenged areas of Burkina Faso. *Trop. Anim. Health Prod.* 45, 497–503. doi: 10.1007/s11250-012-0248-4
52. Spelman, R. J., Ford, C. A., McElhinney, P., Gregory, G. C., and Snell, R. G. (2002). Characterization of the DGAT1 gene in the New Zealand dairy population. *J. Dairy Sci.* 85, 3514–3517. doi: 10.3168/jds.S0022-0302(02)74440-8
53. Sved, J. A., Cameron, E. C., and Gilchrist, A. S. (2013). Estimating effective population size from linkage disequilibrium between unlinked loci: theory and application to fruit fly outbreak populations. *PLoS ONE* 8:e69078. doi: 10.1371/journal.pone.0069078
54. Tenesa, A., Navarro, P., Hayes, B. J., Duffy, D. L., Clarke, G. M., Goddard, M. E., et al. (2007). Recent human effective population size estimated from linkage disequilibrium. *Genome Res.* 17, 520–526. doi: 10.1101/gr.6023607
55. Thaller, G., Krämer, W., Winter, A., Kaupe, B., Erhardt, G., and Fries, R. (2003). Effects of DGAT1 variants on milk production traits in German cattle breeds. *J. Anim. Sci.* 81, 1911–1918.
56. Toro, M. A., Fernández, J., and Caballero, A. (2009). Molecular characterization of breeds and its use in conservation. *Livest. Sci. Special Issue: Anim. Genet. Resour.* 120, 174–195. doi: 10.1016/j.livsci.2008.07.003
57. Uimari, P., and Tapio, M. (2011). Extent of linkage disequilibrium and effective population size in Finnish Landrace and Finnish Yorkshire pig breeds. *J. Anim. Sci.* 89, 609–614. doi: 10.2527/jas.2010-3249
58. Van Raden, P. M., O'Connell, J. R., Wiggans, G. R., and Weigel, K. A. (2011a). Genomic evaluations with many more genotypes. *Genet. Sel. Evol.* 43:10. doi: 10.1186/1297-9686-43-10
59. Van Raden, P. M., Olson, K. M., Null, D. J., and Hutchison, J. L. (2011b). Harmful recessive effects on fertility detected by absence of homozygous haplotypes. *J. Dairy Sci.* 94, 6153–6161. doi: 10.3168/jds.2011-4624
60. Zhou,L.,Heringstad,B.,Su,G.,Guldbrandtsen,B.,MeuwissenT.H.E.,Svendsen, M.,etal.(2014).Genomic predictions based on a join reference population fortheNordicRedcattlebreeds. *J.DairySci.* 97,4485–4496.doi:10.3168/jds. 2013-7580

CHAPTER 5

Sexual Dimorphism of miRNAs Secreted by Bovine *In vitro*-produced Embryos

Nicole Gross, Jenna Kropp and Hasan Khatib*

Department of Animal Sciences, University of Wisconsin, Madison, WI, USA

Sexual dimorphism of bovine blastocysts has previously been observed through differences in development, cell death, metabolism, telomere length, DNA methylation, and transcriptomics. However, dimorphism in the secretion of miRNAs to culture media has not yet been evaluated. The objectives of this study were to determine if sex-specific blastocyst miRNA secretion occurs and to further investigate the role these miRNAs may have in the interaction between a blastocyst and the maternal environment. *In vitro* embryo culture was performed and media from male and female blastocysts was collected into sex-specific pools. Profiling of 68 miRNAs revealed a total of eight miRNAs that were differentially expressed between female and male-conditioned media. Validation by qPCR confirmed higher expression of miR-22 ($P < 0.05$), miR-122 ($P < 0.05$), and miR-320a ($P < 0.05$) in female media for three additional biological replicates. To examine the potential roles of secreted miRNAs to the media in communication with the maternal environment, miR-22, miR-122, and miR-320a were each supplemented to four replicates of primary bovine endometrial epithelial cell culture. Uptake of miR-122 ($P < 0.05$) and miR-320a ($P < 0.05$) was detected, and a trend of uptake was detected for miR-22 ($P > 0.05$). Further, expression of the progesterone receptor transcript, a predicted target of all three miRNAs, was found to be upregulated in the cells following supplementation of miR-122 ($P < 0.05$) and miR-320a ($P < 0.05$), and a trend upregulation of the transcript was observed following miR-22 ($P > 0.05$) supplementation. This work demonstrates that male and female conceptuses are able to differentially secrete miRNAs at the blastocyst stage and that these miRNAs have the ability to induce a transcriptomic response when applied to maternal cells. This knowledge builds on the known dimorphic differences in conceptuses at the blastocyst stage and demonstrates a role for blastocyst-secreted miRNAs in cell–cell communication.

Keywords: microRNA, sexual dimorphism, embryo, culture media, bovine

1. INTRODUCTION

Mammalian embryos exhibit sexual dimorphism in development, genetics, and epigenetics (Laguna-Barraza et al., 2013). *In vitro* production systems employed to generate bovine embryos, have reported that male embryos develop faster (Avery et al., 1992) and have increased blastocyst rates (Camargo et al., 2010; Ghys et al., 2016) as well as higher total cell numbers (Xu et al., 1992; Ghys et al., 2016; Oliveira et al., 2016) compared to female embryos. On day 7 of development, *in vitro*-produced female bovine blastocysts show a higher incidence of cell apoptosis than male blastocysts (Ghys et al., 2016), with the earliest detection of this effect on day 6 (Oliveira et al., 2016). Dimorphism in terms of metabolic strategy is also a distinct; this is demonstrated by the differential uptake of amino acids between male and female embryos (Sturmey et al., 2010), and was shown in a study by Green et al. (2016) in which alteration of glucose availability in culture medium induced a stronger skew in sex ratio of bovine blastocysts.

When cultured under the same conditions, embryos exhibit sexual dimorphism in up to one-third of actively expressed genes (Bermejo-Álvarez et al., 2010). Further, when exposed to adverse culture conditions, male and female conceptuses exhibit differentially altered transcriptomes (Heras et al., 2016). Additionally, mitochondrial distribution, telomere length, and DNA methylation have shown sexual dimorphism in bovine embryos (Bermejo-Álvarez et al., 2008). Though differences have been detected between male and female blastocysts, a better understanding of the role these changes play in communication to the maternal environment is warranted.

To fully understand successful pregnancy, it is important to elucidate the signaling mechanisms, or dialog, between an embryo and the mother. A potential signaling mechanism involved in regulation of embryo development is the secretion of miRNAs into the extracellular environment. miRNAs are a family of short, single-stranded, non-coding RNAs that are approximately 19–25 nucleotides in length. These small molecules modulate sequence-specific mRNA transcription, leading to regulation of gene expression (Wahid et al., 2010). Sexually divergent miRNA expression may contribute to observed differences between male and female embryos. As shown on the miRbase archive[1] (release 21; Kozomara and Griffiths-Jones, 2014), the Bos taurus Y chromosome has no known miRNAs, whereas the X chromosome encodes 61 known miRNAs. miRNAs can be secreted out of a cell through extracellular vesicles, via apoptotic bodies, or by being bound to AGO proteins (Turchinovich et al., 2013). In cattle, the presence of circulating miRNAs has been reported in bodily fluids such as milk (Sun et al., 2015), blood, and follicular fluid (Noferesti et al., 2015) and in the *in vitro* culture media of blastocysts (Kropp et al., 2014; Kropp and Khatib, 2015). miRNAs have shown involvement in a variety of biological processes including organismal development, cell proliferation, cell death, hematopoiesis, and immunity (Wahid et al., 2010). Evidence that miRNAs modulate signaling networks, such as the purinergic network — an extracellular network mediated by nucleosides such as adenosine and ATP (Ferrari et al., 2016) — supports the potential of embryo-secreted miRNAs to serve as important pathway regulators.

Embryo-secreted miRNAs have demonstrated potential to affect endometrial transcriptomes. For example, miR-30b transfected into human endometrial epithelial cells induced transcriptomic changes in the cells (Ye et al., 2015). miRNA-661 has been shown to decrease maternal protein expression, and correlate with implantation outcome (Cuman et al., 2015). The ability of these miRNAs to alter uterine cell transcriptomes and protein expression further warrants a more comprehensive investigation of embryonic miRNAs as signaling molecules to the mother. However, it is not yet known whether miRNAs present in the culture medium of *in vitro*-produced embryos differ between male and female embryos. It is unknown if these miRNAs serve a function in the dialog between the embryo and the mother. We hypothesized that sexual dimorphism exists in the profiles of miRNAs secreted by blastocysts, and that these miRNAs serve as a signaling mechanism for male and female conceptuses to alter the gene expression profile in maternal endometrial cells.

2. MATERIALS AND METHODS

2.1 Ethics Statement

This study is exempt from approval of the institutional and national requirements of Animal Care and Use Committee because cows used for *in vitro* fertilization were not handled at the University of Wisconsin facilities. Ovaries used for embryo production were purchased from Applied

Reproductive Technology, LLC (Monona, WI, USA), and permission was granted by the company to perform *in vitro* fertilization using ovaries.

2.2 Experiment I: miRNA Profiling in Media

2.2.1 *In vitro* Embryo and Media Production

Embryos and media were procured as described by Kropp and Khatib (2015). Ovaries were obtained from Applied Reproductive Technology, LLC (Monona, WI, USA), and follicles were aspirated to recover oocytes. Oocytes were washed in Vigro TL-Hepes (Bioniche, Pullman, WA, USA) supplemented with 3% bovine serum albumin, sodium pyruvate, and gentamicin and placed in cohorts of 10 oocytes per 50 μl drop of maturation media for 24 h. Maturation media consisted of M-199 media supplemented with gonadotropins (FSH and LH), estradiol, sodium pyruvate, 10% fetal bovine serum and gentamicin. Following 24 h of maturation, cumulus-oocyte complexes were washed in Vigro TL-Hepes (Bioniche) and transferred in cohorts of 10 into a 44 μL drop of fertilization media supplemented with fatty acid-free bovine serum albumin (FAF-BSA), sodium pyruvate, and gentamicin. Sperm was prepared using a Percoll discontinuous gradient as described by Parrish et al. (1995), where the final concentration was adjusted to 1 million sperm per mL and 2 μL was added per drop. Additionally, 2 μl each of penicillamine-hypotaurine-epinephrine and heparin were added to each fertilization drop.

Gametes were co-cultured with sperm for 20 h at which point presumptive zygotes were stripped of their cumulus cells and washed in supplemented Vigro TL-Hepes (Bioniche). The presumptive zygotes were placed 25 per drop into a 50 μL drop of CR_{1aa} culture media (Rosenkrans et al., 1993; Sagirkaya et al., 2006) supplemented with FAF-BSA, sodium pyruvate, amino acids, and gentamicin. Embryos were then cultured until day 5, whereupon they were morphologically assessed for characteristics of the morula stage as described by Bó and Mapletoft (2013). Morulae were selected based on the appearance of a coalesced/compacted inner cell mass that consumed 60–70% of the perivitelline space. Those deemed as morula stage embryos were washed and placed individually into a 50 μL drop of CR_{1aa} medium lacking FAF-BSA supplementation.

On day 8 of development, individually cultured embryos were morphologically assessed for characteristics of the blastocyst stage of development. Embryos which developed a blastocoel cavity and demonstrated a distinct inner cell mass and outer trophectoderm were deemed blastocysts, whereas those which failed to develop to the blastocyst stage were deemed degenerate embryos. Blastocysts that were at the mid-to-expanded blastocyst stage and quality grades 1 and 2 (Bó and Mapletoft, 2013), as well as each embryo's respective conditioned culture media, were individually collected. Blastocysts were stored in 10 µL TE buffer and media was preserved in aliquots. Both embryos and media were stored at -20°C for subsequent procedures. Two different sires were used for embryo generation.

2.2.2 Sex Determination of Embryos by Nested PCR

Individual embryos were genotyped for sex determination as described by Kirkpatrick and Monson (1993). In brief, PCR was performed using the GoTaq® DNA Polymerase system (Promega Corporation, Madison, WI, USA). Reagents [5X Green GoTaq® Reaction Buffer, dNTPs, magnesium chloride and primers (Supplementary Table 1)], as well as proteinase K, were added to a PCR reaction containing one blastocyst embryo in 10 µl of TE buffer. To lyse each embryo, PCR reactions were incubated at the following temperatures: 50°C for 30 min, 95°C for 10 min, -20°C for 5 min, 95°C for 10 min, and -20°C for 5 min. Taq DNA polymerase was then added to each well and the first PCR was carried out with primers zfx/zfy (Aasen and Medrano, 1990; Kirkpatrick and Monson, 1993) under the following conditions: 95°C for 5 min, followed by 30 cycles of 95°C for 60 s, 55°C for 45 s, and 72°C for 60 s, and a final extension step of 72°C for 7 min. A nested PCR was then performed with primers specific to the X (zfx) and Y (zfy) chromosomes using the initial PCR product as template. The second PCR was cycled under the same conditions as the initial PCR. Embryo sex was confirmed by gel electrophoresis, where a single band was visualized for a female (X chromosome product of 246 base pairs) or two bands were visualized for a male (X chromosome band plus a Y chromosome product of 167 base pairs). A total of three IVF replicates were carried out to procure media samples derived from female and male embryos in which 102 blastocyst embryos were collected. Of the blastocysts collected, 57 blastocyst embryos were successfully genotyped, where 30 were male and 27 were female (Supplementary Table 2).

2.2.3 RNA Extraction from Culture Media and miRNA Profiling

For miRNA differential expression analysis between female and male embryos, a total of three pools of media were generated each for males (each pool consisted of media from 10 embryos) and females (each pool consisted of media from 9 embryos). Embryo pools were determined following IVF completion. Each pool included media from a unique set of individual embryos. Pools were designed to contain media of embryos from all three IVF runs, representing both sires equivalently across pools. RNA was extracted from each pool using a miRNeasy Serum/Plasma kit (Qiagen, Germantown, MD, USA). A total of three extractions were carried out per pool, with an initial input volume of 120–140 µL media sample, consisting of an equal volume of eluent from individual embryos. miRNA profiling was performed using the Firefly® Circulating miRNA Assay Immunology Panel

(ABCAM, Cambridge, MA, USA). The immunology panel includes 68 immune response-related miRNAs chosen by ABCAM based on known functions related to immune response and differential regulation in plasma or serum. We hypothesized that these miRNAs could play a role in the immune response of the mother to the developing embryo. Complementary oligonucleotides which encode hydrogel microparticles were hybridized to selected miRNAs. The oligonucleotide adapter served as a universal PCR priming site, allowing for fluorescent amplification of the target. Amplified products were then re-hybridized to original oligonucleotide particles, and an EMD Millipore Guava 8HT Flow Cytometer (Merck, Darmstadt, Germany) was used to quantify the hybridization. Measurements were performed in triplicates.

Statistical analysis of differentially expressed miRNAs was performed using the Firefly® Analysis Workbench Software (ABCAM) in which miRNA expression levels of male and female embryos were compared to each other. An unpaired *t*-test was used to determine significant differential expression ($P < 0.05$) of miRNAs between male and female embryo groups.

2.2.4 Validation of Differentially Expressed miRNAs Using Quantitative Real-time PCR (qRT-PCR)

For the validation of differentially expressed miRNAs detected by Firefly® particle technology, three miRNAs (miR-22, miR-122, and miR-320a) were chosen for further analysis by qRT-PCR. A total of five additional IVF replicates were performed to generate 146 embryos derived from one sire and different dams. Fertilization, embryo grading, media collection, and embryo genotyping were performed as described above. A total of 101 blastocysts embryos were successfully genotyped for validation, resulting in 51 male and 50 female blastocysts (Supplementary Table 2). Male and female pools of media, with volumes of 40–90 μL, were created and RNA was extracted as described above. The three male pools were derived from 17 embryos each, while three female pools were derived from 17, 16, and 16 embryos. Total RNA was reverse-transcribed using a miScript II RT kit (Qiagen) with HiSpec Buffer following the manufacturer's instructions. The qRT-PCR method was used to determine relative fold difference in miRNA expression between male and female pools. A miR-39 spike-in control (Qiagen) was added at the time of extraction to serve as an internal control. The miScript SYBR Green Kit (Qiagen), and specific primers (Supplementary Table 3) corresponding to each mature miRNA sequence (Qiagen) were used. Cycling conditions were carried out in a Bio-Rad iCycler real-time PCR machine as follows: 95°C for 15 min followed by 40 cycles of 94°C for 15 s, 55°C for 30 s, and 70°C for 30 s. C_T values > 33 were considered beyond the threshold for detection. The $2^{-\Delta\Delta C_T}$ method by Livak and Schmittgen (2001) was used to determine mean fold change in miRNA gene expression, where $\Delta\Delta C_T = (C_{T,\text{Target miRNA}} - C_{T,\text{miR-39}})_{\text{female media}} - (C_{T,\text{Target}}$

$_{miRNA}$ - $C_{T,miR-39})_{male\ media}$. The ΔC_T values were evaluated for significance using an unpaired *t*-test for each miRNA. The TargetScan software[2] (Lewis et al., 2005) was used to determine candidate target genes of validated miRNAs. Each miRNA was searched by name, under both cow and human for species selection. Top predicted targets were viewed. Those which were targeted in both cow and human, and additionally had known roles in embryo development were manually identified.

2.3 Experiment II: Uptake of miRNAs into Bovine Endometrial Cells

2.3.1 Primary Cell Culture of Bovine Endometrial Cells

To investigate the potential of validated miRNAs as signaling molecules to maternal tissues, a primary cell culture system was implemented. Bovine endometrial epithelial cells (BEECs) (Cell Applications Inc., San Diego, CA, USA) were cultured for 6–8 passages and seeded in a 96-well culture plate at 7000 cells per cm^2 per manufacturer instruction. Cells were cultured in Bovine Endometrial Epithelial Cell Growth Medium (Cell Applications Inc.) and passaged using Hank's Balanced Salt Solution, Trypsin/EDTA and Trypsin Inhibitor (Cell Applications Inc.), as directed.

2.3.2 miRNA Supplementation and Collection of Primary Endometrial Epithelial Cells

Synthetic mimics for the miRNAs miR-22, miR-122, and miR-320a (Qiagen) — specific to those found to be differentially expressed between male and female embryos — were supplemented to BEECs to assess whether these miRNAs modulate maternal gene expression. At 24 h post-passage of cells, 50 nM mimic miRNA was added to a well of a 96-well plate containing BEECs. Notably, no transfection reagent was used in order to more closely simulate an *in vivo* setting. A control of non-treated cells was simultaneously cultured. A total of four biological replicates were produced for each miRNA mimic treatment. After 24 h of co-culture with the miRNA mimic, the medium was aspirated from each well, cells were washed twice in 200 μL PBS, and lifted with trypsin. Cells were pelleted at 300 g for 5 min, and the excess supernatant was aspirated from each tube and discarded.

To evaluate the effect of miRNAs on gene expression in the endometrial epithelial cells, an additional experiment was performed, in which cells were treated with 1 μL of Lipofectamine 2000 (Thermo Fisher Scientific, Madison, WI, USA) in conjunction with 50 nM of miRNA and Opti-MEM medium, as specified by the manufacturer. A control of lipofectamine-only treated cells was cultured simultaneously.

2.3.3 Primary Endometrial Epithelial Cell miRNA and mRNA Extraction and Quantification

Total RNA was extracted from the BEECs using a MiRNeasy Mini Kit (Qiagen) to assess miRNA uptake by the maternal cells and expression of genes targeted by supplemented miRNAs, respectively. Reverse transcription and quantification of uptake of miRNAs in cells were performed as described for media. Reverse transcription of mRNA was carried out using an iScript cDNA kit (Bio-Rad, Hercules, CA, USA) and qRT-PCR to evaluate the expression of candidate mRNA targets was performed using iTaq Universal SYBR Green Supermix (Bio-Rad). The β-actin gene was selected as an internal control according to its stability across samples in comparison to GAPDH. Intron-spanning primers were designed for each target mRNA to avoid amplification of genomic DNA (Supplementary Table 4). The CFX Connect Real-Time PCR Detection System was used for mRNA quantification with the following cycling conditions: 95°C for 30 s, followed by 40 cycles of 95°C for 5 s, and 60°C for 30 s. Gene expression analysis was performed using the $2^{-\Delta\Delta C_T}$ method (Livak and Schmittgen, 2001), with a control group of non-treated cells. To assess the effects of supplementation and mRNA expression changes, biological replicates were standardized as previously described (Willems et al., 2008), maximally reducing inter-sample variation between cells. Following log transformation, mean centering, and autoscaling, an unpaired *t*-test was performed to assess significance ($P < 0.05$).

3. RESULTS

3.1 Embryo Genotyping and miRNA Profiling

Out of 248 blastocysts collected through eight rounds of IVF, 158 embryos were successfully genotyped, resulting in a total of 81 males and 77 females. Thus, no significant sex bias was observed in our IVF system ($P > 0.05$). A total of 68 miRNAs were profiled using Firefly® technology, of which miRNAs miR-122 (*P*-value = 0.048), miR-16 (*P*-value = 0.042), miR-30b (*P*-value = 0.029), miR-320a (*P*-value = 0.042), miR-15b (*P*-value = 0.089), miR-16-2 (*P*-value = 0.061), miR-17 (*P*-value = 0.072), and miR-22 (*P*-value = 0.069) were found to be upregulated in female-conditioned embryo media (Figure 1). A subset of miRNAs found to be differentially expressed in media of female embryos compared to male embryos (miR-122, miR-22, and miR-320a) was selected for validation using qRT-PCR analysis. The qRT-PCR for miR-122, miR-22, and miR-320a was performed on three biological replicates of media for embryos produced from new dams.

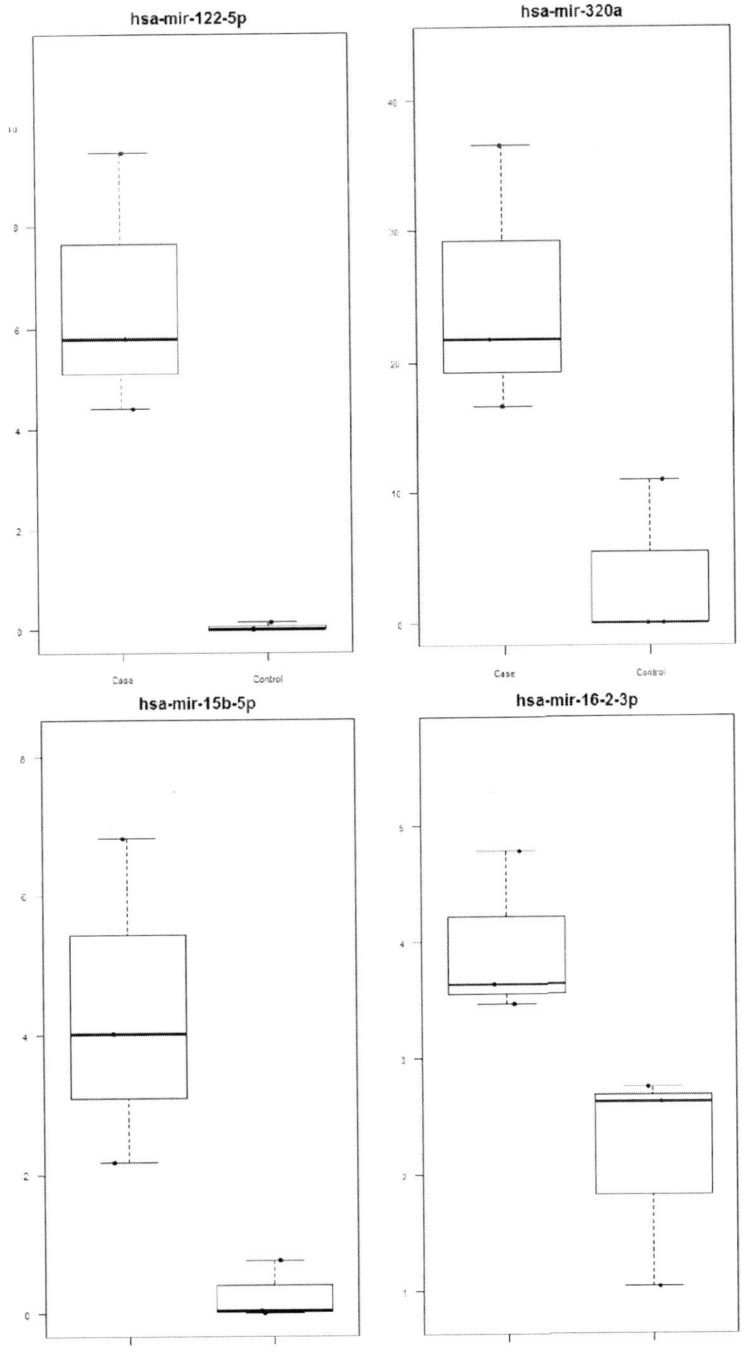
hsa-mir-122-5p
Case
Control
hsa-mir-320a
Case
Control
hsa-mir-15b-5p
Case
Control
hsa-mir-16-2-3p
Case
Control

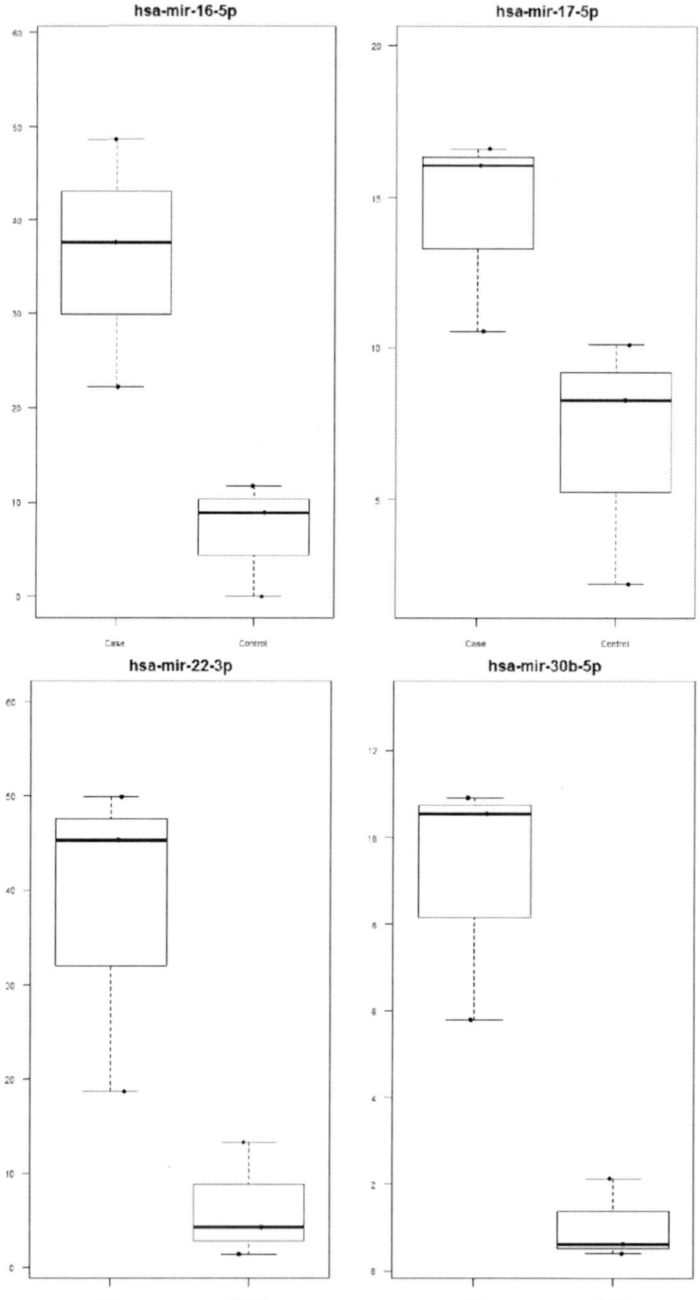

Figure 1. Significantly upregulated miRNA in female versus male media as determined by Firefly®Circulating miRNA Assay Immunology Panel of 68

miRNAs (ABCAM, Cambridge, MA, USA). Data are represented as signal values for each sample. miRNAs miR-122 (*P*-value = 0.048), miR-16 (*P*-value = 0.042), miR-30b (*P*-value = 0.029), miR-320a (*P*-value = 0.042), miR-15b (*P*-value = 0.089), miR-16-2 (*P*-value = 0.061), miR-17 (*P*-value = 0.072), and miR-22 (*P*-value = 0.069) were found to be significant. Female media is represented as case and male media is represented as control.

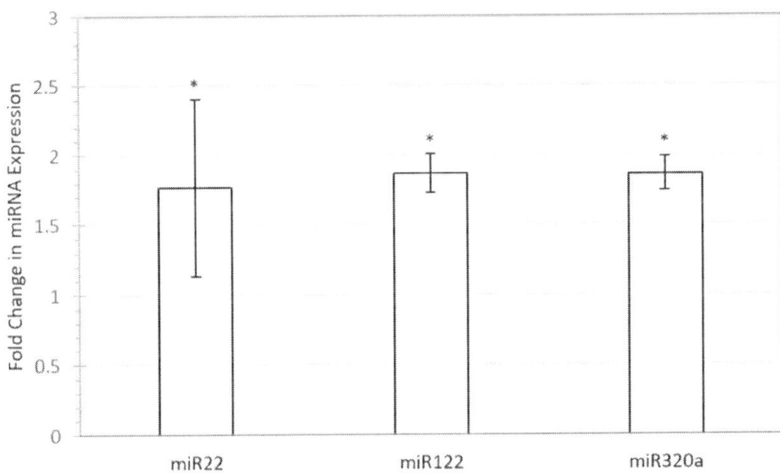

Figure 2. Validation of selected miRNAs with quantitative real-time PCR (qRT-PCR). Fold difference of expression for miR-22 ($P < 0.05$), miR-122 ($P < 0.05$), and miR-320a ($P < 0.05$) in female versus male-conditioned culture medium. Error bars represent SE for the mean fold change of the expression range. Asterisk denotes significant difference between males and females.

Expression analysis revealed upregulation of all three miRNAs in the media of females, with overall fold changes of 1.77, 1.86, and 1.86 for miR-22 ($P < 0.05$), miR-122 ($P < 0.05$), and miR-320a ($P < 0.05$), respectively (Figure 2).

3.2 miRNA Supplementation and Induction of Gene Expression Changes in Primary Cells

To test whether embryonic miRNAs are taken up by maternal endometrial cells, synthetic miRNA mimics corresponding to miR-22, miR-122, and miR-320a were supplemented to primary endometrial epithelial cells. Figure 3 depicts the relative fold difference in expression of miRNA within BEECs following co-culture with the miRNA mimic in comparison to untreated control cells. MiR-122 ($P < 0.05$) and miR-320a ($P < 0.05$) were significantly higher in expression compared to control non-supplemented cells, with fold changes of 28.61 and 44.96, respectively. The miR-22 ($P > 0.05$) expression did not significantly differ between supplemented and control cells, though uptake was observed, with a fold change of 2.49.

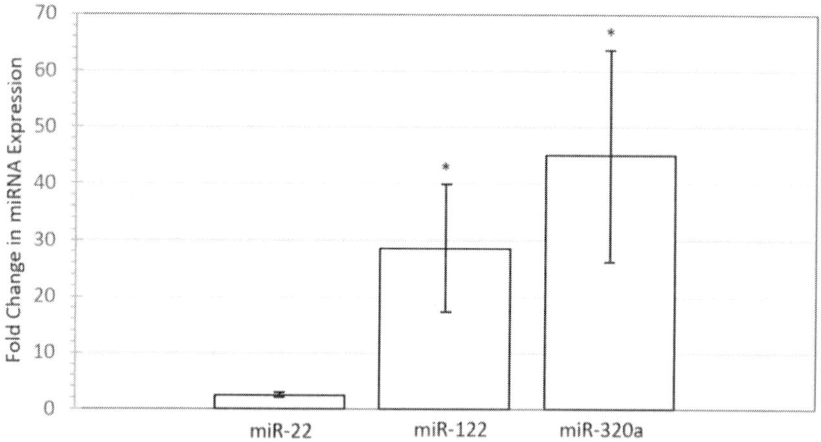

Figure 3. Increased expression of miRNAs in Bovine Endometrial Epithelial Cells following supplementation with 50 nM miRNA mimic for miR-22 ($P > 0.05$), miR-122 ($P < 0.05$), and miR-320a ($P < 0.05$) compared with untreated control for four biological replicates. Error bars represent the SE for the mean fold change of the expression range. Asterisk denotes significant difference between treated and control cells.

To assess the impact of miRNA supplementation on gene expression changes in the BEECs, we used TargetScan[3] (Lewis et al., 2005; Agarwal et al., 2015) to search for target genes of miR-22, miR-122, and miR-320a that could have important roles in fetal-mother communication. Among several candidate genes, the progesterone receptor (*PGR*) gene was selected because it can be targeted by all three embryonic miRNAs, and is involved in the mediation of the effects of progesterone, a fundamental hormone for maintenance of pregnancy (Spencer and Bazer, 2002). Figure 4 shows the fold difference in expression of *PGR* in endometrial cells supplemented with miRNAs miR-122, miR-22, and miR-320a compared to control cells. Interestingly, qRT-PCR analysis of a *PGR* transcript targeted by two miRNAs revealed a significant upregulation of its expression in all four biological replicates for each treatment, with fold changes of 2.17 and 4.18 for miR-122 (P-value = 0.002) and miR-320a (P-value = 0.004), respectively. A trend for upregulation of expression was also seen for cells treated with miR-22 (P-value = 0.077), with a fold change of 1.48. Additionally, lipofectamine-transfected cells exhibited upregulation of *PGR* gene expression for all three miRNA treatments, with fold changes of 1.51 (miR-22), 1.50 (miR-122), and 1.41 (miR-320a) when compared with a lipofectamine-only treated control (Supplementary Table 5).

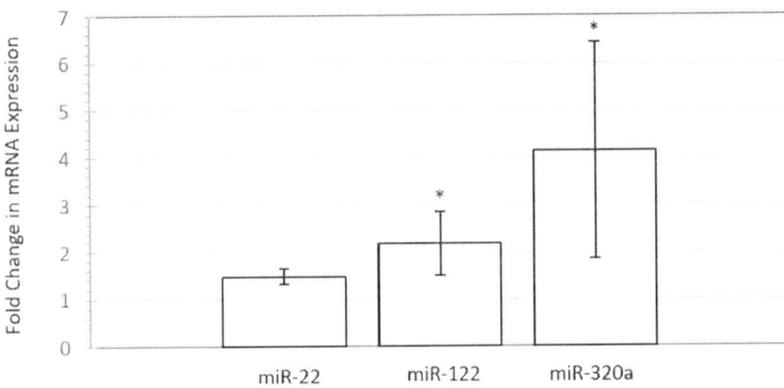

Figure 4. Progesterone receptor (*PGR*) gene expression is increased following miRNA mimic (50 nM) supplementation to bovine endometrial epithelial cells for miR-22 ($P < 0.08$), miR-122 ($P < 0.05$), and miR-320a ($P < 0.05$) compared with untreated control for four biological replicates. Error bars are represented as SE of the mean fold change for the expression range. Asterisk denotes significant difference between treated and control cells.

4. DISCUSSION

Blastocyst-secreted miRNAs may play a remarkable role in the communication of sexually dimorphic states of conceptuses to the mother. Our findings show for the first time that male and female blastocysts differentially secrete miRNAs into culture media. Furthermore, we demonstrate that these miRNAs are taken up by maternal endometrial cells. Two miRNAs targeted *PGR* and induced upregulation of this gene, and a trend of upregulation of *PGR* was observed for the third. The observed sexually dimorphic signaling establishes a purposeful role of miRNAs in regulating the implantation process in bovine embryos.

4.1 Sexual Dimorphism of miRNA in the Culture Media of Preimplantation Embryos

No difference in the ratio of male to female blastocysts was observed in this study, which is in agreement with other reports of *in vitro*-produced bovine embryos (Bermejo-Álvarez et al., 2010; Oliveira et al., 2016). Conversely, other studies have observed a higher proportion of male to female blastocysts (Camargo et al., 2010; Ghys et al., 2016). These differences in development reported across studies may reflect susceptibility of female conceptuses to environmental factors specific to the IVF system employed. For example, production of a higher proportion of male blastocysts was associated with a higher concentration of glucose in culture media (Kimura et al., 2008).

Epigenetic sexual dimorphism has previously been reported through observations of differential DNA methylation (Bermejo-Álvarez et al., 2008) and gene expression (Bermejo-Álvarez et al., 2008, 2010; Forde et al., 2016). Our study is the first to show sexual dimorphism in the secretion of specific miRNAs from blastocysts into the culture medium. This finding is particularly important to understanding the mechanisms of signaling within the preimplantation stage of development, as it plausibly indicates differential physiological needs of an individual embryo that can be communicated to the dam through miRNA secretion. Moreover, differential miRNA secretion could be developed into tools that serve as a method for monitoring embryonic health.

In the present study, a total of eight miRNAs were found to be upregulated in the conditioned culture media of female embryos. Indeed, patterns observed during development indicate the secreted miRNAs upregulated in female media may serve as non-invasive biomarkers for embryo growth. The observation that miR-122 decreases from hours 0 to 22 of maturation in bovine oocytes (Abd El Naby et al., 2016) and the detection of miR-320a in human embryo culture medium at the cleavage, morula, and blastocyst stages (Capalbo et al., 2016) indicate these miRNAs could serve as continuous markers of embryo developmental checkpoints. A study by Feng et al. (2015) identified that presence of miR-320 in the human follicular fluid is decreased in patients using intracytoplasmic sperm injection to conceive, and knockdown of miR-320 in mouse oocytes negatively affects developmental potential of embryos through inhibition of the Wnt signaling pathway. Additionally, miR-22 functions in human cells during apoptosis — a process which is known to be differentially regulated between male and female blastocysts (Ghys et al., 2016; Oliveira et al., 2016), suppressing tumor growth through inhibition of ATP citrate lyase (Xin et al., 2016). Further, miR-22 has been observed in media of bovine embryos which degenerate prematurely, failing to successfully form blastocysts (Kropp and

Khatib, 2015). Both miR-30b (Ye et al., 2015) and miR-15 (Zhang et al., 2015) are involved in regulation of the epithelial-mesenchymal transition in cancer cells, which bears many similarities to the epithelial-mesenchymal transition important for successful implantation and gastrulation (Kalluri and Weinberg, 2009). miR-16 has been shown to inhibit angiogenesis, through targeting of vascular endothelial growth factor, serving as a potential cause of recurrent spontaneous abortions (Zhu et al., 2016). Taken together, the miRNAs found in this study may serve as non-invasive markers for the differential needs of male and female embryos.

4.2 miRNAs as Signaling Molecules for Early Development

Significant uptake of miR-122 and miR-320a, as well as a trend of uptake for miR-22, were observed in this study. Internalization of miRNAs by endometrial cells exhibits the ability for these molecules to interact with the maternal environment and indicates their potential to function as signaling molecules. This study showed endometrial cell uptake of miRNA at half the concentration used previously by Cuman et al. (2015) on human endometrial cells, and without need of a transfection reagent. Lipofectamine 2000 was used in this study to show miRNAs (with aided transport into cells) can affect expression of *PGR*. Indeed, internalization of miRNAs, as well as increase in *PGR* was observed compared with the lipofectamine-only control.

Mechanisms through which endometrial cells take up various miRNAs have yet to be demonstrated, though the uptake of two out of three distinct miRNAs indicates passage of miRNAs into cells may be somewhat selective. It is unknown whether endometrial cells perform uptake through various modes, given that transport of extracellular miRNAs can occur through binding of the miRNA to the AGO protein, apoptotic bodies, or extracellular vesicles (microvesicles and exosomes) (Turchinovich et al., 2013). Successful transport of miRNAs could impact developmental success of the early embryo, and should thus be further interrogated.

Interestingly, the miRNAs investigated in this study have been shown to affect processes critical to the establishment of an interface between the embryo and the mother. Angiogenesis, a process necessary for embryo viability, is inhibited by miR-320 in rat myocardial microvascular endothelial cells (Wang et al., 2009). Normal rat ovarian function is linked to the roles of miR-122, which stimulates activation of the sterol response binding protein pathway, leading to LH receptor downregulation (Menon et al., 2015). The estrogen receptor ERα was shown to be downregulated by transfection with miR-22 in endometrioid carcinomas, where continuous stimulation by estrogen is considered a risk factor for tumorigenesis (Li and

Yang, 2013). Thus, miRNAs found in our study have the potential to regulate cell function in the mother.

4.3 Involvement of Progesterone Receptor in Conceptus-to-Mother Communication

Computational prediction revealed that all three miRNAs (miR-22, miR-122, and miR-320a) collectively target *PGR*, and further functional analysis showed a trend of upregulation of the *PGR* transcript following supplementation with these miRNAs individually. Interestingly, upregulation of *PGR* was correlated with the level of uptake of each miRNA into endometrial cells (Figures 3, 4). Most reported mechanisms for miRNA targeting of transcripts involve degradation of the transcript or inhibition of translation (Wahid et al., 2010). However, upregulation of transcripts by miRNAs has been described to occur through direct and indirect mechanisms, as well as in relation to cell state and transcript composition (Rusk, 2008; Vasudevan, 2012; Orang et al., 2014). Particularly, miR-122 has shown the ability to upregulate a transcript for hepatitis C virus by providing a scaffold for binding of essential factors to transcription, such as RNA polymerase, and enhancing binding of the 40s ribosomal subunit (Orang et al., 2014). Additional modes of upregulation include stabilization of the 5′ end of transcripts, cell cycle state-specific regulation, and protection of AU-rich regions of transcripts (Rusk, 2008; Vasudevan, 2012; Orang et al., 2014), as well as various undefined mechanisms. In the present study, the exact mechanism through which this upregulation of *PGR* occurs was not investigated. Future studies should expand on understanding the modes of upregulation which could lead to these effects on *PGR*.

Further, the substantial upregulation of the *PGR* transcript by all three miRNAs demonstrates a complex collective targeting network of miRNAs working toward regulation of a common gene. TargetScan[4] (Lewis et al., 2005; Agarwal et al., 2015) analysis of targeting sites (data not shown) indicates all three miRNAs could bind to separate regions of the *PGR* 3′ UTR, demonstrating the possibility that several miRNAs could also simultaneously induce effects on the *PGR* transcript. This gives cause for future investigation of the potential networks conceptus-produced miRNAs may form. Interpreting these interactions may better elucidate signaling mechanisms between the conceptus and mother during preimplantation development.

The most well-studied aspect of *PGR* is its interaction with progesterone, a hormone essential to the maintenance of pregnancy and thus, upregulation of *PGR* expression could have implications for the understanding of mechanisms of cow fertility prior to and during implantation. Establishment of uterine receptivity to the conceptus has been shown to require a loss of *PGR* expression in the luminal epithelium and then later from the glandular epithelium as a result of elevated progesterone (Bazer et al., 2009). Progesterone presence in the endometrium is involved in stimulation of blastocyst growth and elongation as well as upregulation of a number of genes critical for adhesion as well as amino acid transport (Spencer et al., 2007). Studies evaluating artificial supplementation of progesterone have demonstrated an association of its administration with induction of gene expression profiles in the endometrium which are similar to expression profiles of normal endometrium interacting with more advanced conceptuses later in normal development (Forde et al., 2009). Embryos do not necessarily need to be present in the uterus at times of elevated progesterone in order to benefit from its effects (Clemente et al., 2009; O'Hara et al., 2014) and artificial elevation of progesterone early in the estrous cycle has been seen to cause decreased size of the corpus luteum, which could induce luteolysis and lead to early embryo loss (O'Hara et al., 2014). Therefore, it is possible that miRNA secretion may serve as a way for embryos to maintain a local level of control over *PGR* in order to mediate effects of progesterone on the endometrium. Although cyclic and non-cyclic cows show little divergence in endometrial transcriptomes until day 16, termed the recognition point of bovine pregnancy (Forde et al., 2011), local effects at early stages of development remain a challenge to measure and may still exist. It is unknown whether the upregulation of *PGR* in the endometrium during the preimplantation period is favorable for successful embryo development and implantation.

Our study was based on *in vitro* investigation of pooled media of blastocysts. *In vivo* models have not been employed to evaluate preimplantation embryo-secreted miRNAs. This is due partially to the challenge of developing precise techniques to measure expression changes induced by embryos within maternal tissues. For example, one study determined that transcriptomic changes occur in the epithelial cells in the oviduct as a result of embryo transfer, but individual embryo influences on the cells were undetectable and were only found to occur with the transfer of 50 embryos (Maillo et al., 2015). Further investigation of these secreted miRNAs should also lead to a better understanding of the differences between individual blastocysts in order to uncover the biology behind this secretion and to develop ways to monitor individual conceptuses throughout development.

Although this study was limited to a panel of miRNAs chosen specifically for their impacts on immunity as opposed to a high-throughput sequencing approach, the results provide a basis for identifying larger-scale dimorphisms of secreted miRNAs. A comprehensive understanding of the exact roles for sexually dimorphic miRNA production has not yet been attained, but future research should focus on understanding larger-scale interactions and collective targeting of subsets of miRNAs on mRNA transcripts in order to better elucidate the roles these molecules may play in maternal cells and to accommodate optimal development of blastocysts relative to their sex.

5. CONCLUSION

The current study demonstrates the potential for blastocysts of differing sex to produce dimorphic signals in the form of miRNAs, which can collectively impose changes in transcripts of maternal cells. Endometrial cell internalization of miRNAs, and subsequent upregulation of *PGR* is a remarkable display of the potential for miRNAs to function as signaling molecules during preimplantation development and allows for a new perspective on understanding the preparation of maternal cells for interaction with blastocysts, which already contain inherent differences due to their sex. Male and female embryo dimorphism in miRNA production allows for future discovery of invaluable biomarkers of embryo signaling, and understanding the potential roles of miRNAs in preimplantation development blastocyst-to-mother communication.

AUTHOR CONTRIBUTIONS

NG carried out the experiments, data analysis and drafting of the manuscript. JK participated in the *in vitro* production of embryos and editing the manuscript. HK conceived the study and participated in the design of experiments and drafting of the manuscript.

FUNDING

This work was supported by USDA Formula Funds (MSN168911).

ACKNOWLEDGMENTS

The authors would like to thank John Parrish and Rick Monson for guidance in the production of embryos.

REFERENCES

1. Aasen, E., and Medrano, J. (1990). Amplification of the ZFY and ZFX genes for sex identification in humans, cattle, sheep and goats. *Biotechnology* 8, 1279–1281.
2. Abd El Naby, W. S., Hagos, T. H., Hossain, M. M., Gad, A. Y., Rings, F., Cinar, M. U., et al. (2016). Expression analysis of regulatory microRNAs in bovine cumulus oocyte complex and preimplantation embryos. *Zygote* 21, 31–51. doi: 10.1017/S0967199411000566
3. Agarwal, V., Bell, G. W., Nam, J. W., and Bartel, D. P. (2015). Predicting effective microRNA target sites in mammalian mRNAs. *Elife* 4, 1–38. doi: 10.7554/eLife.05005
4. Avery, B., Jorgensen, C. B., Madison, V., and Greve, T. (1992). Morphological development and sex of bovine in vitro fertilized embryos. *Mol. Reprod. Develop.* 32, 265–270.
5. Bazer, F. W., Wu, G., Spencer, T. E., Johnson, G. A., Burghardt, R. C., and Bayless, K. (2009). Novel pathways for implantation and establishment and maintenance of pregnancy in mammals. *Mol. Hum. Reprod.* 16, 135–152. doi: 10.1093/molehr/gap095
6. Bermejo-Álvarez, P., Rizos, D., Rath, D., Lonergan, P., and Gutierrez-Adan, A. (2008). Epigenetic differences between male and female bovine blastocysts produced in vitro. *Physiol. Genomics* 32, 264–272. doi: 10.1152/physiolgenomics.00234.2007
7. Bermejo-Álvarez, P., Rizos, D., Rath, D., Lonergan, P., and Gutierrez-Adan, A. (2010). Sex determines the expression level of one third of the actively expressed genes in bovine blastocysts. *Proc. Natl. Acad. Sci. U.S.A.* 107, 3394–3399. doi: 10.1073/pnas.0913843107
8. Bó, G. A., and Mapletoft, R. J. (2013). Evaluation and classification of bovine embryos. *Anim. Reprod.* 10, 344–348.
9. Camargo, L. S. A., Freitas, C., de Sa, W. F., de Moraes Ferreira, A., Serapiao, R. V., and Viana, J. H. M. (2010). Gestation length, birth weight and offspring gender ratio of *in vitro*-produced Gyr (*Bos indicus*) cattle embryos. *Anim. Reprod. Sci.* 120, 10–15. doi: 10.1016/j.anireprosci.2010.02.013
10. Capalbo, A., Ubaldi, F. M., Cimadomo, D., Noli, L., Khalaf, Y., Farcomeni, A., et al. (2016). MicroRNAs in spent blastocyst culture medium are derived from trophectoderm cells and can be explored for human embryo reproductive competence assessment. *Fertil. Steril.* 105, 225.e1-3–235.e1-3. doi: 10.1016/j.fertnstert.2015.09.014
11. Clemente, M., De La Fuente, J., Fair, T., Al Naib, A., Gutierrez-Adan, A., Roche, J. F., et al. (2009). Progesterone and conceptus elongation in cattle: a direct effect on the embryo or an indirect effect via the endometrium? *Reproduction* 138, 507–517. doi: 10.1530/REP-09-0152
12. Cuman, C., Van Sinderen, M., Gantier, M. P., Rainczuk, K., Sorby, K., Rombauts, L., et al. (2015). Human blastocyst secreted microRNA regulate endometrial epithelial cell adhesion. *EBioMedicine* 2, 1528–1535. doi: 10.1016/j.ebiom.2015.09.003
13. Feng, R., Sang, Q., Zhu, Y., Fu, W., Liu, M., Xu, Y., et al. (2015). MiRNA-320 in the human follicular fluid is associated with embryo quality in vivo and affects mouse embryonic development in vitro. *Sci. Rep.* 5:8689. doi: 10.1038/srep08689
14. Ferrari, D., Bianchi, N., Eltzschig, H. K., and Gambari, R. (2016). MicroRNAs modulate the purinergic signaling network. *Trends Mol. Med.* 22, 905–918. doi: 10.1016/j.molmed.2016.08.006
15. Forde, N., Carter, F., Fair, T., Crowe, M. A., Evans, A. C. O., Spencer, T. E., et al. (2009). Progesterone-regulated changes in endometrial gene expression contribute to advanced conceptus development in cattle. *Biol. Reprod.* 81, 784–794. doi: 10.1095/biolreprod.108.074336
16. Forde, N., Carter, F., Spencer, T. E., Bazer, F. W., Sandra, O., Mansouri-Attia, N., et al. (2011). Conceptus-induced changes in the endometrial transcriptome: How soon does the cow know she is pregnant? *Biol. Reprod.* 85, 144–156. doi: 10.1095/ biolreprod.110.090019
17. Forde, N., Maillo, V., O'Gaora, P., Simintiras, C. A., Sturmey, R. G., Ealy, A. D., et al. (2016). Sexually dimorphic gene expression in bovine conceptuses at the initiation of implantation. *Biol. Reprod.* 95, 92. doi: 10.1095/biolreprod.116.139857

18. Ghys, E., Dallemagne, M., De Troy, D., Sauvegarde, C., Errachid, A., and Donnay, I. (2016). Female bovine blastocysts are more prone to apoptosis than male ones. *Theriogenology* 85, 591–600. doi: 10.1016/j.theriogenology.2015.09.050
19. Green, M. P., Harvey, A. J., Spate, L. D., Kimura, K., Thompson, J. G., and Roberts, R. M. (2016). The effects of 2,4-dinitrophenol and d-glucose concentration on the development, sex ratio, and interferon-tau (IFNT) production of bovine blastocysts. *Mol. Reprod. Dev.* 83, 50–60. doi: 10.1002/mrd.22590
20. Heras, S., De Coninck, D. I. M., van Poucke, M., Goossens, K., Bogado Pascottini, O., Van Nieuwerburgh, F., et al. (2016). Suboptimal culture conditions induce more deviations in gene expression in male than female bovine blastocysts. *BMC Genomics* 17, 72. doi: 10.1186/s12864-016-2393-z
21. Kalluri, R., and Weinberg, R. A. (2009). Review series: the basics of epithelial-mesenchymal transition. *J. Clin. Invest.* 119, 1420–1428. doi: 10.1172/JCI39104.1420
22. Kimura, K., Iwata, H., and Thompson, J. G. (2008). The effect of glucosamine concentration on the development and sex ratio of bovine embryos. *Anim. Reprod. Sci.* 103, 228–238. doi: 10.1016/j.anireprosci.2006.12.014
23. Kirkpatrick, B. W., and Monson, R. L. (1993). Sensitive sex determination assay applicable to bovine embryos derived from IVM and IVF. *J. Reprod. Fertil.* 98, 335–340. doi: 10.1530/jrf.0.0980335
24. Kozomara, A., and Griffiths-Jones, S. (2014). MiRBase: annotating high confidence microRNAs using deep sequencing data. *Nucleic Acids Res.* 42, 68–73. doi: 10.1093/nar/gkt1181
25. Kropp, J., and Khatib, H. (2015). Characterization of microRNA in bovine in vitro culture media associated with embryo quality and development. *J. Dairy Sci.* 98, 6552–6563. doi: 10.3168/jds.2015-9510
26. Kropp, J., Salih, S. M., and Khatib, H. (2014). Expression of microRNAs in bovine and human pre-implantation embryo culture media. *Front. Genet.* 5:91. doi: 10.3389/fgene.2014.00091
27. Laguna-Barraza, R., Bermejo-Álvarez, P., Ramos-Ibeas, P., De Frutos, C., López-Cardona, A. P., Calle, A., et al. (2013). Sex-specific embryonic origin of postnatal phenotypic variability. *Reprod. Fertil. Dev.* 25, 38–47. doi: 10.1071/RD12262
28. Lewis, B. P., Burge, C. B., and Bartel, D. P. (2005). Conserved seed pairing, often flanked by adenosines, indicates that thousands of human genes are microRNA targets. *Cell* 120, 15–20. doi: 10.1016/j.cell.2004.12.035
29. Li, H., and Yang, B. B. (2013). Friend or foe: the role of microRNA in chemotherapy resistance. *Acta Pharmacol. Sin.* 34, 870–879. doi: 10.1038/aps.2013.35
30. Livak, K. J., and Schmittgen, T. D. (2001). Analysis of relative gene expression data using real-time quantitative PCR and. *Methods* 25, 402–408. doi: 10.1006/meth.2001.1262
31. Maillo, V., Gaora, P. O., Forde, N., Besenfelder, U., Havlicek, V., Burns, G. W., et al. (2015). Oviduct-embryo interactions in cattle: two-way traffic or a one-way street? *Biol. Reprod.* 92, 144–144. doi: 10.1095/biolreprod.115.127969
32. Menon, B., Gulappa, T., and Menon, K. M. J. (2015). MIR-122 regulates LH receptor expression by activating sterol response element binding protein in rat ovaries. *Endocrinology* 156, 3370–3380. doi: 10.1210/en.2015-1121
33. Noferesti, S. S., Sohel, M. M. H., Hoelker, M., Salilew-Wondim, D., Tholen, E., Looft, C., et al. (2015). Controlled ovarian hyperstimulation induced changes in the expression of circulatory miRNA in bovine follicular fluid and blood plasma. *J. Ovarian Res.* 8, 81. doi: 10.1186/s13048-015-0208-5
34. O'Hara, L., Forde, N., Carter, F., Rizos, D., Maillo, V., Ealy, A. D., et al. (2014). Paradoxical effect of supplementary progesterone between day 3 and day 7 on corpus luteum function and conceptus development in cattle. *Reprod. Fertil. Dev.* 26, 328–336. doi: 10.1071/RD12370
35. Oliveira, C. S., Saraiva, N. Z., de Lima, M. R., Oliveira, L. Z., Serapião, R. V., Garcia, J. M., et al. (2016). Cell death is involved in sexual dimorphism during preimplantation development. *Mech. Dev.* 139, 42–50. doi: 10.1016/j.mod.2015.12.001

36. Orang, A. V., Safaralizadeh, R., and Kazemzadeh-Bavili, M. (2014). Mechanisms of miRNA-mediated gene regulation from common downregulation to mRNA-specific upregulation. *Int. J. Genomics* 2014:970607. doi: 10.1155/2014/970607
37. Parrish, J. J., Krogenaes, A., and Susko-Parrish, J. L. (1995). Effect of bovine sperm separation by either swim-up or Percoll method on success of in vitro fertilization and early embryonic development. *Theriogenology* 44, 859–869. doi: 10.1016/0093-691X(95)00271-9
38. Rosenkrans, C. F., Zeng, G. Q., MCNamara, G. T., Schoff, P. K., and First, N. L. (1993). Development of bovine embryos in vitro as affected by energy substrates. *Biol. Reprod.* 49, 459–462. doi: 10.1095/biolreprod49.3.459
39. Rusk, N. (2008). When microRNAs activate translation. *Nat. Methods* 5, 122–123. doi: 10.1038/nmeth0208-122a
40. Sagirkaya, H., Misirlioglu, M., Kaya, A., First, N. L., Parrish, J. J., and Memili, E. (2006). Developmental and molecular correlates of bovine preimplantation embryos. *Reproduction* 131, 895–904. doi: 10.1530/rep.1.01021
41. Spencer, T. E., and Bazer, F. W. (2002). Biology of progesterone action during pregnancy recognition and maintenance of pregnancy. *Front. Biosci.* 7, d1879–d1898.
42. Spencer, T. E., Johnson, G. A., Bazer, F. W., Burghardt, R. C., and Palmarini, M. (2007). Pregnancy recognition and conceptus implantation in domestic ruminants: roles of progesterone, interferons and endogenous retroviruses. *Reprod. Fertil. Dev.* 19, 65–78. doi: 10.1071/RD06102
43. Sturmey, R. G., Bermejo-Álvarez, P., Gutierrez-Adan, A., Rizos, D., Leese, H. J., and Lonergan, P. (2010). Amino acid metabolism of bovine blastocysts: a biomarker of sex and viability. *Mol. Reprod. Dev.* 77, 285–296. doi: 10.1002/mrd.21145
44. Sun, J., Aswath, K., Schroeder, S. G., Lippolis, J. D., Reinhardt, T. A., and Sonstegard, T. S. (2015). MicroRNA expression profiles of bovine milk exosomes in response to *Staphylococcus aureus* infection. *BMC Genomics* 16, 806. doi: 10.1186/s12864-015-2044-9
45. Turchinovich, A., Samatov, T. R., Tonevitsky, A. G., and Burwinkel, B. (2013). Circulating miRNAs: cell-cell communication function? *Front. Genet.* 4:119. doi: 10.3389/ fgene. 2013.00119
46. Vasudevan, S. (2012). Posttranscriptional upregulation by microRNAs. *Wiley Interdiscip. Rev. RNA* 3, 311–330. doi: 10.1002/wrna.121
47. Wahid, F., Shehzad, A., Khan, T., and Kim, Y. Y. (2010). MicroRNAs: synthesis, mechanism, function, and recent clinical trials. *Biochim. Biophys. Acta* 1803, 1231–1243. doi: 10.1016/j.bbamcr.2010.06.013
48. Wang, X. H., Qian, R. Z., Zhang, W., Chen, S. F., Jin, H. M., and Hu, R. M. (2009). MicroRNA-320 expression in myocardial microvascular endothelial cells and its relationship with insulin-like growth factor-1 in type 2 diabetic rats. *Clin. Exp. Pharmacol. Physiol.* 36, 181–188. doi: 10.1111/j.1440-1681.2008.05057.x
49. Willems, E., Leyns, L., and Vandesompele, J. (2008). Standardization of real-time PCR gene expression data from independent biological replicates. *Anal. Biochem.* 379, 127–129. doi: 10.1016/j.ab.2008.04.036
50. Xin, M., Qiao, Z., Li, J., Liu, J., Song, S., Zhao, X., et al. (2016). miR-22 inhibits tumor growth and metastasis by targeting ATP citrate lyase: evidence in osteosarcoma, prostate cancer, cervical cancer and lung cancer. *Oncotarget* 7, 44252–44265. doi: 10.18632/oncotarget. 10020
51. Xu, K. P., Yadav, B. R., King, W. A., and Betteridge, K. J. (1992). Sex-related differences in developmental rates of bovine embryos produced and cultured in vitro. *Mol. Reprod. Dev.* 31, 249–252. doi: 10.1002/mrd.1080310404
52. Ye, Z., Zhao, L., Li, J., Chen, W., and Li, X. (2015). miR-30d blocked transforming growth factor beta1-induced epithelial-mesenchymal transition by targeting snail in ovarian cancer cells. *Int. J. Gynecol. Cancer* 25, 1574–1581. doi: 10.1097/IGC.0000000000000546
53. Zhang, W. L., Zhang, J. H., Wu, X. Z., Yan, T., and Lv, W. (2015). miR-15b promotes epithelial-mesenchymal transition by inhibiting SMURF2 in pancreatic cancer. *Int. J. Oncol.* 47, 1043–1053. doi: 10.3892/ijo.2015.3076

54. Zhu, Y., Lu, H., Huo, Z., Ma, Z., Dang, J., Dang, W., et al. (2016). MicroRNA-16 inhibits feto-maternal angiogenesis and causes recurrent spontaneous abortion by targeting vascular endothelial growth factor. *Sci. Rep.* 6:35536. doi: 10.1038/srep35536

CHAPTER 6

Begait Goat Production Systems and Breeding Practices in Western Tigray, North Ethiopia

Hagos Abraham[1,2*], Solomon Gizaw[3], Mengistu Urge[2]

[1]*Department of Animal Science, Bule Hora University, Bule Hora, Ethiopia*

[2]*School of Animal and Range Sciences, Haramaya University, Dire Dawa, Ethiopia*

[3]*LIVES Project, International Livestock Research Institute (ILRI), Addis Ababa, Ethiopia*

ABSTRACT

A study was undertaken to characterize production system, breeding practices and production constraints of Begait goat in Kafta humera district, northwestern lowlands of Ethiopia. Personal observations, focus group discussions and a detailed structured questionnaire were used to collect data from 150 respondents (99 small-scale and 51 large-scale farmers). Mixed crop-livestock production system was found to be the dominant farming system in the study district. The farmers kept a variety of livestock species including goats, sheep, cattle, chickens and donkeys. Small ruminants are the most numerous and are of great importance in the area. Goats are herded with sheep in the open fields by the young boys in the family/hired grazers. They have a number of roles though mainly kept as a source of cash income. Breeding was generally uncontrolled. Size, growth rate and libido were the most frequently reported traits in selecting bucks, whereas does were selected based on size, twining ability and milk yield. Effective population size and rate of inbreeding were calculated by considering random mating and flocks were not mixed. The major constraints hampering goat production in the area were water scarcity, feed and grazing land shortages and insufficient veterinary services among the others. Therefore, addressing these constraints is important to design a successful genetic improvement scheme in the area for goat.

Keywords: Farming System, Husbandry Practice, Marketing, Production Constraints

1. INTRODUCTION

Goats (Capra hircus) are of the most beneficial animals. Ethiopia is one of the top ten goat producing countries in the world. As per the livestock census carried out in 2016, Ethiopia owns approximately 30 million goats; of which 71.57% are females and the rest 28.43% are males. Majority of the national goat population is found in the lowlands. Nearly all (99.99%) of the goats are indigenous types [1] [2] , which have become adapted to a wide agro-ecological zones of the country [3] due to natural selection. The environmental adaptation of indigenous breeds facilitates livestock production in a range of agro-ecological conditions and constraints [4].

Goats are deeply embedded in almost every African culture and are true friends to the rural poor [5]. For instance, goats in Ethiopia are important to the subsistence, economic and social livelihoods of their owners [6] [7] [8]. On top of this, Ethiopia generates hard currency through exporting live goats, goat meat and goat skins. At optimum off-take rates, the country has a potential to deliver 1.13 and 2.00 million goats to national and international markets per annum, respectively [7].

Despite the large goat population is found in the country and its contributions to household income and to the country's national economy, productivity per unit of animal is low. Genetic improvement is one way to increase productivity of the goat genetic resources in the country. According to [9], development of animal genetic improvement plans will only be successful when accompanied by good understanding of the production system and when simultaneously addressing constrains. However, there is little or no adequate information on majorities of indigenous goat breeds/types in Ethiopia to design appropriate breeding plans, suggesting a low level of priority for this area of research. Specifically, the study aims at:

1) Characterizing production system and breeding practices of Begait goat;
2) Identifying the major constraints facing Begait goat production in its home tract.

2. MATERIALS AND METHODS

2.1. Description of the Study Area

This research work was done in Kafta and Setit humera districts of western zone of Tigray National Regional State, northwest lowlands of Ethiopia. Setit humera town, a special district, is the administrative center of Kafta humera district and western zone of the region. In this study, the district has been integrated as part of Kafta humera district. The study area covers a total area of 71.80 km^2 of which 54.20% is cropland.

The study area is situated 1372 km away from the administrative center of Addis Ababa city to northwest direction. It is bordered by Sudan to the west, Tahtay Adyabo district to the east, Eritrea to the north and Wolkayit and Tsegedei districts to the south. Geographically, it is located in between the latitudes 13°14'N - 14°27'N and longitudes 36°27'E - 37°32'E. The study area comprises kolla (lowland) and weinadega (midland) agro-climatic zones with an altitude ranging from 560 - 1849 m above sea level. The annual precipitation in the lowlands and midlands is 448.8 and 1102.5 mm, respectively [10] occurring between Jun and September. The mean minimum and maximum temperatures in the lowlands are 25°C and 27.5°C, respectively while the corresponding values in the midlands are 20°C and 25°C. The hottest months are between April and June with temperatures rising up to 42°C.

2.2. Farming System

The agricultural production system in the study area is mixed farming. There are two types of farming systems in the district viz. small-scale and large-scale based farming systems. As a result, farm activities predominantly crop production and livestock rearing are performed by small-scale and large-scale farmers. According to [11] , above half (58%) of the total cropland is cultivated by large-scale farmers while small-scale farmers cultivate the rest 42%. Crop production is a rain fed production system without irrigation practice. Sesame and sorghum are the major staple food and cash crops grown in the area. Other crops like cotton, millet, maize and teff are well grown.

Livestock production is similar in both farming systems, and it is a traditional type which is characterized by minimal inputs as is in other areas of Ethiopia. [12] Demonstrated that low input production system is found in all livestock production systems prevailing in the country except in Peri-urban and urban system. The most dominant farm animal species of the area are cattle and small ruminants. Animal feeding is free grazing.

2.3. Sampling and Data Collection

Prior to site selection, livestock experts in the district were contacted for brief discussions. In addition, secondary data were reviewed to better understand the current production systems and the area dominated by Begait goat. Based on this information, four rural kebeles and the special district were selected as study sites. A total of 150 Begait goat owning farmers (99 small-scale and 51 large-scale farmers) were selected at random.

Interviews were conducted at the farmers' residences using structured questionnaire with the assistance of development agents. The questionnaires covered information on key household characteristics, livestock possession, flock structure, purposes of keeping goat, mating system, marketing system and production constraints. To complement the survey work, focus group discussions were held with a total of 8 - 10 key informants per selected study site. In addition, general information about the district was obtained from secondary data sources.

2.4. Data Analysis

The survey data were analyzed using SPSS software (SPSS for windows, release 16.0, 2007). F test was applied when required to test statistical significances. Indices were calculated for all ranking data using the formula: Index = S [3 for rank 1 + 2 for rank 2 + 1 for rank 3] given for an individual reason divided by S [3 for rank 1 + 2 for rank 2 + 1 for rank 3] for all reasons [13] . By considering the number of breeding males (Nm) and breeding females (Nf), the effective population size (Ne) and rate of inbreeding (ΔF) were also estimated using the equations of , $Ne = 4(Nm \times Nf)\ Nm + Nf$ and $\Delta F = 1\ 2Ne$, respectively [14].

3. RESULTS AND DISCUSSION

3.1. Household Information

Table 1 presents some key characteristics of the respondents across the two farming systems. The overall average age of the sampled household head was 50 ± 10.44 years, implying that the respondents were adults with a good experience in goat farming. The overall mean family size was 4.85 ± 0.89, which is comparable with the national estimate of 4.80 persons per household [15].

Of the total sampled household heads, 98.67% were males. The rest (1.33%) households were female headed. Indeed, it is unlikely to find a female headed household unless she is widowed or divorced. Literacy rate among the household heads was 69.67%. Similar results were reported by [16] in Dale district and [17] in Goma district. The better educational background obtained in this study would be a good opportunity for further animal genetic improvement programs in the study area, since literate communities are more likely adopt and practice new technologies. The study also revealed that most (98%) of the respondents were married. The percentage of unmarried and widowed was 0.67% and 1.33%, respectively. The present findings indicate that goat farming can be performed by every social class of the community regardless of their background characteristics, viewing the significant importance of goat for its owners.

3.2. Livestock Holding and Flock Structure

Table 2 demonstrates flock structure and livestock possession across the two farming systems. Overall, sampled farmer owned 43.67 goats, 42.90 sheep, 10.19 cattle, 9.70 chickens and 0.96 donkeys. There was a slight increase in mean livestock holding of respondents from small-scale to large-scale farming even if the difference is not statistically significant except for cattle. None of the respondents owned mules and horses. This finding supports the fact that mules and horses could not adapt the hot environmental conditions of the area.

The mean flock size of goats in the present study was close to the work of [18] and [19] , who reported that 44.0 and 48.5 heads per household in Siti zone of Somali region and Abergelle district of Amhara region, respectively. In contrast, the mean flock size recorded in this study is higher than the previous average flock size of 21.20 Begait goats in the area [20] . This difference might be due to random sampling error or due to changes in goat farming system over time in the area.

Table 1. Background characteristics of the respondents by farming system.

Descriptor variables	Small scale farming (n = 99)	Large scale farming (n = 51)	Overall (N = 150)
	Mean ± SD	Mean ± SD	Mean ± SD
Age(years)	48.24 ± 9.88	53.10 ± 10.85	50.00 ± 10.44
Family size	4.91 ± 0.94	4.75 ± 0.80	4.85 ± 0.89
Sex	Percent	Percent	Percent
Male	98.33	100.00	98.67
Female	1.67	-	1.33
Educational level			
Illiterate	27.30	37.30	30.33
Literate	72.70	62.70	69.67
Marital status			
Married	97.00	100.00	98.00
Unmarried	1.00	-	0.67
Widowed	2.00	-	1.33

N: Number of interviewed households; SD: Standard deviation.

Table 2. Mean livestock holdings and flock structure across the two farming systems.

Livestock species	Small-scale farming	Large-scale farming	Overall
	(mean ± SD)	(mean ± SD)	(mean ± SD)
Goat	42.21 ± 18.42[a]	46.43 ± 15.94[a]	43.67 ± 17.49
Kids	11.92 ± 5.22[a]	13.16 ± 4.33[a]	12.34 ± 4.96
Buckling	3.18 ± 1.28[a]	3.51 ± 1.07[a]	3.29 ± 1.22
Doeling	5.18 ± 2.29[a]	5.71 ± 1.88[a]	5.36 ± 2.17
Bucks	1.89 ± 0.92[a]	1.99 ± 0.81[a]	1.94 ± 0.89
Does	19.54 ± 8.41[a]	21.47 ± 6.99[a]	20.19 ± 7.98
Castrates	0.67 ± 0.47[a]	0.76 ± 0.43[a]	0.70 ± 0.46
Sheep	41.68 ± 17.86[a]	45.41 ± 15.07[a]	42.90 ± 17.09
Cattle	9.62 ± 4.68[b]	11.29 ± 3.98[a]	10.19 ± 4.51
Chicken	9.45 ± 3.69[a]	10.16 ± 0.70[a]	9.70 ± 3.57
Donkey	0.97 ± 0.28[a]	0.94 ± 3.34[a]	0.96 ± 0.20

Means with the same letter within the same row are not significantly different ($p > 0.05$); SD: standard deviation.

The number of males and females in the sample population and their ages were often used as an indicator of a traditional management system in Africa [21] . The present finding revealed that the proportion of does represents the largest class followed by suckling kids, while castrates represent the lowest proportion (Table 2). This is in good agreement with previous findings in Ethiopia [17] [22] [23] [24] and elsewhere in Africa [25] [26] [27] . The higher proportion of does than other age groups suggests that adult females stay in a flock for breeding purposes and/or milk production. However, the less number of buckling as compared to doeling is because of marketing and slaughtering of buckling. There was a high level of elimination of the majority of born male kids.

Small ruminants were the predominant livestock species in the area (Table 3) and this shows that their wide acceptability and ease of adaptation in the area. Sheep and goat numbers are growing fastest in the mixed farming systems [28] as subsistence farmers prefer small stock because the risk of losing large ruminants is high [29] . In rural areas of Ethiopia, because of their subsistence and economic reasons, goats and sheep have been described as bank reserve which can be drawn upon when cash money is required. This confirms a study by [30] , who reported that keeping livestock, especially small ruminants plays role as safety net that enables households to get quick income to settle urgent financial needs.

3.3. Purposes of Keeping Goat

Knowledge of reasons for keeping animals is a prerequisite for deriving operational breeding goals [6] [31] . Table 4 summarizes reasons of keeping goats and the ranking of these purposes by owners. The most frequently reported reason for keeping goat was cash income generation followed by milk and meat production for home use. This implies that sale of goats to generate cash constitute the primary purpose among the other benefits of keeping goats in the area. This was also observed in similar studies [32] [33] [34] [35], who reported that cash income as the most important reason in goat farming. However, the purpose of keeping goats in rural areas of South Africa is mainly for traditional purposes and meat consumption [36].

3.4. Land Holding and Goat Production

Mixed crop-livestock production system takes place as an important source of livelihood in the area. Huge ruminant animals such as goats are kept and a vast area of land is cultivated. Overall cropland holding of small-scale and large-scale farmers was observed to be 4.10 and 86.90 hectares, respectively. This indicates that there is huge crop residue production in the area, a considerable source of goat feed. Normally, crop residues are grazed or stockpiled for animal feeding. In some cases, the stored residues are sold during the critical period of feed scarcity. Sorghum crop residues are by far the most abundant feed resource in the area.

Table 3. Percentage of households owning livestock species across the two farming systems.

Livestock species	Small-scale farming		Large-scale farming		Overall	
	n = 99	%	n = 51	%	N = 150	%
Goat	99	100.00	51	100.00	150	100.00
Sheep	80	80.80	38	74.50	118	78.67
Cattle	66	66.67	33	64.70	99	66.00
Chicken	69	70.00	30	58.82	99	66.00
Donkey	67	67.67	29	56.86	96	64.00

Table 4. Reasons for keeping goats as ranked by respondents in the area.

Purpose	Rank			Index
	Rank 1	Rank 2	Rank 3	
Cash income	119	17	12	0.448
Meat home consumption	11	10	118	0.190
Milk	20	123	7	0.348
Skin	0	0	3	0.003
Saving value	0	0	10	0.011

Index = Σ [3 for rank 1 + 2 for rank 2 + 1 for rank 3] for particular purpose divided by Σ [3 for rank 1 + 2 for rank 2 + 1 for rank 3] for all purposes. The highest index value indicates the highest importance of the trait.

4. GOAT HUSBANDRY PRACTICES

4.1. Feeds and Feeding Management

The different feed resources reported in the area were natural pasture, browse species, crop residue and crop aftermath. Private grazing land was not common in the area. It was reported that communal grazing was the most abundant feed source for goat in the area though grazing drastically reduces in the dry season. Indigenous browse species are available feed resources for goats mainly in the dry season and the drought periods (March-June). It was indicated that Acacia mellifera, Acacia seyal, Anogeissus leiocarpus, Balanite aegyptica, Dichrostachys cinerea, Sclerocarya birrea, Combretum hertmannianum, Rhus nataliensis, Zizyphus spina-cristi were the predominant plant species widely utilized as goat feed source.

Crop residue and crop aftermath grazing were the other vital feed resources of goat during crop harvesting and after harvesting has occurred. Animals feed on the residue in two ways. The harvested crop residue is stacked near to homestead and fed to selected group of animals in the dry season. Otherwise, standing residue is left for grazing. However, the expansion of cropland, increase in livestock population and area enclosure are negatively affecting feed resource in the area.

4.2. Water Sources and Watering Management

Water resource is pertinent and vital for the subsistent life of livestock and livestock owners. In the study area, river, pond and borehole form the major source of water for domestic use and livestock watering. Indeed, water points are limited and large number of animals watered at the same water points leading to high chances of spreading diseases and land degradation. There were only two government constructed water points. Besides, private water points were very few.

Distance to water points varied with season and was invariably longer during the dry season. This posed challenges to breeders and sometimes limited access to water. A large proportion (45.33%) of the respondents travel 6 - 10 km to water their animals in the dry season. While in the wet season 67.33% of the total respondents watered their animals at a distance of less than 1 km. Shorter watering frequencies were used to water animals in the area. The proportions of the respondents that watered their goats once a day and twice in a day were 82% and 18%, respectively. Watering with more than a day frequency is not reported.

4.3. Housing System

Based on personal observations and interviewed households' information, goats were housed with sheep in nights. None of the respondents had provision for accommodating different age groups. Only sick ones and newly born kids/lambs were retained alone in sheds. Sheds were constructed of locally available materials. Sheds were not permanent as animals migrate from place to place in search of feed and water. However, farmers in urban areas used permanent houses separated from or attached to their homesteads.

4.4. Herding Practice

The common herding management in the area was free grazing. All the interviewed farmers replied that they herded goats with sheep. One or two herdsman and at least one herding dog used to accompany flocks for grazing. There is also the possibility of mixing with other goats and sheep flocks in the nearby vicinity of villages. Flocks of small size of less than 30 were pooled together to a maximum size of 60 to 120 heads for grazing. And then the flocks owned by different farmers were taken together by hired grazers.

4.5. Common Goat Diseases and Health Management

Diseases are a major constraint to the improvement of livestock industry in the tropics as they decrease production and increase the morbidity and mortality [37] . The most commonly prevailing diseases which hamper goat production in the study area are presented in Table 5. Respondents indicated that diseases affect all age groups of goats and mortalities from disease are high. More than half (53%) of the sampled respondents were able to separate between the different diseases. Diseases were named according to their symptoms.

Vaccination services were provided by Bureau of Agriculture and Rural Development of the district at seven clinics. In addition, farmers use modern drugs either from government clinics or open markets and treat their animals themselves. Some farmers also claimed that veterinary officers were called to treat their sick animals. The most extensively used drug in the area was oxytetracycline and locally known as leshlesh. Likewise, branding was applied when animals get sick and for identification purpose.

4.6. Origin, Distribution and Current Status of Begait Goat

Indeed, there is no documented information on the exact origin of Begait goat. According to [20], it is believed that the breed was derived from Nubian type goats. In the area the breed is also known as Barka/Hassan. From individual interviews, 37% of sampled farmers agreed that the breed's name "Barka" is an indicator of the breed's origin that is Barka area from Eritrea while most (63%) of the respondents believed that the breed's origin could be Eritrea and Sudan. From the focus group discussions, it became obvious that Begait goat is primarily kept by Beni Amir ethnic group who lives in Eritrea and Sudan bordering Ethiopia. This observation is supported by [38], who documented that Beni Amir ethnic group occupies the borders between much of Eritrea's Barka valley and eastern Sudan.

Table 5. Local names and scientific equivalents of the common goat diseases in the area.

Scientific name	Local name (*Tigrigna*)
PPR (pest des petits ruminants)	Gulhay
Foot rot	Mojelle
Orf/Contagious Ecthyma/Sore mouth	Af-gumed
Pasteurellosis	Mi'eta
Small pox	Enfrir
Anthrax	Megerem

Despite the exact date of immigration remains unclear, Begait goat is extensively distributed across northwestern lowlands of Ethiopia. It is estimated that >110,000 Begait goats are found in northwestern and western zones of Tigray region. Based upon the results of focus group discussions and individual interviews, population size of the breed is at increasing trend. The possible reason reported for this trend was the involvement of more farmers in Begait goat production as the breed is more prolific and resistant to drought conditions with better milk yield and meat quality. At present, Bureau of Agriculture and Rural Development of Tigray region has established a ranch, at Mai Woyni kebele, western zone of the region, for all Begait animal species aiming at conservation, breeding and improvement.

4.7. Breeding Practice and Selection Criteria

In the study area, mating was random. There was no report of controlled mating. Bucks run with does throughout the year and castration was an uncommon practice in the area. This resulted in indiscriminate and uncontrolled breeding. The productive life span of buck was reported on average 5 years. Most (83%) of the respondents had own buck. Farmers who had no breeding males, purchase buck from local markets or got buck service from their neighbors. This finding is in line with previous findings [14] [39] [40].

The most common way of selecting goats as parents for the coming generations is to use the offspring of a chosen parent (buck and/or doe). Linear index is the best strategy for selecting replacements in the livestock industries [41]. Selection criteria for buck and doe with corresponding index values are presented in Table 6.

The respondents prioritized body size, growth rate and libido as selection criteria for buck with the indices of 0.329, 0.232 and 0.183, respectively. While the most important selection criterion for does were body size (index = 0.330), twining ability (index = 0.219) and milk yield (index = 0.166). Others like mothering ability, kid survival, kidding interval and age at puberty were also reported as criteria but with lower rankings.

Table 6. Selection criteria for buck and doe in the study area as ranked by owners.

Class and selection criteria	Rank			Index
	Rank 1	Rank 2	Rank 3	
Buck				
Body size	82	18	14	0.329
Drought resistance	18	18	28	0.131
Growth rate	17	70	18	0.232
Libido	12	31	67	0.183
Coat color	15	8	17	0.087
Horn	6	5	6	0.038
Doe				
Body size	85	14	13	0.330
Age at puberty	8	4	3	0.039
Kidding interval	8	5	6	0.045
Twining ability	8	75	23	0.219
Milk yield	14	16	75	0.166
Kid survival	10	15	8	0.076
Mothering ability	17	20	22	0.126

Index = Σ [3 for rank 1 + 2 for rank 2 + 1 for rank 3] for particular trait divided by Σ [3 for rank 1 + 2 for rank 2 + 1 for rank 3] for all traits.

4.8. Effective Population Size and Level of in Breeding

The effective population size (Ne) and level of inbreeding (ΔF) were calculated for Begait goat using the averaged breeding males and females under closed breeding practice. Majority of the bucks were originated from their respective flocks which lead to inbreeding. Communal herding in contrast increases genetic diversity by rising Ne [6], which is a common practice in the study area. The obtained results for Ne and (ΔF) were 7.08 and 0.07, respectively. The rate of inbreeding was equivalent to the maximum acceptable value of 0.063 [42].

4.9. Reproductive Performance

According to respondents, the age at puberty in Begait goat was 7.41 ± 1.85 months for males and 8 ± 1.78 months for females. The present finding is within the range of age at sexual maturity reported for most of the tropical goat breeds under extensive system of management [20] [43].

The reported mean age at first kidding (AFK) for Begait goat was 14.18 ± 1.24 months. This is in agreement with that reported by [44] for Metema goats that have mean AFK of 13.6 ± 2.44 months. This result is also in agreement with reported age at first kidding for Arsi-Bale goats which is 14.88 ± 0.3 months [16] and shorter than for Somali short-eared goats which is 19.9 ± 7.93 months AFK under traditional management conditions.

Begait goat breed, according to the respondents had an average kidding interval of 8.4 ± 1.37 months. This result was shorter than the reported kidding interval for Abergelle and Central Highland goats which were 11.31 ± 2.21 and 10.3 ± 1.42 months, respectively [45]. Higher longevity under adverse conditions is one of the adaptation traits of tropical livestock. The mean reproductive lifespan of does in the flock was reported 11.44 ± 1.17 years and the average number of kids per doe per lifespan was 15.32 ± 1.76.

4.10. Acquisition and Disposal

The famous methods used to acquire goats were new born, buying and inheritance with the index values of 0.390, 0.343 and 0.186, respectively (Table 7). This is in good agreement with the findings of [33] and [46] in Uganda. Whereas disposal was through selling, slaughtering in holidays and death in that order (Table 7).

Goat marketing in the study area was traditional type. All (100%) of the respondents reported that weighing balance is unknown for selling and/or buying animals; rather visual assessment was used to estimate body condition of animals. Hence, selling price was fixed by negotiation between sellers and buyers. This result agreed with many research reports [16] [47] [48] from Ethiopia. The study results also indicated that higher numbers of goats are sold and better price is fetched in holidays although farmers sell goats at any time of the year depending on their need for money. Goat owners sold their goats primarily for purchasing food items, health expenses, farm input expenses (fertilizer and cropland rent), labor expenses (herdsman and laborers), school expenses for children and to pay back credit.

4.11. Production Constraints

Interviews with the sampled farmers revealed key problems constraining goat production in the area (Table 8). Water shortage, feed and grazing land shortages and inadequate veterinary services were ranked first, second and third, respectively. In line with the present findings previous studies [49] [50] indicated that the major problems of small ruminant farming include inadequate supply of water and pasture mainly in the dry season and problems arising from inadequate veterinary services and infrastructure.

Table 7. Ways of acquiring and disposing goat in the study area as ranked by owners.

Way of acquiring and disposing goat	Rank			Index
	Rank 1	Rank 2	Rank 3	
Acquiring (Entered through)				
Born	82	43	19	0.390
Bought	50	69	20	0.343
Gift	10	11	21	0.081
Inheritance	8	27	89	0.186
Disposing (Exited through)				
Sale	91	28	24	0.392
Death	29	24	78	0.237
Slaughter	17	83	32	0.277
Theft	7	11	11	0.060
Predator	6	4	5	0.034

Index = Σ [3 for rank 1 + 2 for rank 2 + 1 for rank 3] for particular method divided by Σ [3 for rank 1 + 2 for rank 2 + 1 for rank 3] for all methods.

Table 8. Ranked goat production constraints in the study area.

Constraints	Rank			Index
	Rank 1	Rank 2	Rank 3	
Feed and grazing area shortages	28	52	19	0.230
Water shortage	49	24	35	0.256
Inadequate veterinary services	31	32	23	0.200
Labor shortage	9	6	20	0.066
Market problem	20	23	24	0.144
Theft	9	7	17	0.064
Predators	4	6	12	0.040

Index = Σ [3 for rank 1 + 2 for rank 2 + 1 for rank 3] for particular constraint divided by Σ [3 for rank 1 + 2 for rank 2 + 1 for rank 3] for all constraints.

5. CONCLUSION

This study provides insight in to agricultural production system, breeding practices and major production constraints encountered in goat farming in the study area, which are preconditions in developing breeding programs. The present study also revealed that there are two types of farming systems in the area namely small-scale and large-scale farming. Goat production is an integral component of the existing farming systems by providing multifunctional roles to its owners. The high economic significance of goat coupled with its large flock size in the study area would suggest the scope of genetic improvement schemes in the area. However, water scarcity, feed and grazing area shortages, poor veterinary services and market linked problems should be addressed.

ACKNOWLEDGEMENTS

The authors would like to acknowledge farmers for their willingness to share their indigenous knowledge concerning the breed while nursing for the first author during his field stay.

REFERENCES

1. Nether Lands-African Business Council (2010) Livestock in Ethiopia and Opportunity Analyses for Dutch Investment. Fact Sheet, NABC.
2. Central Statistical Agency of Federal Democratic Republic of Ethiopia (2016) Agricultural Sample Survey: Report on Livestock and Livestock Characteristics (Private Peasant Holdings). Statistical Bulletin No. 583.
3. Workneh, A. (1992) Preliminary Survey of Indigenous Goat Types and Goat Husbandry Practices in Southern Ethiopia. M.Sc. Thesis, Alemaya University, Alemaya.
4. Köhler-Rollefson, I. (2000) Management of Animal Genetic Diversity at Community Level. GTZ, Eschborn, Germany. http://hdl.handle.net/10568/3679
5. Peacock, C. (2005) Goats: Unlocking Their Potential for Africa's Farmers. Paper presented at the 7th Conference of Ministers Responsible for Animal Resources, Kigali, 31 October-4 November 2005.
6. Jaitner, J., Sowe, J., Secka-Njie, E. and Dempfle, L. (2001) Ownership Pattern and Management Practices of Small Ruminants in The Gambia-Implications for a Breeding Program. Small Ruminant Research, 40, 101-108. https://doi.org/10.1016/S0921-4488(00)00221-2
7. Adane, H. and Girma, A. (2008) Economic Significance of Sheep and Goats. In: Alemu, Y. and Merkel, R.C., Eds., Sheep and Goat Production Handbook for Ethiopia, Ethiopian Sheep and Goat Productivity Improvement Program (ESGPIP), 1-4.
8. Assen, E. and Aklilu H. (2012) Sheep and Goat Production and Utilization in Different Agro Ecological Zones in Tigray, Ethiopia. Livestock Research for Rural Development, 24. http://www.lrrd.org/lrrd24/1/asse24016.htm
9. Baker, R.L. and Gray, G.D. (2003) Appropriate Breeds and Breeding Schemes for Sheep and Goats in the Tropics: The Importance of Characterizing and Utilizing Disease Resistance and Adaptation to Local Stresses. In: R. Sani, G.D. Gray and R.L. Baker, Eds. https://cgspace.cgiar.org/handle/10568/2933
10. Ethiopian Agricultural Research Organization (2002) An Assessment of the Agricultural Production Base, Technological Packages and Innovation and Intervention Strategies for Commercial farmers in Kafta Humera Wereda of Tigray Regional State.
11. Kafta Humera District Livelihood Report (2007) Disaster Prevention and Preparedness Agency, Ethiopia.
12. Institute of Biodiversity Conservation (2004) The State of Ethiopia's Farm Animal Genetic Resources: Country Report. A Contribution to the First Report on the Stat of the World's Animal Genetic Resources. IBC, May 2004, Addis Ababa.
13. Kosgey, I.S. (2004) Breeding Objectives and Breeding Strategies for Small Ruminant in the Tropics. Ph.D. Thesis, Wageningen University, Wageningen.
14. Falconer, D.S. (1989) Introduction to Quantitative Genetics. 3rd Edition, Longman Scientific and Technical, New York.
15. Central Statistical Agency of the Federal Democratic Republic of Ethiopia (2010) Summary and Statistical Report of the 2007 Population and Housing Census Result, Addis Ababa.
16. Endeshaw, A. (2007) Assessment of Production and Marketing System of Goats in Dale District Sidama zone. M.Sc. Thesis, University of Hawssa, Awassa.
17. Shenkute, B., Getahun, L., Azage, T. and Hassen, A. (2010) Small Ruminant Production in Coffee-Based Mixed Crop-Livestock System of Western Ethiopian Highlands: Status and Prospectus for Improvement. Livestock Research for Rural Development, 22. http://www.lrrd.org/lrrd22/10/shen22186.htm
18. Hulunm, G. (2014) On-Farm Phenotypic Characterization and Performance Evaluation of Bati, Borena and Short Eared Somali Goat Populations of Ethiopia. M.Sc. Thesis, Haramaya University, Haramaya.

19. Solomon, A. (2014) Design of Community Based Breeding Programs for Two Indigenous Goat Breeds of Ethiopia. Ph.D. Thesis, BOKU-University of Natural Resources and Life Sciences, Vienna.
20. Farm-Africa (1996) Goat Types of Ethiopia and Eritrea. Physical description and management systems. Published jointly by FARM Africa and International Livestock Research Institute.
21. Wilson, R.T. (1986) Livestock Production in Central Mali: Long Term Studies on Cattle and Small Ruminants in the Agro-Pastoral System. ILCA Research Report No. 14, 116.
22. Tsedeke, K. (2007) Production and Marketing of Sheep and Goats in Alaba, SNNPR. M.Sc. Thesis, Hawassa University, Awassa.
23. Dhaba, U., Belay, D., Solomon, D. and Taye, T. (2012) Sheep and Goat Production Systems in Ilu Abba Bora Zone of Oromia Regional State, Ethiopia: Feeding and Management Strategies. Global Veterinaria, 9, 421-429.
24. Ahmed, S., Kefelegne, K. and Kefena,E. (2015) Breeding Objective, Selection Criteria and Breeding Practice of Indigenous Goats in Western Ethiopia: Implications for Sustainable Genetic Improvemt. Greener Journal of Agricultural Sciences, 5, 167-176. https://doi.org/10.15580/GJAS.2015.5.072715105
25. Niekerk, W.A. and Pimentel, P.L. (2004) Goat Production in the Smallholder Section in the Boane District in Southern Mozambique. South African Journal of Animal Science, 34, 123-125.
26. Dossa, L.H., Sangaré, M., Buerkert, A. and Schlecht, E. (2015) Production Objectives and Breeding Practices of Urban Goat and Sheep Keepers in West Africa: Regional Analysis and Implications for the Development of Supportive Breeding Programs. SpringerPlus, 4, 281. https://doi.org/10.1186/s40064-015-1075-7
27. Otieno, O.G., Junga, J.O., Badamana, M.S. and Amimo, J.O. (2015) Indigenous Knowledge Used in Breeding and Management of Capra hircus Populations in Kajiado and Makueni Counties, Kenya. Open Journal of Genetics, 5, 111-135. https://doi.org/10.4236/ojgen.2015.53009
28. de Haan, C., Steinfeld, H. and Blackburn, H. (1997) Livestock and the Environment: Finding a Balance. Report of the Commission of the European Communities, the World Bank and the governments of Denmark, France, Germany, The Netherlands, UK and USA.
29. Sölkner, J., Nakimbugwe, H. and Valle Zarate, A. (1998) Analysis of Determinants for Success and Failure of Village Breeding Programmes. Proceedings of the 6th World Congress on Genetics Applied to Livestock Production, Armidale, 273-280.
30. Apori, S.O., Osei, D.Y., Rhule, S.W.A., Doku, C. and Oppong-Anane, K. (2011) Livestock Marketing in the Coastal Savannah Zone of Ghana. Proceedings of the 17th Biennial Conference of the Ghana Society of Animal Production, 20-23 July 2011, 43-45.
31. Rewe, T.O., Indetie, D., Ojango, J.M.K. and Kahi, A.K. (2006) Breeding Objectives for the Boran Breed in Kenya: Model Development and Application to Pasture-Based Production Systems. Animal Science Journal, 77, 163-177. https://doi.org/10.1111/j.1740-0929.2006.00334.x
32. Kugonza, D.R., Bareeba, F.B. and Kirembe, G. (2001) Characterization of Meat Goat Management Systems in Three Districts of Central Uganda. MUARIK Bulletin, 4, 26-29.
33. Semakula, J., Mutetikka, D., Kugonza, R.D. and Mpairwe, D. (2010) Smallholder Goat Breeding Systems in Humid, Sub Humid and Semi Arid Agro-Ecological Zones in Uganda. Global Veterinaria, 4, 283-291.
34. Arse, G., Feyisa, H., Gurmessa U., Merga, M. and Girma, D. (2013) Assessment on Challenges and Opportunities of Goat Farming System in Adami Tulu, Arsi Negelle and Fantale Districts of Oromia Regional State, Ethiopia. African Journal of Agricultural Research, 8, 26-31.
35. Byaruhanga, C., Oluka, J. and Olinga, S. (2015) Socio-Economic Aspects of Goat Production in a Rural Agro-Pastoral System of Uganda. Universal Journal of Agricultural Research, 3, 203-210. https://doi.org/10.13189/ujar.2015.030604
36. Braker, M.J.E., Udo, H.M.J. and Webb, E.C. (2002) Impacts of Intervention Objectives in Goat Production within Subsistence Farming Systems in South Africa. South African Journal of Animal Science, 32, 185-191. https://doi.org/10.4314/sajas.v32i3.3745

37. Mwacharo, J.M. and Drucker, A.G. (2005) Production Objectives and Management Strategies of Livestock-Keepers in South-East Kenya: Implications for a Breeding Programme. Tropical Animal Health and Production, 37, 635-652. http://www.ncbi.nlm.nih.gov/pubmed/16619880 https://doi.org/10.1007/s11250-005-4253-8
38. Chisholm, H. (1911) Beni-Amer. Encyclopædia Britannica 3, 11th Edition, Cambridge University Press, Cambridge.
39. Ndamukong, K.J., Sewell, M.M. and Asanji, M.F. (1989) Management and Productivity of Small Ruminants in Northwest Province of Cameroon. Tropical Animal Health and Production, 21, 109-119. https://doi.org/10.1007/BF02236189
40. Ngere, L.O., Adu, I.F. and Okubanjo, I.O. (1984) The Indigenous Goats of Nigeria. Animal Genetic Resources, 3, 1-9. https://doi.org/10.1017/S1014233900000109
41. Dekkers, J.C. and Gibson, J.P. (1998) Applying Breeding Objectives to Dairy Cattle Improvement. Journal of Dairy Science, 81, 19-35. https://doi.org/10.3168/jds.S0022-0302(98)70151-1
42. Armstrong, J.B. (2006) Inbreeding: Why We Will Not Do It? http://www.parispoodles.com/Inbreeding.html
43. Wilson, R.T. (1991) Small Ruminant Production and the Small Ruminant Genetic Resource in Tropical AFRICA. FAO Animal Production and Health, Rome, Paper No. 88.
44. Tesfaye, T. (2009) Characterization of Goat Production Systems and On-Farm Evaluation of the Growth Performance of Grazing Goats Supplemented with Different Protein Sources in Metema Woreda, Amhara Region, Ethiopia. M.Sc. Thesis, Haramaya University, Haramaya.
45. Belay, D. (2008) Growth, Reproductive Performance and Carcass Characteristics of Abergelle and Central Highland Goat Types under Traditional and Slightly Improved Management in Sekota Woreda. M.Sc. Thesis, Haramaya University, Haramay.
46. Ssewannyana, E., Onyait, O.A., Okwir, W., Ekoi, M., Okello, M., Masaba, J. and Ajibo, G.E. (2004) Characteristics of Rural Goat Production and Marketing in Kumi and Lira Districts, Uganda. Uganda Journal of Agricultural Science, 9, 289-293.
47. Ayele, S., Assegid, W., Jabbar, M.A., Ahmed, M.M. and Belachew, H. (2003) Livestock Marketing in Ethiopia: A Review of Structure, Performance and Development Initiatives. ILRI Socio-Economics and Policy Research Working Paper 52.
48. Gezahegn, A., Mohammad, A.J., Hailemariam, T., Elias, M. and Getahun, K. (2003) Seasonal and Inter-Market Differences in Prices of Small Ruminants in Ethiopia. Journal of Food Products Marketing, 12, 59-77.
49. Egbe-Nwiyi, T.N.C., Igbjekwe, F.N. and Nwosu, C.O. (1999) Gastro-Intestinal Parasitism and Associated Haemotological Changes in Small Ruminants in the Semi Arid Region of North-Eastern Nigeria. Bulletin of Animal Health and Production in Africa, 47, 61-70.
50. Omoike, A. (2006) Prevalence of Diseases among Sheep and Goats in Edo State Nigeria. Journal of Agriculture and Social Research, 6, 23-31.

CHAPTER 7

Congenital Malformations in River Buffalo (*Bubalus bubalis*)

Sara Albarella [1], Francesca Ciotola [1,*], Emanuele D'Anza [1], Angelo Coletta [2], Luigi Zicarelli [1] and Vincenzo Peretti [1]

[1] *Department of Veterinary Medicine and Animal Production, University of Naples Federico II, via Delpino 1, Naples 80137, Italy*

[2] *Associazione Nazionale Allevatori Specie Bufalina–ANASB, Caserta 81100, Italy*

AUTHOR CONTRIBUTIONS

Francesca Ciotola conceived the idea for the manuscript and led the article. Sara Albarella and Emanuele D'Anza performed the bibliographic research and drafted the manuscript. Luigi Zicarelli and Vincenzo Peretti revised critically different versions of the manuscript and added intellectual content. Angelo Coletta provided the new cases reported in this paper. Francesca Ciotola and Sara Albarella reviewed and approved the final version of the manuscript to be published.

SIMPLE SUMMARY

Congenital malformations (due to genetic causes) represent a hidden danger for animal production, above all when genetic selection is undertaken for production improvements. These malformations are responsible for economic losses either because they reduce the productivity of the farm, or because their spread in the population would decrease the total productivity of that species/breed. River buffalo is a species of increasing interest all over the world for its production abilities, as proved by the buffalo genome project and the genetic selection plans that are currently performed in different countries. The aim of this review is to provide a general view of different models of congenital malformations in buffalo and their world distribution. This would be useful either for those who performed buffalo genetic selection or for researchers in genetic diseases, which would be an advantage to their studies with respect to the knowledge of gene mutations and interactions in this species.

ABSTRACT

The world buffalo population is about 168 million, and it is still growing, in India, China, Brazil, and Italy. In these countries, buffalo genetic breeding programs have been performed for many decades. The occurrence of congenital malformations has caused a slowing of the genetic progress and economic loss for the breeders, due to the death of animals, or damage to their reproductive ability or failing of milk production. Moreover, they cause animal welfare reduction because they can imply foetal dystocia and because the affected animals have a reduced fitness with little chances of survival. This review depicts, in the river buffalo (*Bubalus bubalis*) world population, the present status of the congenital malformations, due to genetic causes, to identify their frequency and distribution in order to develop genetic breeding plans able to improve the productive and reproductive performance, and avoid the spreading of detrimental gene variants. Congenital malformations most frequently reported in literature or signaled by breeders to the Department of Veterinary Medicine and Animal Production of the University Federico II (Naples, Italy) in river buffalo are: musculoskeletal defects (transverse hemimelia, arthrogryposis, umbilical hernia) and disorders of sexual development. In conclusion this review put in evidence that river buffalo have a great variety of malformations due to genetic causes, and TH and omphalocele are the most frequent and that several cases are still not reported, leading to an underestimation of the real weight of genetic diseases in this species.

Keywords: congenital malformation; river buffalo; disorders of sexual development; genetic musculoskeletal defects

1. INTRODUCTION

The world's water buffalo population is about 168 million: 95% are in Asia (India, Pakistan, and China), 2% in Africa (Egypt), 2% in South America and the Caribbean, 1% in Europe and Oceania [1] Buffalo bred for milk production belong to the River type, like Murrah and Jaffarabadi (both bred in Brazil, also for meat production), Nili Ravi, Mediterranean River Buffalo, and Mediterranean Italian River Buffalo (MIRB). Swamp buffaloes, bred in South East Asia, are draft animals used extensively in paddy field farming.

Breeding programs are carried out in many countries, such as China, India, and Brazil; however, Italy represents the international point of reference for buffalo breeding as a result of the innovations brought in management, dairy technologies, cheese manufacturing, feeding, and reproduction. Despite the aim of breeding plans being to obtain healthy animals with good morphology and high production, it is not uncommon for animals to be born showing congenital defects due to gene mutations inherited from the parents or caused by new mutations or changes to the DNA. The occurrence of congenital malformations, above all when due to gene variants passed down by parents, in livestock production, causes the slowing of genetic progress, economic loss for the breeders due to the death of animals, or damage to their reproductive and productive ability, such as milk production. Moreover, congenital malformations imply a reduction of animal welfare directly in the affected animals because they have a reduced fitness and a lower adaptability to farming conditions and indirectly because can provoke foetal dystocia. To study the genetic bases of these diseases, and to find strategies to eradicate them in livestock populations is, for breeders, an issue to avoid economic losses [2].

With respect to the worldwide presence of congenital malformations in River buffalo, there are few papers, and a systematic report is still lacking. Purohit et al. [3] reported that 1.78% of dystocic deliveries from fetal causes are due to fetal malformations. In Brazil, Marcolongo-Pereira et al. [4] observed in the Murrah breed the occurrence of congenital defects in 7.54% of buffalo specimens necroscopically examined between 1978 and 2009. The most frequent congenital defects were arthrogryposis (seven animals), myotonia (29 animals), and mechano-bullous genodermatoses (four animals). According to Damè et al. [5] all of these defects were due to the high level of consanguinity of Brazilian buffaloes. In MIRB, transverse hemimelia [6] and sexual developmental abnormalities [7] were mainly reported (114 and six cases, respectively). According to the World Health Organization, congenital diseases are structural or functional anomalies, including metabolic disorders, which are present at the time of birth due to different etiology causes, such as genetic and environmental factors and infections [8].

This paper aims to provides an overview of the congenital malformations reported in river buffalo (*Bubalus bubalis*) in the literature or that have been signaled (and still unpublished) to the Department of Veterinary Medicine and Animal Production of University of Naples Federico II by ANASB and veterinarians, most likely due to genetic causes, according to available information, because of their increasing impact in the world livestock production chain. Moreover, the completion of buffalo genome sequencing and annotation will allow the identification of gene mutations underlying these malformations.

The congenital malformations observed can be classified as: musculoskeletal defects, gastrointestinal defects, craniofacial abnormalities, disorders of sexual development, and other malformations (Table 1).

Table 1. Case number and types of congenital malformations reported in literature in different river buffalo breeds.

Congenital Malformations	Breed	Cases (n)	References
Musculoskeletal defects			
Transverse Hemimelia	MIRB	10	[6]
	MIRB	13	[10]
	Indian Meshana	1	[12]
	MIRB	91	Present study
	unknown	1	[13]
Artrogryphosis	MIRB	1	[9]
	Murrah	6	[20]
	Murrah	1	[21]
Other musculoskeletal defects			
Perosomus elumbis	Nili Ravi	1	[22]
Campylorrhacchis contorta	unknown	1	[26]
Myotonia congenita	Murrah	29	[4]
Megaesophagus	Murrah	9	[5]
Other limb malformations	MIRB	2	Present study
Omphalocele	MIRB	72	[9]
	Pandharpuri	1	[39]
Schistosoma Reflexum	MIRB	2	[9]
	Murrah	1	[44]
Gastrointestinal defects			
Atresia ani	MIRB	2	[9]

Congenital Malformations	Breed	Cases (n)	References
Craniofacial abnormality			
Hydrocephalus	MIRB	3	[9]
	Murrah	7	[4]
	Surti	1	[58]
	–	1	[59]
Polycephaly	MIRB	1	Present study
	–	1	[68]
	–	1	[60]
	–	1	[67]
	Nili Ravi	1	[69]
Other Craniofacial malformations	Murrah	2	[75,76]
	Surti	1	[58]
	MIRB	1	Present study
Disorders of sexual development			
X-monosomy	MIRB	2	[7]
X-Trisomy	MIRB	1	[7]
Sex reversal	MIRB	2	[7]
Cryptorchidism	MIRB	1	Present study
Other malformations			
Meccanobullosus acantholytic dermatosis	Murrah	4	[5]

2. MUSCULOSKELETAL DEFECTS

2.1. Transverse Hemimelia

Transverse hemimelia (TH) is a congenital malformation characterized by total or partial lack of distal limb structures. It is also known as congenital amputation, since the malformed limb is similar to a stump normally developed proximally and distally, but cut off at different points.

In river buffalo, and particularly in MIRB, this is the most frequent congenital malformation diagnosed previously and, similar to other species, all of the cutaneous annexes are present so that in the terminal part of the stump, the sketch of hooves are visible. It was reported for the first time in some farms located in the province of Salerno [9], and then in different farms of Southern Italy [6,10].

At birth buffalo calves with TH are alive and lively, the only symptom is difficulty in moving due to the malformed limb or limbs. At necropsy they show no macroscopic lesions incompatible with life and their karyotype is normal. In Table 2 the cases analyzed in our lab are reported, classified according to phenotypic characteristics; the most frequent lesion is TH of the hock of one hind limb. Anamnestic data and examinations performed on malformed calves and their parents permitted us to suppose the action of one or more autosomal genes at the base of the malformation. Thanks to a genomic sequencing project the genome of 11 affected buffaloes was sequenced and compared with that of 14 healthy controls revealing that variants of *SMARCA4* and *WNT7A* genes are the main drivers for this type of TH malformation. Moreover, the accumulation of homozygous mutations in modifier genes appears to impact the severity of the TH phenotype, resulting in animals that vary from the lack of a single transverse bone in one limb to the complete lack of both hindlimbs with malformation of one forelimb [11], and probably to the lack of all four limbs. Another case that can be classified as TH has been reported by Jethva et al. [12], that observed an Indian Mehsana buffalo male with no hind limbs. More recently, a case of TH has been reported in a buffalo fetus with multiple craniofacial defects [13]. In both cases a genetic base is only supposed, but not supported, by other data.

2.2. Arthrogryposis

Arthrogryposis is a congenital disease characterized by non-progressive joint contractures that can affect upper or lower limbs and/or the vertebral column, leading to various degrees of flexion or extension limitations, evident at birth. It is not a specific diagnosis, but rather a clinical finding of congenital contractures causing severe mobility difficulties in affected individuals. Arthrogryposis has been reported in livestock and pets and is often associated with muscle atrophy or other malformations. It has more than one etiological factor, including physical limitation of in utero movement causing fetal akinesia/hypokinesia syndrome, maternal illness, intrauterine viral infection (Schmallenberg virus, Akabane virus, or Aino virus), toxin exposure, and genetic disorders affecting the fetus [14]. In Horse, this malformation has been associated with trisomy [15] while in sheep a candidate region for arthrogryposis has been identified on the proximal end of OAR5 [16]. In cattle arthrogryposis multiplex congenita has been associated with a genomic deletion encompassing *ISG15*, *HES4*, and *AGRN* genes in the American Angus breed [17]; and in Red Dairy cattle it is associated with a deletion in the first exon of *CHRNB1* [18]; furthermore, Iannuzzi et al., in a Piedmont calf affected by arthrogryposis, found the duplication of the *SMN* gene on BTA20q13.1 [19]. In Southern Brazil Schild et al. [20] found a Murrah buffalo herd in which the disease was transmitted as an autosomal recessive disorder, moreover, in India, Saini et al. [21] reported a case of a Murrah buffalo male fetus born dead with flexions and multiple articular rigidity of joints of all four limbs. Finally, Correale and Consalvo [9] observed, in MIRB, a case of a calf born with an abnormal and permanent ankylosis of limb joints. In the cases described by Saini et al., and Correale and Consalvo, the involvement of all four limbs let us assume gene abnormalities are responsible for the malformation.

Table 2. Phenotypic trait of transverse hemimelia (TH) cases and their frequency in Mediterranean Italian River Buffalo (MIRB) of Italian breeding farms (data from the Department of Veterinary Medicine and Animal Productions, University Federico II, Naples).

Malformed Limb	Anatomic Structures Involved	Frequency (%)
Left hind limb	Proximal epiphysis tibia	6
	Distal epiphysis tibia	2
	Tarsus	10
	Proximal epiphysis metatarsus	8
	Distal epiphysis metatarsus	10
	Proximal epiphysis of first phalanx	8
Right hind limb	Proximal epiphysis tibia	12
	Tarsus	2
	Proximal epiphysis metatarsus	6
	Distal epiphysis metatarsus	6
Both hind limbs	r. distal epiphysis tibia l. proximal epiphysis metatarsus	6
	r. distal epiphysis tibia l. distal epiphysis metatarsus	2
	r. metatarsus l. metatarsus	8
	r. second tarsus bones l. second phalanx	2
	r. knee l. twisted	2
Both hind limbs and one forelimb	r. hind limb second tarsus bones l. hind limb diaphysis tibia l. forelimb loss of the third phalanx	2
	r. hind limb second tarsus bones l. hind limb proximal epiphysis metatarsus r. forelimb hypoplasia	2
	Amelia of both hind limbs r. forelimb longitudinal ectromelia of metacarpus	2
All limbs involved		4

r. = right; l. = left.

2.3. Other Musculoskeletal Defects

A variety of other musculoskeletal malformations have been reported in river buffalo. Mehmood et al. [22] described a case of congenital perosomus elumbis (PE) in a Nili Ravi buffalo characterized by small vertebral column, under-developed pelvis, and deformed hind limbs. PE has also been reported in other animal species, such as cattle, mainly in the Holstein Friesian breed, sheep, pigs, dogs, and horses [23]. The etiology of PE is still not known, but mutations within the homeobox (Hox) gene family are believed the principal causes of this malformation because they are involved in fundamental ontogenetic processes [24,25].

Kumar et al. [26] observed a case of *campylorrhacchis contorta* in a calf born from a primiparous Indian buffalo; the fetus was under-weighed and showed brachignathism/parrot mouth, malformation of the pelvis, lumbar, and sacral regions, and ankylosis of all of the limbs with atrophy of musculature. According to the author, this malformation being a congenital defect is most likely caused by a defective gene or a genetic insult.

Another congenital musculoskeletal defect reported in river buffalo is myotonia congenita; a specific inherited disorder of muscle membrane hyperexcitability. It is caused by reduced sarcolemmal chloride conductance due to mutations in *CLCN1*; the gene coding for the main skeletal muscle chloride channel *ClC-1* [27]. It has been also reported in 29 Murrah buffaloes [4], inherited in an autosomal recessive manner and due to an aberrant splicing of *CLCN1* associated to a synonymous SNP in exon-3 (c.396C>T). All animals showed generalized muscular hypertrophy and action myotonia, especially after being startled with difficulties in rising from recumbency and stiffness at the beginning of the movements [28].

Congenital megaesophagus (ME) consists of esophageal dilation and dysmotility that progressively leads to the inability to drive the food/bolus into the stomach. It has been reported in cats [29], dog [30], horses [31], cattle [32], and buffalo [5]. The etiology of ME is still unclear, and GWAS and whole genome sequencing analyses performed in cats and dogs showed that different genes could be related to this condition, like *COLQ* gene in cats [29], and genes located in a region of CFA12 in the German Shepherd dog breed [30].

Nine cases of ME have been reported in Murrah buffaloes [5] in which genealogical data permitted the establishment that the condition is inherited. Affected animals showed a short tail, delayed growth, chronic bloat, regurgitation, and aspiration pneumonia. They died at 3–9 months of age and, from an anatomophatological point of view, they showed dilated and obstructed esophagus with flaccid musculature, edema, emphysema, and the presence of liquids in the thoracic cavity.

In addition to the previously described congenital malformations involving limbs, at the Department of Veterinary Medicine and Animal Production of the University Federico II of Naples, in MIRB, we have examined a case of forelimb amelia and a calf with a complex of skeletal malformation involving only limbs characterized by hypoplasia of the radius and ulna, the absence of medial femoral condyle, and tibial agenesis (unpublished data). Embrionic limb development is a complex process under the control of several genes, like *SHH*, *GLI3*, *WNT7A*, *Fgfs*, *BMPs*, *ALX4*, *PRMT5*, and *SMARCA4*; thus, they can all be considered candidate genes for these malformations. Recently, a duplication of exon 2 of the *ALX4* has been observed in Galloway Cattle affected by tibial hemimelia syndrome and assumed to be the candidate causative mutation for this malformation [33].

2.4. Umbilical Hernia

Congenital umbilical hernia (omphalocele) is the protrusion, in a newborn, of the peritoneal sac and, eventually, of other abdominal organs that are covered by skin and subcutaneous tissue through the abdominal wall at the umbilical level [34]. It is due to the failure of the closure of the umbilical ring at birth and its severity varies in relation to the extent of the umbilical defect and the amount of the protruding abdominal organs. In small and large animals, it is a common defect and a genetic component is believed to be involved in this malformation. In pigs it is the most relevant congenital disorder and two gene loci have been associated to this malformation: *SWR1928* on SSC7 and *SW830* on SSC10 [35]. In this species, the *CDC73* gene is assumed to be responsible for the malformation [36]. In cattle this condition is the most frequently noted congenital disorder [37] and it seems to be associated with genes located at the centromeric end of BTA8 [38]. Additionally, this is the most common malformation in river buffalo, but only a few cases are well described in the literature, like in the Pandharpuri buffalo breed in which Amle et al. [39] found a male calf with incomplete closure of umbilicus and reducible intestinal hernia associated to cryptorchidism, and in MIRB, in which Correale and Consalvo [9] found 72 cases with different phenotypic traits. Genetic causes in this species have never been studied and it could be interesting to verify if, in river buffalo, responsible genes are located on the long arm of BBU3 homologue of BTA8 [40].

2.5. Schistosoma Reflexum Syndrome

Schistosoma Reflexum or schistosomus reflexus (SR) syndrome is a rare and fatal congenital malformation characterized by exposed abdominal, and sometimes thoracic, viscera [41], failure or incomplete closure of abdominal wall, and marked dorsoflexion and ankylosis of the spine and limbs [42]. Other malformations, like fused ribs and vertebrae, fissure of genitals, and polypodia may also appear. Kumar et al. [43] reported a case of dystocia in a Murrah buffalo due to a fetus with schistosoma reflexum characterized by distinct ventral curvature of the spine, incomplete development of diaphragm attachment and thoracic arc, exposition of abdominal visceral organs, like intestines, omentum, kidney, and liver, and limbs that were touching each other, while the neck and head were in an ankolysed state. In MIRB, Correale and Consalvo [9] reported the birth of two calves with exposition of the abdominal organs and dorsoflexion of the spine; both calvings were at full term pregnancy, but dystocic. In particular, one calf showed a somathopleure dorsal tilt, instead of ventrally, closes on itself, like a bag formed with amnios, a dorsal bag in which limbs were gathered. Moreover, ribs were sternalized on the back, thus resembling a reverse thorax. These cases were diagnosed as SR syndrome. No etiology was supposed in both reports and no genetic analyses has been performed, but it has been reported that an autosomal recessive mode of inheritance for SR exists in pets [44] and in cattle [42].

3. GASTROINTESTINAL DEFECTS

3.1 Atresia Ani

Atresia ani is a congenital malformation of the anorectum due to the failure of the urorectal fold to divide the cloaca completely or of the failure of the perforation of the fetal anal membrane that divides the rectus and anus during fetal development. As a consequence, the anal opening is closed and feces are accumulated in the large intestine. This is the most common defect of the lower gastrointestinal tract reported in mammals [45] and, in particular, in cattle, pigs, and sheep [46], while it has been reported few times in horses and donkeys [47]. In pigs a polygenic inheritance, autosomal recessive in some cases and autosomal dominant transmission in others, has been observed [48] and genome analysis allowed for the identification of candidate loci for atresia ani on SSC1, 9, and 15 [49,50], and *GLI2* as a candidate gene [51]. In cattle, several cases of atresia ani are reported [45,52-55] mainly in Holstein Friesian. In this breed there is some evidence that this condition is inherited in an autosomal recessive manner, but no candidate genes have been found. In MIRB, Correale and Consalvo [9] reported two cases in which the rectum is normally developed but ends in a blind pouch due to the persistence of a complete anal membrane. No data are available about the mode of its inheritance in buffalo or its etiology, but the development of the hindgut is governed by multiple genes in the relevant signaling pathways. Based on the evidence of human genetic studies, murine gene knock-out and teratogenic models, and epidemiological studies, anorectal malformation (ARM) may, most likely, be the result of genetic susceptibility and epigenetic interaction. Genetic factors have been considered important contributing factors in the pathogenesis of ARM [56].

4. CRANIOFACIAL ABNORMALITIES

4.1. Hydrocephalus

Hydrocephalus is a multifactorial, congenital, or acquired disorder characterized by an abnormal accumulation of cerebrospinal fluid (CSF) within the cranial cavity. It is defined as internal when the CFS is accumulated in the ventricles, and external when CFS is accumulated in the subarachnoid space [57]. The increase of CFS is, above all, related to its abnormal reabsorption or defective lymphatic drainage and rarely to its production, and it induces progressive enlargement of the head. Etiology includes: genetic factors, developmental anomalies, intrauterine or prenatal infection, in utero teratogens exposure, dietary deficiencies (vitamin A deficiency in the rabbit), or tumors or bleeding in the brain. Between 1978 and 2009 from 61 autopsies of Murrah buffaloes, Marcolongo-Pereira et al. [4] diagnosed seven cases of hydranencefalia, probably due to hereditary factors, one of which was associated to the condition of albinism. The histopathological examination of this last case revealed the presence of cystic cavities in the thalamus and the substitution of nervous tissue with liquid; moreover, the cortex was thinner. Pandey et al. [58] described a case of hydrocephalus in a male "Bull Dog" calf as shown in Section 4.3. Sharma et al. [59] described the case of dystocia due to hydrocephalic monster as a buffalo calf characterized by an extraordinarily large, football shaped head; on dissection, cranial bones were found markedly thin. Although genetic factors are the most frequent cause of this malformation, the authors have not fully evaluated intra-uterine infections and nutritional factors. Thus, the causes cannot be ruled out in this case. In MIRB Correale and Consalvo [9] have found three cases of hydrocephalus with typical alterations consisting of dilation of cerebral ventricles due to abnormal liquor accumulation and consequent increase of cranium volume. The nervous structures were reduced due to the pressure caused by the increased liquor volume; the calves were born dead or died within a few days after birth. No other malformations are described in those three cases and the possible action of infections and teratogen substances, such as drugs or toxins, is not supposed by the authors. This lets us assume that, in all of the cases, a genetic cause is the most likely cause.

4.2. Polycephaly

Polycephaly is a congenital malformation in which an individual with two (dicephaly) or more heads, probably due to partial or total union of two developing embryos or to a partial duplication of a body, or to the anteroposterior compression of the embryonic disk [60]. It is very rare and the affected individual, typically, is either stillborn or dies very soon after birth. It has been reported in different mammals alone, or associated with other malformations, in cattle [61-64], sheep [65], and goats [66] but, as for other congenital abnormalities, polycephaly is likely to occur more frequently than the veterinary literature would suggest and only in few cases is it possible to hypothesize genetic mutations in the etiology. The most frequent type of polycephalia observed is dicephalia, which is classified in: atlodymus (two complete and separate skull and one neck), iniodymus (two skulls with fusion at the occipital level), and derodymus (two complete and separate skulls with two separate necks [60].

In buffaloes several cases of duplication of the head with different characteristics have been reported: monster with varying degrees of duplication of the face or diprosopus [67] and dicephalus monsters characterized by two full developed heads of similar size fused at the level of atlas [60,68- 70].

In MIRB, we have seen a case of a dicephalic derodymus calf (Figure 1) characterized by complete duplication of cranial structures (two muzzles, four eyes, and four ears) and organs up to the neck. The calf was born dead after a dystocic calving, thus cytogenetic analyses were not performed, while according to genetic analyses performed the calf is not the fusion of two developing embryos (unpublished data).

4.3. Other Craniofacial Malformations

Other congenital craniofacial malformations are very rare in domestic animals and the most frequent reported are brachygnatia, prognathia, micrognathia, clefts of palate, of the face, or of specific facial boneshave. In some cases an inherited condition has been demonstrated [71,72] and sometimes underlying chromosome regions are involved [73], or genetic mutations [74] have been detected.

In India, three cases of "Bull Dog" buffalo calves have been observed; the first one was from a Murrah buffalo fetus of 8.5 months characterized by complete alopecia, broad head, bulging forehead, malocclusions of the jaw, prognathism of the mandible, and pot belly [75]; the second case was a dead male calf from the Surti breed with prognathism, protruding tongue, nose divided by furrows, hydrocephalus, and the whole body covered by hairs [58]. In both cases, thoracic and visceral organs were normal. A third case from a Murrah buffalo has been reported by Singh et al. [76], which was characterized by an enlarged head, bulging of eyes, large ears, short and stumpy legs, short tail, round distended abdomen, hairless body, pot belly, and excessive lipomatous mass in the abdominal cavity. It was supposed to be a simple autosomal recessive defect with some alterations. In MIRB we observed a craniofacial malformation characterized by underdeveloped incisive, maxillary and nasal bones with a consequent tongue prolapse, and lower jaw deformities (Figure 2) that, according to literature [58,75] can be classified as another case of "Bull Dog" calf. The absence of conditions like maternal infections and teratogenic factors exposure assume a genetic cause behind the malformation of this calf.

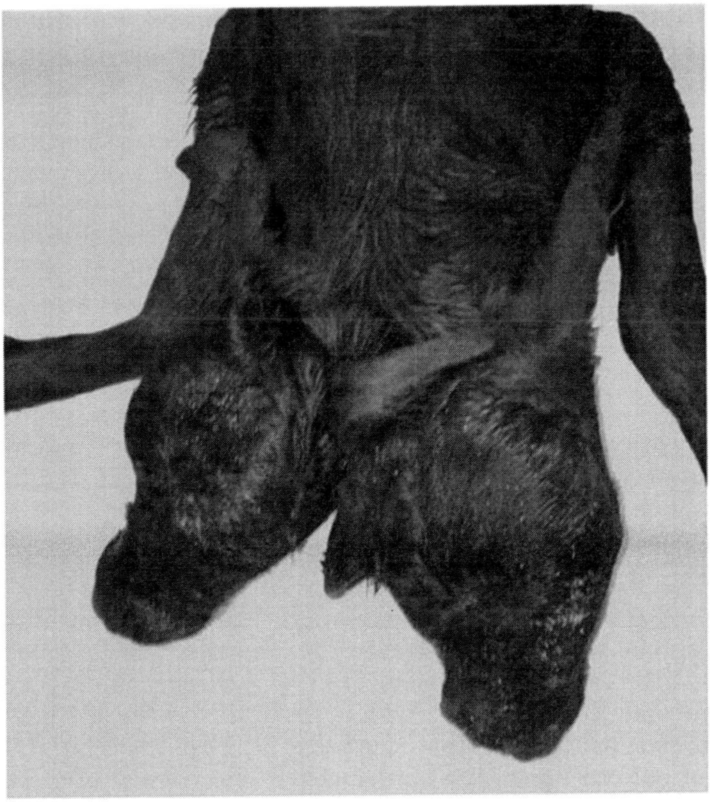

Figure 1. Dicephalic derodymus newborn MIRB calf with complete duplication of cranial structures up to the neck.

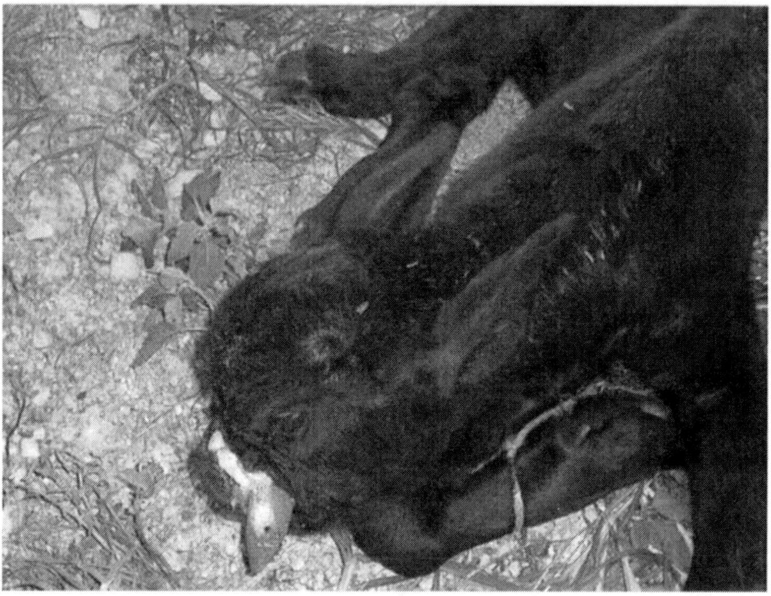

Figure 2. A newborn MIRB calf having underdeveloped incisive, maxillary, and nasal bones with a consequent tongue prolapse and lower jaw deformities.

5. DISORDERS OF SEXUAL DEVELOPMENT

Disorders of sexual development (DSDs) are congenital conditions in which chromosomal, gonadal, or anatomical sex development is atypical [77]. Sexual development is based on three essential steps: (1) the establishment of chromosomal sex which occurs at the time of fertilization; (2) the determination of gonadal sex; and (3) the development of phenotypic sex; the last two steps are under genetic control.

The abnormalities of chromosomal sex are linked to abnormalities in sex chromosome number or structure. In the river buffalo, cases of X-monosomy (Turner's syndrome) [78,79], X-trisomy [80,81], and XXY chromosome complemented with translocation between the two X chromosomes [82] have been reported. It has been estimated that about 20% of MIRB females with reproductive problems and examined with cytogenetic analyses are carriers of sex chromosome abnormalities [83]. In domestic animals the X-monosomy is a rare congenital condition due to the absence of one X-chromosome and has been reported in two Murrah and two MIRB buffaloes [8,79]. Generally the carrier is sterile with serious modifications of the reproductive apparatus, such as an abnormal cervix and uterus, or the absence of gonads. The X-trisomy has been found in two river buffaloes, one Murrah [80] and one MIRB [7]. Both animals were sterile because of abnormalities of internal genitalia (absent gonads). For the first time Patel et al. [82] reported a case of a Mehsana buffalo bull with an XXY chromosome complement and with a translocation between the two X chromosomes. The bull was phenotypically normal with regular reproductive behavior but sterile, due to azoospermia.

The abnormalities of gonadal sex are related to mutation, deletion, or translocation of small parts of chromosome containing one or more genes which regulate the testis and ovarian developing pathways, like *SRY*, *SOX9*, *RSPO1*, and *WNT*. These abnormalities are defined as sex reversal disorders and classified according to the chromosomal complement detected in the individual (male or female) and the presence/absence of *SRY* gene [84]. Therefore, there are four conditions: XX male SRY-negative, XX male SRY-positive, XY female SRY-negative, and XY female SRY-positive [85].

In domestic animals sex reversal disorders are rare and described in horses [84-86], cattle [87], pigs [88], dogs [89], and cats [90]. Up to now, in river buffalo only the XY female condition in two MIRBs with female external genitalia and some phenotypic male traits [7] has been reported. Both animals were sterile due to gonadal dysgenesis; in particular, one was SRY positive and lacking internal sex ducts; and the other one (with an unknown *SRY* condition) showed Müllerian duct atrophy.

The abnormalities of phenotypic sex are related to developmental alterations of internal and external genitalia, which is strictly dependent on secretions from testis in males. In carriers, defined pseudohermaphrodites, the chromosomal and gonadal sex are in agreement, but the internal or external genitalia are ambiguous. The causes of this condition are often unknown because multiple genes are involved in the differentiation of genital apparatus and their roles are still not clear. In dogs, the iatrogenic administration of androgen or progesterone to the dam during gestation could induce masculinization of female fetuses (female pseudohermaphrodites) [89]. In Persistent Müllerian Duct Syndrome (PMDS), a type of canine male pseudohermaphroditism, a mutation of the *MISRII* gene, causes the failure of the regression of Müllerian ducts in males during embryo development [88].

Another disorder of sexual development is cryptorchidism, a condition in which one or both testicles fail to descend into the scrotum and it is an abnormality in the development of phenotypic sex without sexual ambiguity. In addition to environmental factors, like endocrine disruptors, cryptorchidism is at least in part determined by genetic causes and, until now, many candidate genes have been identified [91]. It is very common in livestock and pets, while, in river buffalo, only one case of cryptorchidism has been described in a Pandharpuri calf associated to umbilical hernia. In the affected calf both testicles were located ectopically under the skin of the ventral abdomen alongside the penis [39]. In MIRB, we have examined a male calf with bilateral abdominal cryptorchidism and a penis craniocaudally directed; the calf was born healthy and at the end of gestation (Figure 3). No karyotype abnormalities were found in this calf and gene analyses are still ongoing.

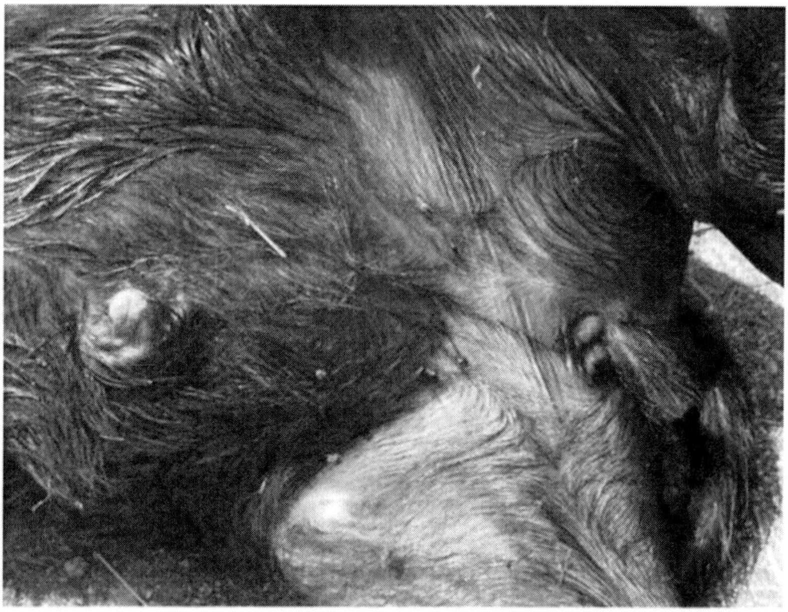

Figure 3. A MIRB male calf with abdominal cryptorchidism and penis craniocaudally directed.

6. OTHER MALFORMATIONS

6.1 Meccanobullosus Acantholytic Dermatosis

This condition is very rare and characterized by a reddish skin that is easy to lacerate, with detachment of the epidermic tissue. At birth the animal is normal, but at 2–3 days of age tufts of hair fall out and the skin starts to lacerate; moreover, injuries to the corneal layer of the hoof may occur. With respect to river buffalo, it has been reported in four Murrah calves transmitted with an autosomal recessive gene and due to the reduced functionality of cadherins [5].

7. CONCLUSIONS

Congenital malformations due to genetic factors represent a significant limit for selection plans and for animal welfare when the underlying gene mutations are unknown and carrier identification is not possible until the mutation is so widespread as to be phenotypically evident.

This is typically due to an increase of the inbreeding rate and the spreading of negative alleles responsible for genetic diseases. To avoid this problem, it is necessary not only to set up breeding plans that avoid the increase in inbreeding, but also to know genetic conditions underlying the diseases and exclude carriers from mating. The completion of buffalo genome sequencing would represent another useful tool to discover those conditions, but in order that these mutations are discovered, a good recording system is necessary and breeders and institutions must be sensitized to the importance of reporting and examining all new cases of malformations, sampling all cases in which genetic causes are assumed. According to the literature and our personal experience, river buffalo have a great variety of malformations due to genetic causes, and TH and omphalocele are the most frequent. However, several cases are still not reported, leading to an underestimation of the real weight of genetic diseases in this species. This theory is confirmed by the fact that, despite 95% of the world buffalo population being bred in Asia, the number of malformations reported in the literature is extremely small, in contrast to what is observed in countries, such as Brazil and Italy, where only 2% of the buffalo world population is bred. Moreover, except for DSDs for which genetic causes have been identified and diagnostic tests are available, for all other malformations, the genetic causes are unknown. In conclusion this report represents a reference point to focus future studies on the most important malformations that affect river buffalo and that compromise the results of genetic improvement plans and the welfare of the animals.

REFERENCES

1. FAO. Water Buffaloes. Available online: http://www.fao.org/agriculture/dairy-gateway/milk-production/dairy-animals/water-buffaloes/en/#.WJsMcLFE_aY (accessed on 15 March 2016).
2. Gholap, P.N.; Kale, D.S.; Sirothia, A.R. Genetic diseases in cattle: A review. *Res. J. Anim. Vet. Fish. Sci.* **2014**, *2*, 24–33.
3. Purohit, G.N.; Kumar, P.; Solanki, K.; Shekher, C.; Yadav, S.P. Perspectives of fetal dystocia in cattle and buffalo. *Vet. Sci. Dev.* **2012**, *2*, 31–42.
4. Marcolongo-Pereira, C.; Schild, A.L.; Pereira-Soares, M.; Vargas, S.F., Jr.; Riet-Correa, F. Congenital defects in ruminants in southern Brazil. *Pesq. Vet. Bras.* **2010**, *30*, 816–826.
5. Damé, M.C.F.; Riet-Correa, F.; Schild, A.L. Hereditary diseases and congenital defects diagnosed in water buffalo (*Bubalus bubalis*) in Brazil. *Pesq. Vet. Bras.* **2013**, *33*, 831–839.
6. Peretti, V.; Ciotola, F.; Albarella, S.; Restucci, B.; Meomartino, L.; Ferretti, L.; Barbieri, V.; Iannuzzi, L. Increased SCE levels in Mediterranean Italian buffaloes affected by limb malformation (transversal hemimelia). *Cytogenet. Genome Res.* **2008**, *120*, 183–187.
7. Di Meo, G.P.; Perucatti, A.; Di Palo, R.; Iannuzzi, A.; Ciotola, F.; Peretti, V.; Neglia, G.; Campanile, G.; Zicarelli, L.; Iannuzzi, L. Sex chromosomeabnormalities and sterility in river buffalo. *Cytogenet. Genome Res.* **2008**, *120*, 127–131.
8. WHO. Congenital Anomalies. Available online: http://www.who.int/ mediacentre/ factsheets/fs370/en/ (accessed on 15 March 2016).
9. Correale, E.; Consalvo, F. About some congenital malformation in buffaloes bred in the Salerno province (Italy). *Bubalus Bubalis* **2003**, *9*, 22–29.
10. Albarella, S.; Ciotola, F.; Dario, C.; Iannuzzi, L.; Barbieri, V.; Peretti, V. Chromosome instability in Mediterranean Italian buffaloes affected by limb malformation (transversal hemimelia). *Mutagenesis* **2009**, *24*, 471–474.
11. Whitacre, L.K.; Hoff, J.L.; Schnabel, R.D.; Albarella, S.; Ciotola, F.; Peretti, V.; Strozzi, F.; Ferrandi, C.; Ramunno, L.; Sonstegard, T.D.; et al. Elucidating the genetic basis of an oligogenic birth defect using whole genome sequence data in a non-model organism, *Bubalus bubalis. Sci. Rep.* **2017**, *7*, 39719.
12. Jethva, P.C.; Patel, S.B.; Manohar, B.K. Amelia of hind limbs in a buffalo calf: A case report. *Indian J. Fields Vet.* **2014**, *9*, 73–74.
13. Honparkhe, M.; Lochan, S.; Shiv, K.; Kumar, N.; Kumar, A. Transversal tetra-hemimelia with multiple craniofacial anomalies in a buffalo calf. *Indian J. Anim. Reprod.* **2016**, *37*, 61–62.
14. Agerholm, J.S.; Hewicker-Trautwein, M.; Peperkamp, K.; Windsor, P.A. Virus-induced congenital malformations in cattle. *Acta Vet. Scand.* **2015**, *57*, 54.
15. Buoen, L.C.; Zhang, T.Q.; Weber, A.F.; Turner, T.; Bellamy, J.; Ruth, G.R. Arthrogryposis in the foal and its possible relation to autosomal trisomy. *Equine Vet. J.* **1997**, *29*, 60–62.
16. Murphy, A.M.; MacHugh, D.E.; Park, S.D.; Scraggs, E.; Haley, C.S.; Lynn, D.J.; Boland, M.P.; Doherty, M.L. Linkage mapping of the locus for inherited ovine arthrogryposis (IOA) to sheep chromosome 5. *Mamm. Genome* **2007**, *18*, 43–52.
17. Beever, J.E.; Marron, B.M. Screening for Arthrogryposis Multiplex in Bovines. U.S. Patent 20110151440 A1, 23 Jun 2011. Available online: http://patft.uspto. gov/ netacgi/nphParser?Sect1=PTO2&Sect2=HITOFF&p=1&u=%2Fnetahtml%2FPTO%2Fsearchbool.html&r=1&f=G&l=50&co1=AND&d=PTXT&s1=%22Screening+arthrogryposis+multiplex+bovines%22.TI.&OS=TTL/ (accessed on 15 March 2016).
18. Agerholm, J.S.; McEvoy, F.J.; Menzi, F.; Jagannathan, V.; Drögemüller, C. A CHRNB1 frameshift mutation is associated with familial arthrogryposis multiplex congenita in Red dairy cattle. *BMC Genom.* **2016**, *17*, 479.

19. Iannuzzi, L.; Di Meo, G.P.; Perucatti, A.; Rullo, R.; Incarnato, D.; Longeri, M.; Bongioni, G.; Molteni, L.; Galli, A.; Zanotti, M.; et al. Comparative FISH-mapping of the survival of motor neuron gene (SMN) in domestic bovids. *Cytogenet. Genome Res.* **2003**, *102*, 39–41.
20. Schild, A.L.; Soares, M.P.; Damé, M.C.; Portianski, E.L.; Riet-Correa, F. Arthrogryposis in Murrah buffaloes in southern Brazil. *Pesq. Vet. Bras.* **2003**, *23*, 13–16.
21. Saini, G.S.; Pandey, A.K.; Chaudhary, R.N.; Kumar, A.; Sharma, S. Arthrogryposis in a Murrah buffalo calf: A case report. *Buffalo Bull.* **2010**, *29*, 318–320.
22. Mehmood, M.U.; Qamar, A.; Raza, S.; Khan, H.; Shahzad, Q.; Sattar, A. Dystocia due to Perosomus Elumbis (Acaudatus) in a buffalo. *Pak. J. Zool.* **2014**, *46*, 1468–1470.
23. Agerholm, J.S.; Holm, W.; Schmidt, M.; Hyttel, P.; Fredholm, M.; McEvoy, F.J. Perosomus elumbis in Danish Holstein cattle. *BMC Vet. Res.* **2014**, *10*, 227.
24. Avedillo, L.J.; Camón, J. Perosomus elumbis in a pig. *Vet. Rec.* **2007**, *160*, 127–129.
25. Gerhauser, I.; Geburek, F.; Wohlsein, P. Perosomus elumbis, cerebral aplasia, and spina bifida in an aborted thoroughbred foal. *Res. Vet. Sci.* **2012**, *92*, 266–268.
26. Kumar, W.P.; Sharma, G.C.; Krishnappa, B.; Rakesh, H.B.; Soni, Y.K. A rare case of Campylorrhacchis contorta in a buffalo: A case report. *Buffalo Bull.* **2014**, *33*, 355–357.
27. Lossin, C.; George, A.L., Jr. Myotonia congenita. *Adv. Gen. San.* **2008**, *63*, 25–55.
28. Borges, A.S.; Barbosa, J.D.; Resende, L.A.; Mota, L.S.; Amorim, R.M.; Carvalho, T.L.; Garcia, J.F.; Oliveira-Filho, J.P.; Oliveira, C.M.; Souza, J.E.; et al. Clinical and molecular study of a new form of hereditary myotonia in Murrah water buffalo. *Neuromuscul. Disord.* **2013**, *23*, 206–213.
29. Gandolfi, B.; Grahn, R.A.; Creighton, E.K.; Williams, D.C.; Dickinson, P.J.; Sturges, B.K.; Guo, L.T.; Shelton, G.D.; Leegwater, P.A.; Longeri, M.; et al. COLQ variant associated with Devon Rex and Sphynx feline hereditary myopathy. *Anim. Genet.* **2015**, *46*, 711–715.
30. Tsai, K.L.; Noorai, R.E.; Starr-Moss, A.N.; Quignon, P.; Rinz, C.J.; Ostrander, E.A.; Steiner, J.M.; Murphy, K.E.; Clark, L.A. Genome-wide association studies for multiple diseases of the German Shepherd Dog. *Mamm. Genome* **2012**, *23*, 203–211.
31. Komine, M.; Langohr, I.M.; Kiupel, M. Megaesophagus in Friesian horses associated with muscular hypertrophy of the caudal esophagus. *Vet. Pathol.* **2014**, *51*, 979–985.
32. Ulutas, B.; Sarierler, M.; Bayramli, G.; Ocal, K. Macroscopic findings of idiopathic congenital megaoesophagus in a calf. *Vet. Rec.* **2006**, *158*, 26.
33. Brenig, B.; Schütz, E.; Hardt, M.; Scheuermann, P.; Freick, M. A 20 bp Duplication in Exon 2 of the Aristaless-Like Homeobox 4 Gene (ALX4) Is the Candidate Causative Mutation for Tibial Hemimelia Syndrome in Galloway Cattle. *PLoS ONE* **2015**, *10*, e0129208.
34. Steinman, A.; Kelmer, G.; Avni, G.; Johnston, D.E. Omphalocele in a foal. *Vet. Rec.* **2000**, *146*, 341–343.
35. Ding, N.S.; Mao, H.R.; Guo, Y.M.; Ren, J.; Xiao, S.J.; Wu, G.Z.; Shen, H.Q.; Wu, L.H.; Ruan, G.F.; Brenig, B.; et al. A genome-wide scan reveals candidate susceptibility loci for pig hernias in an intercross between White Duroc and Erhualian. *J. Anim. Sci.* **2009**, *87*, 2469–2474.
36. Su, Y.; Long, Y.; Ruan, G.; Wu, L.; Zhang, Z.; Xiao, S.; Deng, W.; Lv, X.; Hu, D.; Wu, G.; et al. Identification of susceptibility gene for pig umbilical hernia in different populations using transmission disequilibrium test. *Yi Chuan***2014**, *36*, 995–1005.
37. Čítek, J.; Řehout, V.; Hájková, J. Congenital disorders in the cattle population of the Czech Republic. *Czech J. Anim. Sci.* **2009**, *54*, 55–64.
38. Ron, M.; Tager-Cohen, I.; Feldmesser, E.; Ezra, E.; Kalay, D.; Roe, B.; Seroussi, E.; Weller, J.I. Bovine umbilical hernia maps to the centromeric end of Bos taurus autosome 8. *Anim. Genet.* **2004**, *35*, 431–437.
39. Amle, M.B.; Shelar, R.R.; Thorat, M.G.; Zope, A.N. Congenital umbilical hernia and cryptorchidism in a Pandharpuri buffalo calf. *Buffalo Bull.* **2004**, *23*, 82–83.
40. Iannuzzi, L.; Di Meo, G.P.; Perucatti, A.; Schibler, L.; Incarnato, D.; Gallagher, D.; Eggen, A.; Ferretti, L.; Cribiu, E.P.; Womack, J. The river buffalo (*Bubalus bubalis*, 2n = 50)

cytogenetic map: Assignment of 64 loci by fluorescence in situ hybridization and R-banding. *Cytogenet. Genome Res.* **2003**, *102*, 65–75.
41. Laughton, K.W.; Fisher, K.R.; Halina, W.G.; Partlow, G.D. Schistosomus reflexus syndrome: A heritable defect in ruminants. *Anat. Histol. Embryol.* **2005**, *34*, 312–318.
42. Citek, J. Pedigree analysis of Czech Holstein calves with schistosoma reflexum. *Acta Vet. Scand.* **2012**, *54*, 22.
43. Kumar, S.; Kumar, S.; Sharma, U.; Kushwaha, R.B.; Pandey, A.K. Dystocia due to schistosomus reflexus in a Murrah buffalo. *Indian J. Anim. Reprod.* **2012**, *33*, 84–85.
44. Ozalp, G.R.; Celikler, S.; Simsek, G.; Ozyigit, M.O.; Inan, S. A case of schistosoma reflexum in a cat with chromosomal aberrations. *Reprod. Domest. Anim.* **2011**, *46*, 373–376.
45. Binanti, D.; Prati, I.; Locatelli, V.; Pravettoni, D.; Sironi, G.; Riccaboni, P. Perineal choristoma and atresia ani in 2 female Holstein Friesian calves. *Vet. Pathol.* **2013**, *50*, 156–158.
46. Greber, D.; Doherr, M.; Drögemüller, C.; Steiner, A. Occurrence of congenital disorders in Swiss sheep. *Acta Vet. Scand.* **2013**, *55*, 27.
47. Nelson, B.B.; Ferris, R.A.; McCue, P.M.; Leise, B.S. Surgical management of atresia ani and perineal hypospadias in a miniature donkey foal. *Equine Vet. Educ.* **2015**.
48. Zimmerman, J.J.; Karriker, L.A.; Ramirez, A.; Schwartz, K.J.; Stevenson, G.W. *Diseases of Swine*, 10th ed.; Wiley-Blackwell: Ames, IA, USA, 2012; pp. 119–141.
49. Cassini, P.; Montironi, A.; Botti, S.; Hori, T.; Okhawa, H.; Stella, A.; Andersson, L.; Giuffra, E. Genetic analysis of anal atresia in pigs: Evidence for segregation at two main loci. *Mamm. Genome* **2005**, *16*, 164–170.
50. Wiedemann, S.; Fries, R.; Thaller, G. Genome wide scan for anal atresia in swine identifies linkage and association with a chromosome region on Sus scrofa chromosome 1. *Genetics* **2005**, *171*, 1207–1217.
51. Jin, Q.; Wang, C.; Li, X.; Yu, M.; Zhao, S.H.; Li, X. Molecular characterization and genome-wide mutations in porcine anal atresia candidate gene GLI2. *Mamm. Genome* **2013**, *24*, 500–507.
52. Azizi, S.; Mohammadi, R.; Mohammadpour, I. Surgical repair and management of congenital intestinal atresia in 68 calves. *Vet. Surg.* **2010**, *39*, 115–120.
53. Ghanem, M.; Yoshida, C.; Isobe, N.; Nakao, T.; Yamashiro, H.; Kubota, H.; Miyake, Y.; Nakada, K. Atresia ani with diphallus and separate scrota in a calf: A case report. *Theriogenology* **2004**, *61*, 1205–1213.
54. Ghanem, M.E.; Yoshida, C.; Nishibori, M.; Nakao, T.; Yamashiro, H. A case of freemartin with atresia recti and ani in Japanese Black calf. *Anim. Reprod. Sci.* **2005**, *85*, 193–199.
55. Gülbahar, M.Y.; Kabak, M.; Yarim, M.; Guvenc, T.; Kabak, Y.B. Persistent cloaca, fused kidney, female pseudohermaphroditism and skeletal anomalies in a simmenthal calf. *Anat. Histol. Embryol.* **2009**, *38*, 229–232.
56. Wang, C.; Li, L.; Cheng, W. Anorectal malformation: The etiological factors. *Pediatr. Surg. Int.* **2015**, *31*, 795–804.
57. Thomas, W.B. Hydrocephalus in dogs and cats. *Vet. Clin. N. Am. Small Anim. Pract.* **2010**, *40*, 143–159.
58. Pandey, A.K.; Saini, G.S.; Chander, S.; Chaudhary, R.N.; Jakhar, P.; Singh, M.; Sundar, S.; Yadav, S. Dystocia due to abnormal calf in a buffalo: A case report. *Buffalo Bull.* **2010**, *29*, 315–317.
59. Sharma, S.K.; Joshi, M.; Khosa, J.S.; Singh, D. An unusual case of dystocia due to hydrocephalic monster in a buffalo. *Int. J. Sci. Environ. Technol.* **2015**, *4*, 300–304.
60. El-Sheikh, H.; Hegab, A.O.; Zaabel, S.M. Dicephalicatlodymus monster associated with hydropsamnii in a buffalo cow: A case report. *Vet. Res.* **2010**, *3*, 46–48.
61. Gülbahar, M.Y.; Yüksel, H.; Soygüder, Z.; Erçin, Ö.F. Dicephalus, Arnold-Chiari malformation, spinal dysraphism and other associated anomalies in a Newborn Holstein Calf. *Turk. J. Vet. Anim. Sci.* **2005**, *29*, 565–570.
62. Hind Osman, E.; Badawi, M.N.; Musa, M.B.; Hala Ali, M.; RababAhamed, M. A case of diprosopia in anomalus cross-bred bovine calf. *Vet. Res.* **2007**, *1*, 61–64.

63. Hussein, R.M.N. Congenital anomalies in cattle and buffalo within Mudaina city in Basrah province between period 2007–2009. *Kufa J. Vet. Med. Sci.* **2010**, *1*, 207–218.
64. Wakuri, H.; Mori, T.; Mutoh, K. Arnold-Chiari malformation and associated anomalies in a dicephalic newborn calf. *Okajimas Folia Anat. Jpn.* **1990**, *67*, 339–350.
65. Mazzullo, G.; Germanà, A.; De Vico, G.; Germanà, G. Diprosopiasis in a lamb. A casereport. *Anat. Histol. Embryol.***2003**, *32*, 60–62.
66. Ramadan, R.O. A dicephalic goat with other defects. *Zentral Vet. A* **1996**, *43*, 337–343.
67. Sharma, A.; Sharma, S.; Vasishta, N.K. A diprosopus buffalo neonate: A case report: A case report. *Buffalo Bull.* **2010**, *29*, 62–64.
68. Kumar, P.; Sharma, A.; Singh, M.; Sood, P.; Barman, P. Dystocia due to a dichephalus monster fetus in a buffalo. *Buffalo Bull.* **2014**, *33*, 13–15.
69. Mehmood, M.U.; Abbas, W.; Jabbar, A.; Khan, J.; Riaz, A.; Sattar, A. An Iniodymus Dicephalic Buffalo Neonate. *J. Anim. Plant Sci.* **2014**, *24*, 973–975.
70. Srivastava, S.; Kumar, A.; Maurya, S.K.; Singh, A.; Singh, K. A dicephalus monster in murrah buffalo. *Buffalo Bull.***2008**, *27*, 231–232.
71. Schalles, R.R.; Leipold, H.W.; McCraw, R.L. Congenital defect in cattle. In *Beef Cattle Handbook*; Iowa State University: Ames, IA, USA, 1999; pp. 1–5.
72. Kerkmann, A.; Kuiper, H.; Ganter, M.; Distl, O. Review of literature and results from test matings of East Friesian milk sheep affected with brachygnathia inferior. *Berl. Münch. Tierärztl. Wochenschr.* **2008**, *121*, 292–305.
73. Davies, S.J.; Wise, C.; Venkatesh, B.; Mirza, G.; Jefferson, A.; Volpi, E.V.; Ragoussis, J. Mapping of three translocation breakpoints associated with orofacial clefting within 6p24 and identification of new transcripts within the region. *Cytogenet. Genome Res.* **2004**, *105*, 47–53.
74. Wolf, Z.T.; Leslie, E.J.; Arzi, B.; Jayashankar, K.; Karmi, N.; Jia, Z.; Rowland, D.J.; Young, A.; Safra, N.; Sliskovic, S.; et al. A LINE-1 insertion in DLX6 is responsible for cleft palate and mandibular abnormalities in a canine model of Pierre Robin sequence. *PLoS Genet.* **2014**, *10*, e1004257.
75. Shukla, S.P.; Nema, S.P.; Pandey, A.K.; Garg, U.K. Dystocia due to bull dog calf in a she buffalo. *Buffalo Bull.* **2007**, *26*, 104–105.
76. Singh, G.; Pandey, A.K.; Kumar, S.; Sunder, S.; Kumar, R.; Dutt, R. Lipomatous bull dog calf monster in a Murrah buffalo. *Indian J. Anim. Reprod.* **2012**, *33*, 94–95.
77. Lee, P.A.; Houk, C.P.; Ahmed, S.F.; Hughes, I.A. Consensus statement on management of intersex disorders. *Pediatrics* **2006**, *118*, 488–500.
78. Yadav, B.R.; Kumar, P.; Tomer, O.S.; Kumar, S.; Balain, D.S. Monosomy X and gonadal dysgenesis in a buffalo heifer (*Bubalus bubalis*). *Theriogenology* **1990**, *34*, 99–105.
79. Prakash, B.; Balain, D.S.; Lathwal, S.S. A 49, XO sterile Murrah buffalo (*Bubalus bubalis*). *Vet. Rec.* **1992**, *130*, 559–560.
80. Prakash, B.; Balain, D.S.; Lathwal, S.S.; Malik, R.K. Trisomy-X in a sterile river buffalo. *Vet. Rec.* **1994**, *134*, 241–242.
81. Villagómez, D.A.; Parma, P.; Radi, O.; Di Meo, G.P.; Pinton, A.; Iannuzzi, L.; King, W.A. Classical and molecular cytogenetics of disorders of sex development in domestic animals. *Cytogenet. Genome Res.* **2009**, *126*, 110–131.
82. Patel, R.K.; Singh, K.M.; Soni, K.J.; Chauhan, J.B. Novel cytogenetic finding: An unusual X;X-translocation in Mehsana buffalo (*Bubalus bubalis*). *Cytogenet. Genome Res.* **2006**, *115*, 186–188.
83. Iannuzzi, L. Evolutionary, Clinical and Molecular Cytogenetics in Water Buffalo: An update. *Buffalo Bull.* **2013**, *32*, 244–256.
84. Anaya, G.; Moreno-Millán, M.; Bugno-Poniewierska, M.; Pawlina, K.; Membrillo, A.; Molina, A.; Demyda-Peyrás, S. Sex reversal syndrome in the horse: Four new cases of feminization in individuals carrying a 64,XY SRY negative chromosomal complement. *Anim. Reprod. Sci.* **2014**, *151*, 22–27.
85. Villagómez, D.A.; Lear, T.L.; Chenier, T.; Lee, S.; McGee, R.B.; Cahill, J.; Foster, R.A.; Reyes, E.; St John, E.; King, W.A. Equine disorders of sexual development in 17 mares including XX, SRY-negative, XY, SRY-negative and XY, SRY-positive genotypes. *Sex Dev.* **2011**, *5*, 16–25.

86. Switonski, M.; Chmurzynska, A.; Szczerbal, I.; Lipczynski, A.; Yang, F.; Nowicka-Posłuszna, A. Sex reversal syndrome (64,XY; SRY-positive) in a mare demonstrating masculine behaviour. *J. Anim. Breed. Genet.* **2005**, *122*, 60–63.
87. Bresciani, C.; Parma, P.; De Lorenzi, L.; Di Ianni, F.; Bertocchi, M.; Bertani, V.; Cantoni, A.M.; Parmigiani, E. A Clinical Case of an SRY-Positive Intersex/Hermaphrodite Holstein Cattle. *Sex Dev.* **2015**, *9*, 229–238.
88. Pinton, A.; Pailhoux, E.; Piumi, F.; Rogel-Gaillard, C.; Darré, R.; Yerle, M.; Ducos, A.; Cotinot, C. A case of intersexuality in pigs associated with a de novo paracentric inversion 9 (p1.2; p2.2). *Anim. Genet.* **2002**, *33*, 69–71.
89. Meyers-Wallen, V.N. Review and update: Genomic and molecular advances in sex determination and differentiation in small animals. *Reprod. Domest. Anim.* **2009**, *44*, 40–46.
90. Schlafer, D.H.; Valentine, B.; Fahnestock, G.; Froenicke, L.; Grahn, R.A.; Lyons, L.A.; Meyers-Wallen, V.N. A case of SRY-positive 38,XY true hermaphroditism (XY sex reversal) in a cat. *Vet. Pathol.* **2011**, *48*, 817–822.
91. Urh, K.; Kunej, T. Molecular mechanisms of cryptorchidism development: Update of the database, disease comorbidity, and initiative for standardization of reporting in scientific literature. *Andrology* **2016**, *4*, 894–902.

CHAPTER 8

Changing values of farm animal genomic resources. from historical breeds to the Nagoya Protocol

Sakari Tamminen*

Department of Social Research, University of Helsinki, Helsinki, Finland

The paper reviews the history of Animal genetic resources (AnGRs) and claims that over the course of history they have been conceptually transformed from economic, ecologic and scientific life forms into political objects, reflecting in the way in which any valuation of AnGRs is today inherently imbued with national politics and its values enacted by legally binding global conventions. Historically, the first calls to conservation were based on the economic, ecological and scientific values of the AnGR. While the historical arguments are valid and still commonly proposed values for conservation, the AnGR have become highly politicized since the adoption of the Convention of Biological Diversity (CBD), the subsequent Interlaken Declaration, the Global Plan for Action (GPA) and the Nagoya Protocol. The scientific and political definitions of the AnGRs were creatively reshuffled within these documents and the key criteria by which they are now identified and valued today were essentially redefined. The criteria of "*in situ* condition" has become the necessary starting point for all valuation efforts of AnGRs, effectively transforming their previous nature as natural property and global genetic commons into objects of national concern pertaining to territorially discrete national genetic landscapes, regulated by the sovereign powers of the parties to the global conventions.

Keywords: values, animal genetic resources, convention on biological diversity, interlaken declaration, global plan of action, national genetic landscapes

1. ANIMAL GENETIC RESOURCES AS A GLOBAL MATTER OF CONCERN

Animal genetic resources (AnGRs) have become a topic of renewed interest in international politics taking agricultural species as its object of concern for the last decade. The reasons for this point to a number of intertwined reasons. The first has to do with the unclear legal status and the scope of regulation stemming from biodiversity agreements targeting a wide range of animal species from wild to agricultural, their biological materials and genetic resources within the global politics of late 20th and early 21st century. Here, states, nations and indigenous communities have become new key stakeholders of genetic resources as they have been granted sovereign rights over territorially bound, native "*in situ*" resources within the text of the convention on biological diversity (CBD)[1], signed by over 150 states in 1992. The sovereign rights over GRs have been subsequently re-enforced with the adoption of the Nagoya Protocol on Access and Benefit-sharing[2] entering into force on October 2014. The Nagoya Protocol is the legally binding protocol guiding how to interpret and act upon the genetic resources issues presented in the CBD over 20 years earlier.

The second, and inherently nested interest relates to the way in which access and animal could, or should, be regulated internationally under biodiversity frameworks, including debates on how different kinds of genetic resources from plants to animals and from wild to agricultural species differ from each other and how the difference might have to inform the practical execution of their global governance. The Food and Agriculture Organization of the United Nations (FAO) established a Commission on Plant Genetic Resources in 1983 to deal with policy, access and benefit sharing issues related plant genetic resources. FAO did broaden the mandate of the Commission in 1995 to cover all aspects after the CBD entered into force and after recognizing "that broadening the coverage of the Commission would allow the Organization to deal in a more integrated manner with agrobiodiversity issues" (Food and Agriculture Organization, 1995, p. 66). 2 years later in 1997, the Commission also established separate working groups for animal and plant genetic resources, followed by one expert group for forest genetic resources[3]. All these Committees—specifically established for different types of genetic resources—demonstrate how difficult the policy, ownership and access and benefit sharing issues related to GRs, and especially to AnGRs, are to understand, let alone to manage in practice.

Three examples from international analyses from the last 10 years will clarify some of the difficult aspects related to global agreements and governance on agricultural AnGRs and open up good questions how did the AnGRs become so politically contested objects of agricultural nature.

Consider, for example, a report from 2006 exploring policy options for the "Exchange, Use and Conservation of Animal Genetic Resources," commissioned by the FAO and funded by the Government of the United Kingdom of Great Britain and Northern Ireland, recognized that fundamental tension between the traditional ownership of AnGRs and new global conventions had emerged (Hiemstra et al., 2006) and this tension needs to be resolved on international level. Going through a number of options for AnGR regulation, tellingly the report ended with the summarizing paragraph claiming that *"[c]lassical ownership' of AnGR includes physical ownership and communal 'law of the land' affecting livestock keeping and breeding. There is an increasing tension with developments in the realms of biodiversity law and intellectual property rights protection. Demarcation of these different rights systems and maintaining equity among different stakeholders is crucial to avoiding conflict and increased transaction costs. In this context, it is important to consider the rights of livestock keepers/breeders vis-à-vis national level sovereign rights, as well as obligations between patent holders and breeders/livestock keepers."* (Hiemstra et al., 2006, p. 37).

The report had been commissioned as FAO wanted to clarify the options on how to navigate the world of new political and legal frameworks after CBD for AnGR management. 3 years later another expert report on AnGR raised a concern that relates to the fact that all different types ownership relations faced now a potential disruptive element. *"Private or communal ownership of AnGR, is potentially at least, challenged by national sovereignty over genetic resources. Individual owners may find that their rights to sell breeding animals or other genetic material, particularly across national boundaries, are restricted. Those seeking to buy specific AnGR may find that they are unable to do so, or that they can only do so on terms that are acceptable not only to the owner of the resources but also in compliance with national legislation."*(Commission on Genetic Resources for Food and Agriculture, 2009, p. 29).

Finally, in November 2014, the Intergovernmental Technical Working Group on Animal Genetic Resources for Food and Agriculture concluded after its meeting that more work on AnGR is needed, but that at least the different types of *utilization* of AnGR, the criteria and ways in which the *country of origin* of AnGRs is assessed, and *access and benefit sharing policies* all need further clarification, although at global level a number of internationally binding legal treaties exist (Commission on Genetic Resources for Food and Agriculture, 2014, pp. 19–25).

For a long time, animals, breeds and their genetic resources were solely governed by rights that were based on physical access and use rights to animals as animals and breeds were seen as "wholes," either as living animals or recorded breeds. A mix of private, semi-private and common ownership models for agricultural and farm animals have been in use, and these have also generated much discussion about the forms of entitlement over the life of the animals and the best possible ways to organize these relations (Hardin, 1968; David, 2011). However, as biotechnologies used in animal production have developed—increasing animal growth rates and carcass composition, enhancing disease resistance and improving hair and fiber production (Wilmut et al., 1992; Wheeler et al., 2010; Wheeler, 2013)—the value of individual farm animal, or even the value of a breed, is not only solely calculated in direct relation to its output of agricultural goods (e.g., meat, milk) but also by its value within the social system of breeding. Thus, farm animals and breeds are valuable also because their capacity to produce particular kind of offspring, or to transmit valuable features encoded within the DNA. This capacity can be either codified in rough ideas of maintaining a pure breed type or within the sophisticated algorithms calculating the Estimated Breeding Value in modern farms based on the development of the herdbook, an innovation that enables population management through exact recordings innovated in late 18th and early 19th centuries[4].

Given that the two sources of value in farm animals have been recognized for over 200 years, it is surprising that at the present, in 2015, the global community dealing with AnGRs has ended up in a situation where the access, ownership rights and benefit sharing issues—issues that for a long time remained unchallenged—have become a matter of global concern and source of slowly proceeding political processes where there are no easy resolutions. The issue of "sovereignty" over all types of genetic resources described in the major international agreements gives the signatory states relatively free hands to develop and implement national laws and regulations. In fact, to fulfill their sovereignty over AnGRs, for example, the states must decide what types of entitlements and relationships over AnGRs they should implement, and how this relates to the national rights that farmers have over their animals, for example.

This paper presents two questions and two hypotheses on nature and status of the AnGRs in the post-CBD world:

(1) First, why are the issues of AnGRs for agriculture debated alongside more general issues of biological diversity;

and

(2) Second, how should we understand that idea of sovereignty over genetic resources given to the signatory states of the CBD and the Nagoya Protocol in the context of AnGRs?

The first hypothesis builds on the cultural history of AnGRs movement in the political institutions, most notably in the FAO. I claim that early warnings about the need for the conservation and coordinated management of AnGRs for agriculture did not lead to action and resulted in a failure to mobilize larger communities to action. This, in turn, lead the animal geneticists affiliated with FAO and other interests parties to join forces with environmental conservation movement, especially United Nations' Environmental Programme (UNEP), to gain international support to the issue of conservation then seen as a agenda priority.

Second, and following from the historical reason explained above, the way in which CBD and to certain extent the subsequent Nagoya Protocol defined and understood genetic resources owes much to the world of plant genetic resources (PGR). Defining the right and obligations of signatories through the PGR leads implicitly to the world of plant breeding, which operates differently from animal breeding practices, the key economic relations, and related biological processes. This is also why the key articles and provisions in the biodiversity conventions are couched in strong terms under national governments' sovereign powers.

I claim that this has resulted in a world, where we have moved from a system where animals once were part of a seamless universal nature without political boundaries, to a world that is a collection of discrete "national genetic landscapes" safeguarded by state policies and legal provisions.

2. THE SHORT INSTITUTIONAL HISTORY OF ANGR CONCERNS

The management of farm AnGRs become a topic immediately after the establishment of the United Nation's Food and Agriculture Organization (FAO), however, the concerns related to genetic resources were first identified as a challenge for developing countries. The negative consequences of modern animal production aiming at increasing animal productivity started to raise doubts within the scientific community, and the first calls for genetic conservation followed quickly. Phillips, the first Deputy Director-General for the FAO remembers how he, as the first employeed of the animal section for FAO, had all his career *"already carried out activities relating to animal genetic resources...and the Organization's involvement in this work dates back to 1946"* (Phillips, 1981, p. 5). From early on the worry was about losing local breeds to extinction in developing countries. Local animals were replaced with globally homogenized and more productive breeds that became easily available and were adopted at fast pace. Despite the early warning calls, little to no action aimed at conservation ensued at global level even if FAO produced a number of scientific reports and hosted a series of meetings around the issue between the early 1950s and 1960s.

It was only after the widespread negative impacts of Green Revolution became evident in the 1960's that AnGRs became truly a global matter of concern also for scientists working within developed countries. This was the direct result of unmanaged use of new breeding techniques combined with shrinking and homogenized ecological habitats. For example, in the 1969 regional meeting of the European Association for Animal Production, the issue for "gene pool losses" was already clearly articulated by Maijala (1971) who also identified the root cause for these losses: "*The present era of frozen semen...has reactualized the problem of gene losses...The problem arises mainly from the fact that an effective utilization of the best animals of today automatically means setting aside the poorer animals, strains, breeds and even species*" (Maijala, 1971, pp. 403–444).

In response to these developments, FAO and the United Nations Environment Programme (UNEP) launched a joint project in 1974 with the title of "*Conservation of animal genetic resources.*" It had the key objective to "prepare a list of breeds of farm animals in danger of extinction together with an account of any measures which have been recommended or taken to prevent this extinction" (Mason, 1981, p. 17). A Consultation Report followed in Mason (1981) with a review of the work achieved by the project through the participating regional and national organizations, and made recommendations for future action.

This report was presented in a workshop for animal geneticists working with genetic resources and was framed with Phillip's opening words, that simultaneously exhibited hope and exasperation on the current state of affairs. He proclaimed: " *I am pleased to bid you welcome here, on behalf of the Director—General. It is indeed heartening to see such a distinguished group of animal geneticists assembled to consider the problems of identification, conservation and effective management of animal genetic resources. This is matter critical to man's future, yet it has had little recognition and little real attention* (Phillips, 1981, p. 2)." Phillips' opening speech betrays how, by the early 1980's, the animal scientists had been awakened to the dire straits of genetic resources but the political support of the issue was still weak and more generally, unrecognized as an global political issue. It did not appear in the general global agendas as did other issues related to modernization and increase of production, such as the environmental movement which had started already in 1970s to attract more political attention and gained fast political weight in the international political arenas. As a result, the issue for farm AnGRs did not spur action nor attract funding for conservation efforts (Boyazoglu and Chupin, 1991).

Fast forward a decade into the early 1990's, and one finds more explicit frustration toward the slow progress on conservation efforts and lack of coordinated international action. Explaining the issue and need for AnGRs conservation Hodges (1990), a Senior Officer at FAO wrote that "*the time for technical talk is over. The issues are clear. What is now needed is an effective international decision to provide funds to do what all agree is now necessary the global, regional and national levels*" (Hodges, 1990, p. 153). AnGRs needed more political support but this proved to be hard to gain without rethinking and reframing the issue, and joining forces with other institutional actors. International action did finally follow a few years later in 1992 when the FAO joined forces with the UNEP, and co-organized the Rio Earth Summit in Rio de Janeiro, Brazil. This was also the historical moment for AnGRs. This is the place and the time where genetic resources became newly articulated as parts of nature as they were linked directly to the recently introduced concept of "biodiversity," the key theme of the global meeting for the world's leaders.

In the meeting, UNEP and FAO introduced the global CBD, a convention aimed at saving biodiversity, for larger public and opened it for signatures. It was undersigned by some 160 countries at Rio de Janeiro and over 30 other countries followed suit during the upcoming years. Several of the Articles included in the Convention, addressed the issue of genetic resources directly and introduced an obligation to identify, report and take appropriate actions to conserve genetic resources. The long follow-up work finally resulted in The Global Plan of Action for Animal Genetic Resources (GPA), adopted in 2007, and the Guidelines on the Preparation of national strategies and action plans for AnGRs, published in 2009.

Given the half-century history of AnGR's as a matter of concern for animal geneticists and long and idle wait for political action, the key question is why did the wide scale political traction to save genetic resources only emerge with the introduction of the CBD, leading to the global and national action plans and guidelines specific to AnGRs over a decade later?

3. EARLY FAILURES IN VALUATION

The reason why the FAO and the regional institutions such as the EAAP failed in gaining political traction with their early alarms about the need for conservation measures relates to two shortcomings in the definition and the valuation of AnGRs.

First shortcoming was the lack of consensus in scientific definition, valuation and prioritization of AnGRs that would lead into simple and uniform action recommendations. The question on what is it exactly that needs to be conserved and how to prioritize the required conservation actions was left open, or at best was illustrated through a case of a few particular breeds. The second, and more important shortcoming was the failure of global political and legal identification of the responsible parties and beneficiaries of any value deriving from the costly conservation actions. This, in turn, relates to the fact that up until the CBD, AnGRs were treated as a mixture of "private" and "commons," or as "club commons" (David, 2011) to be shared and used, subject only to individual farmers' and breeding associations' property right regimes and explicit regulations at country level.

After the introduction of the CBD the legal status of AnGRs changed globally as they were politically identified falling under the sovereign power of the signatory parties to the Convention—a major change and complication in access and benefit sharing relations that was later affirmed by the GPA in 2007 and later by the Nagoya Protocol. Understanding the latter is especially important as this understanding exposes the new overarching paradigm under which the value most of the AnGRs today are to be governed.

First, the failure to provide a clear direction for conservation relates to the arguments about the overall role of different kinds of AnGRs in animal production. When the concept of AnGRs were first introduced among the animal scientist, they were framed in and through two different ways (both scientifically informed) of demonstrating the role of AnGRs in animal production. The two ways—the "utilizationist" and "conservationist" standpoints—literally attributed the value of AnGRs in animal production in two incommensurable ways (and to some extent this debate still continues even today). Hodges (1984) report on genetic resources explained the main differences between the two approaches:

"The utilizationist's primary concern is the immediate usefulness of available genetic resources to improve livestock populations...The loss of breeds as distinct identities is not generally a concern, as long as the genes that make these breeds potentially useful are retained in the commercial stocks...The preservationist's primary objective is long-term conservation of genetic resources for future use. This view emphasizes the value of preserving the widest possible spectrum of genetic diversity to be prepared for unpredictable changes of future needs. The greatest possible number of breeds are to be preserved as purebreds." (Hodges, 1984).

The differences of these two views boil down to conserving "the known useful genes" in one form or another versus conserving the "genetic diversity of whole animal breeds" to hedge the uncertainty deriving from unknown future needs. The first approach aims to save the sliced and diced, functionally valuable component of animals regardless of its "breed"; the other also the animal breeds in the purebred form and to maximize diversity as an insurance policy against future unknowns. Although analytically distinct from animals or breeds, the animal scientists first presented the issue of AnGR conservation as a choice between *isolated genetic components* immediately useful in the production of high performance animals or as the maximization of genetic diversity by the conservation of *local breeds in their animal forms*. In these two approaches, AnGR's are conceptually presented as different objects of conservation and seen valuable for different purposes[5].

Second, the failure to identify parties responsible for the conservation irked the conversations as this related directly to the economics of conservation, or more generally, to the political economy of global animal production. The problem was captured in a report produced by the United States' board of Agriculture in its "Managing global livestock genetic diversity":

"The concept of conservation...is complex. One can think of live animals, being preserved in situ, or in some semi-artificial situation; alternatively one may think of cryogenic storage of sperm or fertilized ova or other tissues or gene segments. The economic problems are difficult with both live animals and with haploid or diploid cells. Who is to pay? There are also questions of how many to preserve, for how long, and where" (National Academy of Science, 1993, p. 3).

Although plant varieties and their genetic material were protected by various intellectual property systems since 1930s (see Kloppenburg, 1988), AnGRs were used by farmers and breeder associations alike without generalized and specified rights or restrictions imposed at global level. Since there was no definition on the ownership rights over the genetic materials of animals, the global attribution of conservation responsibility through political processes proved to be impossible without more specific consideration. Yet, for pigs and chicken, ownership and responsibility questions have been more straightforward. This reflects what Tvedt et al., 2007, p. 8) note of the legal protection farm animals and their protection in general, and chicken and pig in particular:

"For farm animals there are strong biological and physical means of protection available: The owner of the animal can more easily than the plant breeder have an overview and control over who is receiving genetic material from his animals or his population. For poultry and pig breeding, however, where farmers often buy hybrids whose genetics are more difficult to reproduce. The sale of hybrids is thus an important strategy for maintaining physical control over the genetic material by physical control over the material. For other breeds, in particular cattle, the physical ownership is often

combined with a register, a herd book that maintains a protocol for the generations of animals fulfilling the criteria for registration."

This is why the different claims about the value of AnGRs and the need for their conservation, made by both the "utilizationists" and "conservationists," rang to deaf ears outside the animal scientist circles. The failure to spur action was not based on the scientific challenge to demonstrate the value of AnGRs in animal production or the lack of consensus in setting the priorities of conservation. Instead, and above all, it was a problem of political economy: who is to pay? And even more importantly, who is to benefit?

4. GLOBAL RE-FRAMING OF ANGRS

Food and Agriculture Organization remained active on AnGRs since the FAO/UNEP consultation program in 1980, established a Committee on Agriculture that kept reminding about the issue at the FAO Council level; designed a FAO expert consultation round on AnGRs in 1989 and in 1992 (Food and Agriculture Organization, 1990, 1999; Steane, 1992). What become clear over the years was that a global binding framework was needed.

Anticipating the global political agreement on AnGRs, de Haen (1992), Assistant Director-General of the Agriculture Department of FAO wrote in 1992 that "*it is clear that there is a greater awareness that a framework for the management of global animal genetic resources must be established. It is most appropriate that this Expert Consultation is taking place now in the context and timing of the Earth Summit, the United Nations Conference on Environment and Development (UNCED) to be held in Brazil in about eight weeks time*" (de Haen, 1992, p. 3).

The first reframing of the AnGR's came in the form of the global CBD a few months later. The Convention had been long in preparation and FAO had been involved in its drafting phases influencing, among other issues, the inclusion of genetic resources to and their definition in the Convention text. There were two important re-framings in the Convention. First was the definition of the genetic resources, as genetic material of "actual or potential value" (CBD, Article 2). This definition bridged the two different views on the valuable material to conserve, or the "utilitizationist" and "preservationist" standpoints. Genetic resources become genetic material that could be attributed with demonstrable or imaginable value. But the question then arises: who has the right to attribute any value claims to AnGRs?

The other reframing answered to this question. Under the definitions of the Article 2 and the Article 15, genetic resources found *"in situ"* within the territory of a signatory were identified as belonging under the sovereign power of signatory states representing the nations of the world, reframing their ownership relationships globally. This is how CBD enacted an important political redefinition of genetic resources: previous problems in the definition of the value of nonhuman life were re-articulated through the politics of nationhood, in the idea of national differences found within the CBD's vision of genetic nature.

With the convention, also AnGRs became tightly nested within the sovereignty of nation-states and their geography. A reversal of the old idea of nations being rooted in natural differences of human populations took place—nonhuman populations, conceptualized as "genetic resources," could now be identified and placed under national or international jurisdiction in terms of their geographical location and the political powers that represented the nationhood that governed that geographical area. A global cartographic demarcation of nonhuman life took place as these novel objects of nature were grafted to the foundations of national sovereignty. They became a new part of the body of nations, a novel form of nonhuman nationhood.

The convention assumes significant amount of power over AnGR and their governance to signatory nation-states. Tvedt et al., 2007, p. 24) interpret the convention and its provisions in the following manner: *"The CBD presupposes the right of a country to exercise sovereign control over its AnGR (accompanied by a number of responsibilities). From the perspective of an exporting country, one of its main concerns is to maintain any property rights it may wish to retain over the AnGR after the resources have left the country. Similarly, it may wish to ensure that the rights of the exporter are respected by the buyer/importer of the AnGR. The most prominent rationale for a country to regulate export of AnGR would be to secure a right over that particular material in the future, including preventing that countries or companies gain control over these resources (e.g., through patenting or other forms of intellectual property rights), which might reduce the value of it in the exporting country."*

This reframing introduced a whole new system where the value of any animal breed will be decided by the nations signatory to the parties but without any common reference what consists legitimate value claim over the material, except the condition of *"in situ."* In the CBD, these are the "conditions where genetic resources exist within ecosystems and natural habitats, and, in the case of domesticated or cultivated species, in the surroundings where they have developed their distinctive properties." (CBD Article 2). This *"in situ"* condition of valuable genetic resources has tremendous effects into how AnGRs are seen within the post-CBD world, especially as AnGRs were now removed from the idea of being freely circulated or tradable objects of nature. They stopped being global commons and instead become subject to the political powers of the Convention parties, many who did not have, and still today do not have, a clear stance what are "valuable" AnGRs to them, and how they will enact their sovereign powers over the access and benefit sharing to the valuable AnGRs. A definitional and legal disorientation followed.

The third re-framing of AnGR's happened as they were presented through ideas derived from the plant and crop worlds. The FAO background study in the CBD on the "Exchange, Use and Conservation of Animal Genetic Resources" acknowledged this as a major problem. It explained:

although current debates regarding agricultural genetic resources have largely had a crop/plant focus, these discussions, and the international instruments or agreements that are emerging have tended to frame the debate for AnGR as well. At first sight plant breeding does not differ much from animal breeding. The genetics of plants and animals are based on the same principles. Plant and animal breeders both need genetic diversity in order to advance and the genetics determine adaptation to particular agro-ecological circumstances, as well as product qualities to a large extent. However, plant varieties can be protected by plant breeder's rights (UPOV), which is not the case for animal breeds/strains. Plant breeders aim at the development of new uniform varieties that are defined by certain phenotypic traits that can identify them from other varieties. Farm animal breeding is largely based on the selection of individuals within populations rather than selection between populations or strains. Farm animal breeders are interested in individual animals (within populations/breeds), while the whole population of a plant variety (clones) is the main focus of plant breeders." (Hiemstra et al., 2006, p. 22).

The third reframing, then, pointed to the difference of animal and plant genetic resources as biological bred resource and legal protected asset: animals might carry interesting genetic traits but it is difficult to exploit one unique genetic characteristic, there are no large international breeding centers but most breeding happens in farms—except for poultry and partly for pigs—and the centers of origin or diversity for AnGR are not as clearly defined as for plants. Most importantly, farmers are not protected by internationally binding rights frameworks—plant breeders, however, are by the International Union for the Protection of New Varieties of plants (UPOV)[6]. The differences between plant and AnGRs make it hard to enforce only system for the two, however, the CBD does exactly this by enforcing the sovereignty of the signatory states as its starting point for rights and obligations via the discourses mostly appropriate to plant genetic resources.

These re-framings of the AnGR dictate much of how global action now unfolds. 15 years after the CBD, in 2007, the state representatives adopted the first "Global Plan of Action for Animal Genetic Resources" (GPA) at the Interlaken Conference held in Switzerland, something that was called a "historical breakthrough" by the FAO Director General Jacques Diouf (Food and Agriculture Organization, 2007, p. iii). The GPA includes the "Interlaken Declaration on Animal Genetic Resources," in which the sovereign rights of states over their AnGRs for food and agriculture was restated (declaration point 2).

5. *IN SITU*, TRANSBOUNDARY, AND DOMESTIC APPLICATIONS

The fact that animals can move across politically established boundaries created a potential problem to these sovereign rights, however, and led to new politically innovated categories of AnGRs, such as "transboundary" species for criss-cross institutionalized country borders. The GPA explained: "Assessing the status of animal genetic resources on a global scale presents some methodo-logical difficulties. In the past, analysis of the Global Databank to identify breeds that are globally at risk was hampered by the structure of the system, which is based on breed populations at the national level. To address this problem…a new breed classification system was developed. Breeds are now classified as either local or transboundary, and further as regional or international transboundary" (Food and Agriculture Organization, 2007, p. 13).

With these political documents not only did animals considered as genetic resources become "national" pertaining to a state, but some of them also became "transboundary," regionally and internationally. The result of this is that political categories infuse with conservation science categories because of the political economy involved in the ownership rights over the actually or potentially valuable genetic resources.

These categories are as much politically informed as they are scientifically true. The definitions of "*in situ*" or "transboundary" are inherently related to the political cartographic demarcation of the natural ecologies of domesticated animals, pointing to the deep connection between politics of value and the science of conservation of farm genetic animal resources. This is what eventually created the incentive for nation-states to act upon the issue of genetic erosion of animal populations, but is now, at the same token, generating new challenges that are beyond the scope of animal scientists or even international organizations to solve.

This complexity is reflected on how the national legislations have been drafted and implemented. Writing about the challenges in the implementation of the CBD legal experts Buck and Hamilton claim that "*[t]he complex subject matter of ABS, its potentially far reaching impact on uses of genetic resources and related information as well as the lack of detail in Articles...have all combined to result in a very low level of domestic implementation by Contracting Parties to the CBD. By 2007, only 39 of the then 189 Contracting Parties had established domestic legislation or were in the process of doing so.*" (Buck and Hamilton 2011, p. 48).

Reporting back on the negotiations of the Nagoya Protocol in Japan in 2010, the protocol that is to meant to clarify the initial CBD, they point out that the key to really "understanding" the real effects of the CBD and Nagoya is dependent on how national governments use their sovereign powers: "*The adoption of the Nagoya Protocol was a major achievement in international biodiversity policy making in 2010...Further international work preparing the entry into force of the Nagoya Protocol will be needed. However, most efforts over the coming years will need to be at domestic level, developing implementing rules to prepare ratification. In all Parties with well-developed or emerging research and development systems this will require significant awareness-raising with stakeholders from research and industry and will result in quite some discussions.*" (Buck and Hamilton, 2011, p. 60).

Most importantly, the national implementation has to take into account that access should take place on "mutually agreed terms" and "be subject to prior informed consent," conditions found in the original CBD and all subsequent treaties. However, other aspects of AnGR can also be regulated, and some of the countries have enacted already requirements for animal genetic material import and export. FAO's Technical Working Group on AnGR Access and Benefit sharing issues explained in its recent report in 2014 that "*[t]he sovereign right of states to determine access to genetic resources should not be confused with other categories of entitlement, such as the private ownership of an animal. A farmer's ownership of an animal may be conditioned by certain laws. For example, animal welfare legislation may regulate the handling, husbandry and transport of the animal. Other laws may require the animal to be vaccinated against specific diseases, and so on. In a similar way, ABS measures may require that, even though an animal is the private property of a farmer or the collective property of a community, certain conditions (e.g., related to the need for "prior informed consent") must be met before it can be provided to a third party for research and development*" (Commission on Genetic Resources for Food and Agriculture, 2014, Item 18).

Indeed, some of the countries have already exercised their sovereign rights. For example China has adopted a set of rules to AnGR, or "*Measures of examination and approval of the entry and exit of animal genetic resources and the research in cooperation with foreign entities in their utilization,"* in 2008. These include a set of import and export rules, such as prohibition *"on the export of newly discovered and unverified"* AnGR in cooperation with *"any foreign institution of individual."* Also, any research and use of AnGR involving foreign collaborators requires permission from the Chinese authorities. South-Africa, on the other hand, now requires a *"genetic impact assessment"* before the import of new breeds. These studies need to be prepared by reputable South African animal scientists and submitted to the relevant authorities (see Commission on Genetic Resources for Food and Agriculture, 2009, p. 34). The national implementation of the sovereign rights over genetic resources can happen in many ways, not only by regulating of access or benefit sharing but also by use and impact, as the examples from China and South Africa demonstrate.

6. CONCLUSION

The challenges are located now within the realm of national politics where the *in situ* condition of genetic resources are turning animals into collections of nationally valuable animals, governed not by the previous ideals of global commons but by the logic of "actual and potential" value, by innovated political re-categorizations of natural beings, and by national restrictions to the access, use and benefit sharing of AnGRs. We have moved from a world where animals once were part of a seamless universal nature without boundaries to a world that is a collection of discrete "national genetic landscapes."

Over the course of the short history of AnGR conservation, the natural identities of farm animals have been refashioned from being objects of breeding to boost the productivity of individual animals and breeds to objects that can be defined actually or potentially valuable as nationally recognized genetic resources. The change in their identity is a creative outcome product of the animal breeding and conservation sciences that have argued the value of animals on the basis of scientific evidence as well as the global politics surrounding the ownership rights over genetic resources considered valuable. AnGRs, including farm animals, are now as much political as they are scientific, as much "cultural" than they are "natural" by their essence.

Table 1 above summarizes the key changes in the conceptualization and valuation of AnGRs before and after the introduction and ratification of the CBD, the Global Plan of Action and the Nagoya Protocol.

Table 1. The historical changes in the value system for animal genetic resources.

Animal genetic resources	Pre-CBD/GPA/Nagoya	Post-CBD/GPA/Nagoya
Nature of animal genetic resources	• Natural breeds or functional genetic components • Identification based on scientific definition and evaluation of conservation need	• Natural-Cultural objects • Identification based on the politically agreed "*in situ*" condition • Genetic material found within the geographically bounded territories of the nation-states, with the exception of politically innovated "transboundary" conditions and categories of animals
Access and benefit sharing	• Private access, or club commons • Freely usable for those who have physical access and local permission to use • No enforced benefit sharing	• National sovereignty to decide • Usable only under the rule of the sovereign party to the conventions • Local political decision on access and benefit sharing principles.
Criteria for valuation	• Actual or potential value, no consensus in general	• Actual AND potential value • Based on territorial *in situ* condition and local political valuation of important national genetic landscapes to conserve

What becomes clear while looking at the key changes in the value system of AnGRs is that AnGRs have become increasingly complex objects for breeders, scientists and politicians alike, with no easy answers to how the balancing of rights, responsibilities and benefit sharing in the near future. While AnGRs have finally become a global issue with high political priority and action, so have the political conditions under which the animals live become inherently global entanglement of science and politics, culture and nature.

At the same time, the status of AnGRs that reside outside of the CBD system—either owned by private companies or breeding societies before the entry into force of the CBD in 1993—is unclear. Although they are not objects of CBD's articles, they might still be affected by and become targets of legal interventions by way in which for example China and South Africa have applied the sovereignty over genetic resources within their respective AnGR regulations. This makes the global system even more complicated, and most likely with a number of unforeseen challenging cases in the future.

The CBD, the Global Plan of Action and the Nagoya Protocol present a global value system framing AnGRs in a way that is finally generating conservation action at national level. But on the global level, the system is more muddled than ever calling for a great deal of conceptual, political and legal analysis to bring more clarity to the current condition that requires the generation of discrete genetic landscapes and marks AnGRs with their nationally correct *in situ* location as their political condition of existence. Given the complex history of AnGRs as a global matter of concern, creating clarity to the present situation will not be easy.

At least three key questions need to be clarified with regards to AnGR and the different claims laid over them in order to move on in the global politics, in the creation and implementation of legal frameworks at national level, and in the reflection of the true impact of the CBD and Nagoya Protocol.

1) What is the true scope of the CBD and Nagoya Protocol in terms of AnGR types? Are there types of AnGRs that remain totally unaffected by and reside outside of the scope of the global conventions?

2) Do the signatory parties (nation-states) have prototypical reactions to—or at least broadly identifiable patterns in—the implementation of their sovereign powers over AnGR?

3) If the signatory parties do exhibit identifiable patterns, a guiding typology of CBD and Nagoya Protocol implementation at national level would help to make sense of how governments are adopting the global agreements (e.g., types of entitlement claims, access regulations etc) at large.

Addressing these three points would already give a much richer and much more coherent overview of AnGRs' status in the post-CBD and post-Nagoya Protocol world than is currently available for public. We do suggest that the institutions driving the global framework on genetic resources provide it soon.

FOOTNOTES

1. https://www.cbd.int/
2. http://www.cbd.int/abs/
3. The names of these expert bodies are: the Intergovernmental Technical Working Group on Animal Genetic Resources for Food Agriculture, the Intergovernmental Technical Working Group on Plant Genetic Resources for Food and Agriculture to deal with specific matters in their areas of expertise, and the Intergovernmental Technical Working Group on Forest Genetic Resources.
4. General herdbooks emerged in Europe in the late eighteenth—and early nineteenth century—the first one for cattle was the "*Short-Horned Cattle Herd*" book, published in 1822 in England. Elsewhere, general registries were published in France (1855), Germany (1864), Holland (1874), and Denmark (1881) (Derry, 2003, p. 8; Walton, 1999, p. 153; Ritvo, 1995, p. 420). The idea of "Estimated Breeding Value" is based on this herdbook keeping but introduces a more refined statistical modeling into the calculation of breeding value of a individual animal at the end of C21th.
5. There are a number of ways to maximize diversity. Conserving a sum of isolates of pure inbred populations will allow saving rare genetic combinations adapted to specific environmental conditions but might result in losing overall diversity. Other options, such as maintaining a large out bred population resulting from crossbreeding, would also provide a large diversity but is not usually the overall aim of conservation programs. It is today generally recognized that a combination of ex-situ and in-situ measures are complementary strategies.
6. http://www.upov.int

REFERENCES

1. Boyazoglu, J., and Chupin, D. (1991). *"Editorial," Animal Genetic Resources Information Bulletin.* Rome: Food and Agriculture Organization of the United Nations.
2. Buck, M., and Hamilton, C. (2011). The nagoya protocol on access to genetic resources and the fair and equitable sharing of benefits arising from their utilization to the convention on biological diversity. *Reciel* 20, 47–61. doi: 10.1111/j.1467-9388.2011.00703.x
3. Commission on Genetic Resources for Food and Agriculture. (2009). *The Use and Exchange of Animal Genetic Resources for Food and Agriculture.* Rome: FAO.
4. Commission on Genetic Resources for Food and Agriculture. (2014). *Report of the Eighth Session of the Intergovernmental Technical Working Group on Animal Genetic resources for Food and Agriculture.* Report CGRFA/WG-angr-8/14/REPORT. Rome: FAO.
5. David, P. A. (2011). "Breaking anti commons constraints on global scientific research: some new moves in "Legal Jujitsu"," in *Designing the Microbial Research Commons: Proceedings of an International Symposium*, ed. P. F. Uhlir (Washington, DC: National Academies Press), 13–24.
6. de Haen, H. (1992). "Opening statement," in *The Management of Global Animal Genetic Resources. Proceedings of an FAO Expert Consultation*, ed. J. Hodges (Rome: FAO), 23–25.
7. Derry, M. E. (2003). *Bred for Perfection: Shorthorn Cattle, Collies, and Arabian Horses Since 1800.* Baltimore: Johns Hopkins University Press.
8. Food and Agriculture Organization. (1990). "Animal genetic resources: a global programme for sustainable development," in *Animal Production and Health Paper No. 80. Proceedings of an FAO Expert Consultation*, Rome.
9. Food and Agriculture Organization. (1995). *Major Trends and Policies in Food and Agriculture.* Rome: FAO.
10. Food and Agriculture Organization. (1999). *The Global Strategy for the Management of Farm Animal Genetic Resources.* Rome: FAO.
11. Food and Agriculture Organization. (2007). *Global Plan of Action for Animal Genetic Resources and the Interlaken Declaration. Commission on Genetic Resources for Food and Agriculture.* Rome: FAO, Commission on Genetic Resources for Food and Agriculture.
12. Hardin, G. (1968). The tragedy of the commons. *Science* 162, 1243–1248. doi: 10.1126/science.162.3859.1243
13. Hiemstra, S. J. A. G., Drucker, M. W., Tvedt, N., Louwaars, J. K., Oldenbroek, K., Awgichew, S., et al. (2006). *Exchange, Use and Conservation of Animal Genetic Resources Policy and Regulatory Options.* CGN Report 2006/06. Wageningen University and Research Centre: Centre for Genetic Resources, the Netherlands (CGN).
14. Hodges, J. (1984). "Review of the fao/unep programme on animal genetic resources conservation and management," in *FAO Animal Production and Health Paper 44/1: Animal Genetic Resources Conservation by Management, Data Banks and Training Rome.* 2–9.
15. Hodges, J. (1990). Animal genetic resources. *Impact Sci. Soc.* 40, 143–154.
16. Kloppenburg, J. R. (1988). *First the Seed: The Political Economy of Plant Biotechnology, 1492–2000.* Cambridge: Cambridge University Press.
17. Maijala, K. (1971). Need and methods of gene conservation in animal breeding. *Ann. Fenet. Sel. Anim.* 2, 403–415.
18. Mason, I. L. (1981). "Cooperative work by FAO and UNEP on the conservation of animal genetic resources," in *FAO Animal Production and Health Paper 24: Animal Genetic Resources Conservation and Management. Proceedings of the FAO/UNEP Technical Consultation* (Rome: FAO), 16–29.
19. National Academy of Science. (1993). "Preface," in *Managing Global Livestock Genetic Resources Committee on Managing Global Genetic Resources: Agricultural Imperatives Board on Agriculture National Research Council.* ed. Barbara J. Rice (Washington, DC: National Academy Press), 1–20.
20. Phillips, R. W. (1981). "The identification, conservation and effective use of valuable animal genetic resources," in *FAO Animal Production and Health Paper 24: Animal Genetic*

Resources Conservation and Management. Proceedings of the fao/unep Technical Consultation in 1980. Rome: FAO, 2–6.

21. Ritvo, H. (1995). "Possessing Mother Nature. Genetic Capital in Eighteenth-Century Britain," in *Early Modern Conceptions of Property*, eds J. Brewer and S. Staves (London and New York: Routledge), 413–426.
22. Steane, D. E. (1992). Note on the fao expert consultation on management of global animal genetic resources rome, 7-10 APRIL 1992. *Anim. Genet. Resour. Informat.* 9, 3–6.
23. Tvedt, M., Hiemstra, S. J., Drucker, A. G., Louwaars, N., and Oldenbroek, K. (2007). *Legal Aspects of Exchange, Use and Conservation of Farm Animal Genetic Resources.* FNI Report 1/2007. Lysaker, FNI.
24. Walton, J. (1999). Pedigree and productivity in the British and North American cattle kingdoms. *J. Hist. Geogr.* 25, 441–462. doi: 10.1006/jhge.1999.0161
25. Wheeler, M. B. (2013). Transgenic animals in agriculture. *Nat. Educ. Knowl.* 4, 1.
26. Wheeler, M. B., Monaco, E., Bionaz, M., and Tanaka, T. (2010). The role of existing and emerging biotechnologies for livestock production: toward holism. *Acta Sci. Vet.* 38, 463–484.
27. Wilmut, I., Haley, C. S., and Woolliams, J. A. (1992). Impact of biotechnology on animal breeding. *Anim. Reprod. Sci.* 28, 149–162. doi: 10.1016/0378-4320(92)90101-I

CHAPTER 9

Sheep Production and Breeding Systems in North Western Lowlands of Amhara Region, Ethiopia: Implication for Conservation and Improvement of Gumz Sheep Breed

Yohannes Dagnew[1], Mengistu Urge[2], Yosef Tadesse[2], Solomon Gizaw[3]

[1]*Department of Animal Science, Bule Hora University, Bule Hora, Ethiopia*

[2]*School of Animal and Range Science, Haramaya University, Dire Dawa, Ethiopia*

[3]*International Livestock Research Institute, Addis Ababa, Ethiopia*

ABSTRACT

A questionnaire survey and focused group discussions were conducted to characterize sheep production systems and traditional breeding practices of Gumz sheep as an essential step for designing the Gumz sheep breed conservation and improvement programme. A total of 240 households were included in the study during the period of November 2016 to March 2016.

Smallholder farmers kept sheep as a source of income, meat and wealth with an index value of 0.43, 0.19 and 0.15, respectively. Important purposes of large scale farmers for keeping sheep were source of cash income (0.57) and investment opportunity (0.22). Average sheep flock sizes were 17.25 ± 0.68 and 90.63 ± 5.38 at smallholder and large scale production systems, respectively. Smallholder sheep flocks' composition was 7.47 ± 0.28 Gumz, 3.68 ± 0.25 Rutana and 4.52 ± 0.22 Rutana-Gumz crossbred sheep. The corresponding figures for large scale farms were 21.2 ± 1.4, 30.7 ± 1.69 and 24.7 ± 1.47 sheep. The average numbers of rams in the smallholder sheep flocks were 0.31 ± 0.04 Gumz, 0.39 ± 0.04 Rutana and 0.24 ± 0.03 crossbreds, respectively. The corresponding figures for large scale farms were 3.10 ± 0.13 Rutana and 0.90 ± 0.14 crossbreds. The Gumz sheep was large in number, but farmers prefer Rutana and its crosses than Gumz in both systems. Diseases, stock theft and labour shortage were reported as the main constraints of sheep production. Although the pure Gumz breed meets the multifaceted roles of sheep for the smallholder farmers, the Gumz sheep population was declining in number in the study area due to uncontrolled breeding with Rutana and high preference of farmers for crossbreds over Gumz sheep. It is observed that there is a risk of dilution and loss of genetic diversity of the Gumz sheep. Hence, we recommend designing of conservation-based breeding program to conserve the locally adapted Gumz breed as well as to improve sheep production and productivity in the area through rational utilization of all the three genotypes.

Keywords: Breeding Practices, Conservation, Gumz, Production System

1. INTRODUCTION

Ethiopia is endowed with diverse agro-ecological zones and corresponding large and diverse sheep resources numbering about 28.3 million heads [1]. Sheep are important sources of cash income and living banks to their owners [2] [3]. Sheep also play an important role in providing export commodities such as live animals and skins/leather to earn foreign currency [4]. Sheep account for 19% and 95% of the total livestock and small ruminant live animal export, respectively [5]. Ethiopia's lowland sheep and goat breeds are in high demand in neighboring and Middle East countries. Lowland breeds are more preferred than highland breeds by exporters [6]. The performance of the sheep industry has been poor compared with other African countries due to inadequate feed and nutrition, widespread disease and health problem, poor management and marketing system [7] [8] . The current sheep production may not be meeting the increasing domestic demand due to increased human population and urbanization [9]. There is also increasing demand for sheep and sheep products in the neighbouring countries like Sudan and North African countries, which is an opportunity for producers. Northwestern lowland of Amhara region is among the high potential areas for livestock production and for the supply of live animals to the Sudan market [10] and used either for local consumption or for re-exporting to Egypt, Libya and Yemen. The major sheep breeds kept in this area are Gumz, Rutana and their crossbred. The local Gumz sheep are prolific, adaptable to the hot environmental condition of the area and known for their tasty meat [11]. However, farmers in the study areas are reported to be more interested in the Rutana and their crossbred with Gumz sheep which are more preferred than the indigenous Gumz breed in the export market. High human population growth and improvement of living standard increase the demand for animal protein, which consequently leads to bulk supply of the best animals and crossing with exotic breeds to meet the demand for large number and best animals in the market. This is seriously threatening the existence of indigenous populations. Hence, the current drive for rapid livestock development through crossbreeding and the threat status of indigenous sheep [12] requires research to design and implement suitable breeding strategies so as to improve productivity and conserve indigenous breeds. Absence of adequate baseline information about the production system is considered as one of the bottlenecks for development of strategy for breed improvement and conservation in most developing countries. [13] also underlined the need for comprehensive description of the production system in order to understand the special adaptative character of breeds and designing of breeding programmes. Thus, characterization of production system, breeding practices and major production constraints help to ensure sustainable use and development of sheep genetic resource. This information is also useful for policy makers and extension services to plan improvement and conservation strategies for the sheep breed. This study was therefore undertaken to generate information on the sheep breeds and breeding systems, to characterize the production system, to describe the production objectives and breeding practices of the sheep keepers, to provide baseline information for designing conservation-based breeding programs for Gumz sheep and rational utilization of the Rutana sheep genetic resource.

2. MATERIALS AND METHODS

2.1. Study Areas

The study was conducted in the lowlands of north western Ethiopia, Amhara region, along the Sudanese border, in Metema and Quara districts (Figure 1). Metema and Quara districts were selected for conducting the study due to accessibility and distribution of the sheep breeds studied purposively. In both of the districts, agricultural production is characterized by a mixed crop-livestock production system.

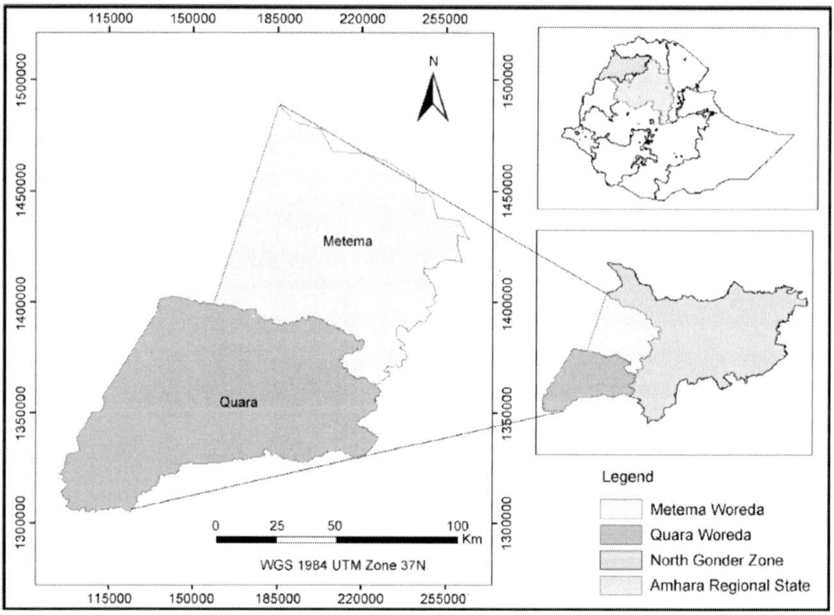

Figure 1. Map of study districts in North Gondar Zone of Amhara region

The major crops grown are sesame and cotton, which are the major export commodities of the country [14] . Metema is located between 12°40'N and 36°8'E. It is situated at an altitude ranging from 500 to 1608 m.a.s.l. It has a uni-modal pattern of rainfall (June to September); with annual rainfall ranging from 850 to 1110 mm with a minimum and maximum average temperature of 22°C and 43°C, respectively [11] . The production system of Metema district is characterized as a small scale and large scale mixed crop-livestock system. Sheep production is a common practice. Quara is located between 11°47'N and 12°21'N latitude and 35°16'E and 35°47'E longitude.

The district is situated at an altitude ranging from 528 to 654 m.a.s.l. The annual rainfall range from 600 mm to 1200 mm, and the average daily temperature is 31°C with 41.1°C and 19.2°C maximum and minimum temperature, respectively [15] . Sheep production in the area is practiced both by smallholder and large-scale farmers.

2.2. Sampling Strategy and Data Collection Procedures

In the north western lowlands, the study areas were classified into two production systems (smallholder and large scale crop-livestock production systems) based on sheep flock size, farmers sheep production practice, their breed preferences, level of input use and marketing systems. Multi-stage stratified sampling was employed to select Kebeles purposively based on sheep breed distribution, sheep population and accessibility. Three Kebeles were selected from each district representing smallholder production systems. Similarly, three investment sites were selected based on sheep breed distribution, sheep population and accessibility from each district representing large-scale production systems. Accordingly, a total of six kebeles (3 from each district) representing smallholder systems and six investments site (3 from each districts) representing large-scale production systems were selected. To calculate the total sample size, the following parameters were considered: 95% level of confidence (CL), 5% desired level of precision and about 20% of the farmers kept both Gumz and Rutana sheep as confirmed by [11] in the study areas that are known for keeping Gumz sheep; thus, the sample size is determined using the single population proportion formula as described by [16] : Therefore, based on the above formula, the total sample size was 240 households. 180 of them would be sheep owners from smallholder production systems, and the rest 60 respondents from large scale production systems. A total of 240 households (30 from each kebele and 10 from each investment sites) were randomly sampled for the survey. Focus group discussions were held with development agents, key informants and elders before the commencement of the actual survey. These discussions were used to obtain information about the history of sheep breeds, population trends (declining, stable or increasing), farmers' reasons for keeping the various breeds and crossbreds, herding practice, grazing/communal land use, mobility and major sheep production constraints. A set of detailed structured questionnaire and check list were prepared and used to collect information from the 240 sheep owners of which 180 were farmers and 60 were large scale farmers in different sites by guided interviews. The questionnaire was prepared by ILRI (International Livestock Research Institute)-OADB (Oromiya Agricultural Development Bureau) for survey of livestock breeds in Oromiya and found to be effective for such studies [17] and adopted for the current research with some modification to fit in the study areas situation. This questionnaire was interpreted in local language to make it first understandable. The questionnaire was pre-tested to check clarity and appropriateness of the questions. Based on the questionnaire information on breeding systems, production objectives, flock structure, production constraints, feeding management, diseases prevalence and marketing system were gathered. Observational studies were also applied in the study areas in addition to historical information about the production system collected from direct questioning of the sheep owners. Primary and secondary data were the sources of information used for the study. Primary data sources were the key informants during guided interviews. The households were selected based on sheep possessions and willingness to be part of the survey. The method of data collection employed was a single-visit formal survey [18]. For group discussion, 10 sheep breeders with long experience in sheep rearing were selected from each districts and interviewed to gather reliable information on the history of sheep breeds, farmers' reasons for keeping the various breeds and major sheep production constraints. The questionnaire were designed to obtain information on general household characteristics, livestock and flock structure, flock management, breeding practices, disease prevalence, production objectives, feeding management and production constraints. Also information on the general composition of the flock, breed, age, sex, purpose of keeping sheep, way of herding and market systems were collected for this study.

2.3. Data Analysis

The data were analyzed using descriptive statistics (SPSS 20.0, 2008). Farmer's preference rankings were summarized into index as weighted averages. The indices were calculated as suggested by [19] ; Index = Sum of (3 X number of household ranked first + 2 X number of household ranked second + 1 X number of household ranked third) given for an individual reason, criteria or preference divided by the sum of (3 X number of household ranked first + 2 X number of household ranked second + 1 X number of household ranked third for overall reasons, criteria or preferences.

3. RESULTS AND DISCUSSION

3.1. Sheep Production Systems

A good understanding of the production system under which indigenous sheep breed are raised is essential in planning suitable improvement and conservation programs for breeds [20] [21] . The existence of smallholder and large scale sheep production systems in the present study area is in accordance to the previous studies [14] . The smallholder production system is characterized by small-scale holding and traditional rain fed system, which mainly rely on family labor, cultivate 1 to 10 ha of land, grow sesame and cotton as cash crops and sorghum as food crop. Cattle, goat, sheep and donkey were the livestock species reared by smallholders. In large scale systems, cultivated land range from 50 to 600 ha. Farmers produced sesame, cotton and sorghum as cash crops. Large capital investment, predominantly mechanized rain fed system, larger livestock holding and substantial number of hired labor were some of the distinguishing characteristics of the large scale system. Cattle, sheep and goat are the main livestock species kept by large scale farmers. The characteristics of the two production systems should be considered in sheep development programs in the study areas. Moreover, a separate sheep breeding programs may be required to capture the advantage of the production systems.

3.2 Land and Livestock holding

Average land holding per farmer was 5.9 and 118 ha for smallholder and large scale farmers, respectively. According to [22] the smallholder holding is greater than the average landholdings of the highland districts of Debark (1.66 ha) and Layarmachiho (2.03 ha) districts. Only 8.6% and 11.5% of the land owned by smallholder and large scale farmers was used for grazing while most of the remaining land was used for growing crops. Out of the total crop land, about 65% and 35% were used for cash crops and staple foods, respectively. The trend in land use was similar to the result reported by [14] . Grazing land has been decreasing due to an ever increasing human population and current greater use of communal grazing lands for crop production [14]. According to [23] the feed supply of the natural pasture in the Amhara National Regional State of Ethiopia is decreasing for similar reasons. Since the farmers were those having sheep, hundred percent of the respondents practiced sheep rearing (**Table 1**). Large scale farmers had the largest sheep flock sizes as compared to the smallholder farmers in both study districts. The smallholder sheep holding reported for Metema district is comparable to that reported by [11], which was 16.02 ± 14.1 sheep per household. The large flocks kept by the large scale farmers in the study area is due to viable cross border trade with Sudan and availability of uncultivable free grazing areas. The smallholder's sheep holding in the present study area is also larger than that in the highland regions of Ethiopia where land is in short supply due to human population pressure. The difference exist in animal holding should be considered in designing breeding programs in order to increase the chance of successful genetic improvement since flock size is a major determinant factor in designing effective genetic improvement. The dominance of sheep as compared to goat in the large scale production systems (**Table 1**) might be due to the high price and year round export market demand for sheep in Sudan and other Arabian countries. Dominance of goat and sheep over cattle in the smallholder production systems might be because of the fact that small ruminants are the major source of cash income, capital saving and food for smallholder farmers. The same trend was observed in Afar lowlands of Ethiopia where sheep and goat flock size was relatively higher than other livestock species [4] [24] .

Table 1. Average livestock size per household (Mean ± Standard error) in different production systems.

Livestock species	Production systems				Over all	
	Smallholder		Large-scale		Small-holder	Large scale
	Metema (90)*	Quara (90)	Metema (30)	Quara (30)		
Cattle	16.24 ± 1.77	14.98 ± 0.85	43.28 ± 5.43	35.07 ± 4.15	15.61 ± 0.98	41.20 ± 3.87
Sheep	16.36 ± 1.02	18.14 ± 0.91	92.10 ± 8.5	85.76 ± 5.48	17.25 ± 0.68	90.63 ± 5.38
Goat	13.87 ± 0.82	14.56 ± 0.77	12.41 ± 1.62	14.41 ± 0.99	14.21 ± 0.56	13.53 ± 0.92
Donkey	1.47 ± 0.07	1.62 ± 0.07	---	---	1.54 ± 0.05	---

*Numbers in brackets are number of respondents.

3.3 Relative contribution of sheep farming

Table 2 shows ranking of the contribution of sheep farming, other livestock enterprises and crop production for the family consumption and cash income. The smallholder systems give priority to crops as a source of food followed by cattle and goat. Whereas, for income generation cash crops contributed more than any other farming activities followed by sheep. In large scale systems, food crops ranked first as a means of food source followed by sheep and cattle. Large scale farmers ranked cash crops and sheep as first and second important source of income generation. This study showed that the smallholder and large scale farmers directly or indirectly depend on livestock mainly sheep as means of income generation and food source for the family. This use function of sheep should be taken into account during designing of breed improvement and conservation strategy. The contribution of sheep in the smallholder system was highly comparable to cash crops such as sesame, which are high value export commodities. The contribution of sheep production as a source of family income and food source (meat) recorded in the present study area is in agreement with reports from other parts of Ethiopian like Somali and West Gojjam [25] . The contribution of the off-farm employment (self and formal employment) also accounts for some portion of cash earned.

Table 2. Small scale and large scale farmers' ranking (index) of importance of major farming activities for the supply of food and income.

Importance	Food source			Income source		
	Smallholder I	Large scale I	Overall I	Smallholder I	Large scale I	Overall I
Food crop	0.41	0.32	0.36	0.09	0.10	0.10
Cash crop	0.07	0.08	0.08	0.33	0.36	0.35
Cattle	0.21	0.21	0.21	0.12	0.14	0.13
Sheep	0.11	0.23	0.17	0.30	0.28	0.29
Goat	0.20	0.16	0.18	0.10	0.12	0.11
Off farm	0.0	0.0	0.0	0.06	0.0	0.03

I = index; I = sum of [3 for rank 1 + 2 for rank 2 + 1 for rank 3] given for an each farming activity divided by sum of [3 for rank 1 + 2 for rank 2 + 1 for rank 3] summed for all the farming activity.

3.4 Sheep production objectives

Table 3 show the ranks of sheep production objectives by smallholder and large scale farmers. Results of earlier studies [20] indicated the knowledge of the reasons for keeping small ruminants is a prerequisite for setting operational breeding goals. The result of the present study show sheep to have multipurpose functions to support the livelihoods of smallholder farmers whereby their financial, source of food and wealth function ranked as paramount importance. However, intangible benefits of sheep, example, savings, insurance mainly against cop failure, socio-cultural and ceremonial purpose and prestige, were also considered important in the smallholder production system [2] [5] [19] . Similarly, tangible and intangible benefits of sheep, which show the multiple functions of indigenous local breeds in low input systems, were also reported in other studies [4] [7] [11] [26] [27] . Successful breeding program could be mainly achieved through including the intangible benefits (cultural, social and environmental) in the breeding objectives for sheep under smallholder production systems in the tropics. Therefore, considering the different socio-cultural perspective of the communities is important in the adoption of any breeding programme. The need to include intangible benefits in breeding goal definition for low-input production system has been discussed by [19] . This study noted higher economic value for litter size, lambing frequency and 12 month live weight to be important when intangible returns were included in the breeding objectives in smallholder production systems. In large scale systems, the primary reason of keeping sheep was income generation and investment opportunities indicating demand driven production objectives. The variation in production objectives between production systems observed in the current study are likely to be related to differences in social and economic contexts. In spite of all the multifaceted advantages of Gumz sheep mentioned by smallholder farmers, this study showed declining flock sizes in Gumz sheep, which implies the need for breeding strategies that target genetic improvement and conservation of the breed.

Table 3. Purpose of keeping sheep and the ranking of the importance of these purposes by production system.

Purposes	Production systems								Overall
	Smallholder				large scale systems				
	1	2	3	I	1	2	3	I	I
Income	81.7	6.7	1.7	0.43	83.3	25.0	41.7	0.57	0.50
Consumption	6.1	35	28	0.19	0.0	10.0	0.0	0.04	0. 12
Saving	2.8	28.3	23.3	0.14	0.0	18.3	15.0	0.09	0.11
Wealthy	9.4	25.6	31.7	0.15	0.0	15.0	20.0	0.08	0.12
Investment	0.0	0.0	0.0	0.0	6.7	31.7	23.3	0.22	0.11
Insurance	0.0	4.4	40.6	0.08	0.0	0.0	0.0	0.0	0.04

I = index; I = sum of [3 for rank 1 + 2 for rank 2 + 1 for rank 3] given for an individual purpose divided by sum of [3 for rank 1 + 2 for rank 2 + 1 for rank 3] summed for all the purpose.

3.5. Sheep Breeding Systems

The breeding systems practiced in the study area were both pure-breeding and crossbreeding. The most important sheep breeds found in the study areas were the indigenous Gumz and the exotic Rutana and few highland breeds. In addition, there is a crossbred sheep between Gumz and Rutana. Farmers reported that they managed to keep purebred Gumz for pure breeding and still practice crossbreeding between Gumz and Rutana breed. The proportion of the three genotypes (Gumz, Rutata and crossbreds) varies across the two production systems. The smallholder farmers primarily keep the indigenous Gumz sheep followed by the crossbreds (Table 4). Whereas, the most represented sheep breed in the large scale production systems was the introduced Rutana sheep followed by the crossbreds. The average sheep flock size reported for smallholder system in the present study was higher than that reported at different districts by [11] and [28] for Gumz (16.02) and Horro (11.3), respectively. Larger flock size (24.0 to 19.0/heads per household) than in the present study was also reported by [29] and [30] in the central highland and lowlands of Ethiopia. Variation in flock sizes of sheep and goats among the production system across Ethiopia was also confirmed by the study of [12] . This variation is attributed to factors such as role of livestock, availability of land and feed and reliability of crop production. Hence, the suitability and availability of uncultivable large extensive grazing land for sheep production and year round demand of sheep at Sudan market could have contributed to owning of larger sheep flock size.

The higher smallholder flock size is advantages for designing village-based breeding programs. Also larger flocks in commercial production are suitable to introduce selective breeding within each farm. There was a difference in proportion among breeds and sexes in the sheep flocks. Both smallholders and commercial farmers owned higher number of Gumz ewes and ewe lambs than ram lambs and rams. This results in a larger number of lambs, which can speed up the effectiveness of selective breeding because of the high selection intensity. The highest proportion of breeding ewes in the current study was similar to that reported in earlier studies [20] [25] indicating the practice of retaining ewes for breeding purposes and as a manifestation of wealth and prestige. The removal of males for sale is another possible factor contributing to the high proportion of ewes per flock. Since the main purpose of sheep production in the study area is for cash income and home consumption, male animals are either sold or slaughtered and their number remains small in the flock. Large scale farmers do not keep Gumz rams rather they use Rutana as sire breed to produce crossbreed. The small size and lower carcass weight of Gumz as compared to Rutana and crossbred sheep is the reason of the practice. As a result large scale farmers do not have the willingness to get involved in pure Gumz sheep rearing. This practice may reduce the population of the Gumz sheep breed and require better breeding strategy to conserve indigenous Gumz breed. The large scale farmers also keep small number of crossbred rams with the purpose of maintaining pure Rutana female, which may lead to reduced Gumz blood level of the next generation of lambs with a consequence of dilution and long term replacement of the Gumz with Rutana sheep breed.

Table 4. Mean of sheep composition by age and sex group per household in the two production system.

		Smallholders systems						Large scales systems					
		Gumz		Rutana		Crossbred		Gumz		Rutana		Crossbred	
Age	Sex	Mean	SE	Mean	SE	Mean	SE	Mean	SE	Mean	SE	Mean	SE
0 - 6	F	0.70	0.06	0.57	0.07	0.81	0.06	-	-	3.38	0.31	5.60	0.35
	M	1.17	0.09	0.61	0.08	1.04	0.08	-	-	4.02	0.43	6.40	0.58
6 - 12	F	1.30	0.09	0.66	0.07	0.95	0.07	-	-	3.73	0.30	3.82	0.27
	M	0.87	0.07	0.61	0.07	0.87	0.06	-	-	4.17	0.48	4.80	0.38
>12	F	3.00	0.13	0.88	0.06	0.62	0.07	21.2	1.4	12.3	0.69	3.22	0.53
	M	0.31	0.04	0.39	0.04	0.24	0.03	-	-	3.10	0.13	0.90	0.14
Overall means		7.47	0.28	3.68	0.25	4.52	0.22	21.2	1.4	30.6	1.69	24.7	1.47

SE = Standard Error; M = Male; F = Female.

3.6 Herding Practices

Herding practice have an implication for designing genetic improvement programs and introducing improved sheep management such as strategic health interventions at village level [31] . The flock herding practices of the smallholder and large scale farmers reflects the breeding managements and has an impact on the flock size. The major type of herding practiced in the study areas was free grazing. About 51%, 31.7% and 17.2% of the smallholder farmers herd sheep together with goat, separate and either separate or sometimes with goat depending on the availability of hired labor, respectively. Because of their feeding habit, farmers prefer to manage sheep separately, but the majority of the smallholder farmers keep sheep with other livestock because of the shortage of labor [4] [24] reported that sheep and goat are herded together in Afar pastoral and agro pastoral system. In the contrary about 62% of the large scale farmers herd sheep separately and 38% rear sheep and goat flock together. According to [29] , sheep in the central highlands of Ethiopia are herded separately for grazing all year round using family members. (**Table 5**)

3.7. Constraints to Sheep Production

Major constraints associated with sheep production as ranked by respondents are presented in **Table 6**. Diseases, labor shortage, stock theft, drought, predator and conflict between farmers were identified as the most important constraints limiting sheep production (**Table 6**). This is in agreement with that reported by [5] in Southern Region of Ethiopia. Disease and parasite were the most important constraints hindering sheep production by causing high mortalities. Similar to the present study, [32] attributed the decline in livestock population in northern Ethiopia to the high disease and parasite prevalence and shortage of feed. Labour shortage as a major constraint to sheep production was also reported in different parts of Ethiopia [7] [24] . The high theft incidences reported in the study area was ascribed to proximity of the districts to Sudan making it easier for sheep rustlers to trek the animals across the border. The marketing channels of stock, which involve towns and its surrounding townships where there is a ready market for the mutton meat and made it easier to trade the stolen stocks [33] . To reduce sheep thefts, law enforcement agents should augment farmers' efforts by carrying out regular day and night patrols in the village. Formation of neighborhood watch committees to patrol the villages during the evening can also help to reduce the risk [34] . Feed shortage especially in the long dry season was critical problem in both production systems. Predator and conflict were minor concerns in both systems of sheep production.

Table 5. Herding practices across production systems in the study areas.

	Production System	
Sheep herding	Smallholder systems (%) (n = 180)	Large Scale systems (%) (n = 60)
Herded separately	31.7	61.7
Herded with goat	51.1	38.3
Separately or together with goat	17.2	--

3.8 Feed resources

In the wet season, natural pasture grazing on communal and private land was the major sources of feed for sheep in both production systems (**Table 7**). Dry seasons feed resource comprises natural pasture, trees and shrubs, stubble grazing and crop residues with variable magnitude of uses of the feed types between the two seasons. Higher private natural pasture use by large scale farmers than small holders is due to the large proportion of uncultivated lands in this system. Feed conservation practices and crop residue utilization are poor in both production systems. The natural pasture provides larger proportion of the feed resource in the wet season, but it hardly satisfies the quantity and quality required by the animal in dry season. Moreover, availability of other feed resources is less [35] , indicating the common phenomenon of feed shortage during the dry season.

Table 6. Sheep production constraints ranked by respondents and priority indices in different production systems.

Constraints	Production systems								
	Smallholders				Large scales				Overall I
	1	2	3	I	1	2	3	I	
Diseases	54.4	35.0	20.0	0.42	43.3	31.7	25.0	0.36	0.39
Feed shortage	8.3	6.7	8.3	0.08	5.0	11.7	10.0	0.08	0.08
Labour shortage	13.3	20.6	31.7	0.19	21.7	20.0	20.0	0.21	0.20
Theft	17.2	30.0	25.6	0.23	18.3	15.0	21.7	0.18	0.21
Predator	6.7	7.8	13.9	0.08	8.3	13.3	10.0	0.10	0.09
Conflict	0.0	0.0	0.5	0.0	3.3	8.3	13.3	0.07	0.04

I = index; I = sum of [3 for rank 1 + 2 for rank 2 + 1 for rank 3] given for an individual constraint divided by sum of [3 for rank 1 + 2 for rank 2 + 1 for rank 3] summed for all the constraints.

Table 7. Sheep feed resources ranked by respondents and priority indices in smallholder and large scale production systems.

	Smallholder		Large scale systems		Overall	
	Wet	Dry	Wet	Dry	Wet	Dry
Feed source	I	I	I	I	I	I
Communal grazing	0.75	0.38	0.67	0.33	0.71	0.36
Private grazing	0.17	0.22	0.2	0.28	0.19	0.25
Crop residues	0.05	0.11	0.00	0.12	0.03	0.11
Trees and shrubs	0.03	0.16	0.13	0.17	0.07	0.17
Crop after math	0.00	0.13	0.00	0.08	0.00	0.11

I = index; I = sum of [3 for rank 1 + 2 for rank 2 + 1 for rank 3] given for an individual feed divided by sum of [3 for rank 1 + 2 for rank 2 + 1 for rank 3] summed for all the feeds.

The present study confirmed the finding of the previous studies [10] [22] who noted reduced pasture productivity in time and getting dry and scarce in the dry season. The scarcity of feed is more exacerbated by large number of seasonal transhumant livestock movement from the neighboring districts to the study areas in search of feed during the main rainy season and the poor feed conservation practices in the area [10] [14] [22] . To bridge the dry season feed shortage, a high percentages of both smallholder and large scale farmers' use fed supplements such as crop by products (sesame seed cake and wheat bran) and sorghum grains. Feed supplementation target the critical feed shortage period (March through to May), and mainly offered to sick animals. Improvement of extensive sheep production system through introduction of better feed development and conservation mechanism is very important. Among the various suggested strategies, supplementing the fibrous low quality basal diet with more nutritious and digestible feedstuffs could be one of the solutions to alleviate this situation and to improve sheep productivity. In this regard, [35] noted that sheep offered noug seed cake as supplement to hay showed higher body weight gain performance and conversion efficiency than those consumed only hay. Therefore, interventions that improve the productivity of sheep are important in creating wealth and improving the standard of living of resource-poor farmers [36] .

Table 8 presents the percentages of household grazing management and ways of herding across the production systems in the study areas. Sheep production is based on free grazing in the study areas. All of the respondents in the large scale systems reported that sheep flock was herded alone without mixing with other flocks. whereas 42.5% of the smallholder farmers herding their sheep flocks separately and 55.9% herding with their neighboring and the rest 1.7% herding their flocks with Sudan pastoralists in the communal rangelands. There is the possibility of mixing flocks with other neighbouring and transhumance Sudan pastoralist's sheep flocks. In Afar, pastoralist (64.8%) explained that their animals mix with others flocks during the rainy season [4].

Table 8. Percent of respondents who used different management type for grazing and herding of sheep across production system in the study areas.

Management type	Smallholders (N = 180)	Large Scale (N = 60)	Over all (N = 240)
Grazing management			
Free grazing	100	100	100
Way of herding			
Herded separately	42.5	100	71.3
Herded with neighbouring flocks	55.9	-	27.9
Herded with pastoralists	1.7	-	0.8

N = number of obeservation.

Although, mixing of flocks reported by more than half of the smallholder farmers has importance of using rams in the communal grazing areas and reduces the risk of inbreeding and it might be create uncontrolled mating between the Gumz and Rutana sheep breed.

3.9 Sheep health problems

Disease prevalence are often regarded to be major factors limiting the productivity of sheep raised by most rural farmers in the tropics and sub tropics [20] and the farmers are not able to achieve the expected amount of benefit from sheep production. Foot rot, pasteurellosis and respiratory disease are the major sheep diseases which affect sheep productivity in the smallholder production systems of the present study area and veterinary health care is also poor. Vaccination was provided for few common diseases and the service was provided by the government only during seasonal outbreak. Some smallholder farmers indicated that they sourced help from government veterinary services, while most of them treat the animals themselves by buying drugs from private drug suppliers. Sheep pox, pasteurellosis and external parasites comprise the major reported diseases in the large scale systems. In contrast, employed veterinarians treat the sick animals in most of the large scale farmers. Types of disease and limited veterinary services reported in this study were in agreement with earlier reports [4] [11] . Similarly, these findings agree with previous reports [27] [37] that identified these factors as constraints to sheep production in different parts of the country as well as in the tropics [20] . Thus, previous and current findings indicate that healthcare to be one of the important packages that shall be considered during genetic improvement and designing of conservation programs. Application of train and visit methods of veterinary extension to farmers would result in improved disease diagnosis and ultimately improve animal health husbandry. Wider utilization of indigenous breeds tolerant to disease could be another way in alleviating some of the problems. Farmers should also be made aware that they need to dip and vaccinate their sheep in order to reduce disease incidences. The advantage of using locally adapted breeds as the way of combating diseases hence improving the livelihood of the rural Poor has been discussed elsewhere in tropics [38] . (**Table 9**)

3.10. Marketing Systems

Sale of live animals is the primary purpose of sheep keeping followed by house consumption. Sheep were sold throughout the year for export market since there is always demand for sheep in Sudan. Both legal and informal sheep trade are operating at different magnitudes in the Ethio-Sudan cross-border. [10] [39] also reported the existence of trans-boundary trade between Ethiopia and Sudan which operate under legal and informal routs. About 68% of the large scale farmers primarily target legal traders who supply sheep for export market, while 25% targeted hired traders. On contrary, 56% of smallholder farmers target informal hired traders who supply sheep for export market, while 29% of smallholder target primarily household consumptions. [39] reported that hired traders come from Sudan to north western lowlands and make deal and transfer the animal through illegal routs. The practice of informal trade might be due to the high tariff rate charged in legal trade. Smallholder farmers indicated that higher price is received when they sold for informal traders compared to legal traders. According to interview result, 78.1% of smallholder farmers offer sheep for sale to meet their urgent needs at any time during the year. This supports the results of other authors who indicated sales of animals in order to meet emergencies of cash need [39] . About 21.9% of smallholder farmer's sold during major holidays of Christian and Muslims. On contrary, 86% of the large scale farmers indicated that they supply sheep to export market throughout the year. This is probably due to the perennial demand of sheep and attractive price in Sudan market. While, 14% of the large scale farmers sold sheep during major holidays of Muslims. This could be an added advantage to implement breed improvement program. Many breeding plan have failed since they do not consider market demand elsewhere in the world.

Table 9. Major sheep disease ranked by respondents and priority indices in different production systems.

Diseases	Smallholders				Large scale				Over all I
	1	2	3	I	1	2	3	I	
Foot rot	17.8	20.6	25.0	0.20	13.3	10.0	21.7	0.13	0.17
Internal parasites	15.6	12.8	17.2	0.16	11.7	18.3	13.3	0.14	0.15
ORF	16.7	11.1	15.6	0.14	6.7	6.7	10.0	0.07	0.11
Pasteurellosis	18.3	22.8	12.8	0.19	15.0	25.0	11.7	0.18	0.19
Sheep pox	7.8	9.4	3.9	0.08	26.7	8.3	11.7	0.18	0.13
Respiratory disease	15.6	5.0	15.6	0.16	10.0	20.0	13.3	0.14	0.15
External parasites	8.3	5.6	10.0	0.07	16.7	11.7	18.3	0.15	0.10

I = index; I = sum of [3 for rank 1 + 2 for rank 2 + 1 for rank 3] given for an individual disease divided by sum of [3 for rank 1 + 2 for rank 2 + 1 for rank 3] summed for all the diseases.

Animals are sold at an early age and males of 3 - 6 months in smallholder (34.2%) and 6 - 12 months in large scale systems (48.1%) contributed the highest percentages of sold animals indicating young males are the most demanded and required class by sheep exporters and also fetched higher price as compared to other age group of sheep in Sudan market. However, this practice will have an impact on inbreeding and loss of genetic material since males that are rapidly growing are picked and sold before used for mating. Earlier work also reported negative selection practices due to the sale of best performing young animals [4] [7] [40] [41] noted that complete lack or insufficient number of breeding males in sheep flocks result in delayed services and negatively affect the flocks' reproductive performance and overall productivity. Breeding ewes were sold (3.6%) to local market very rarely in large scale systems than smallholder systems (14.4%). This may be due to the fact that large scale sheep owners kept females for breeding and only sold aged and un-productive ewes. Most common sheep breed sold to export market throughout the year was the Rutana sheep and its crosses. Rutana and its crosses fetched higher prices than Gumz sheep. The Gumz sheep was more preferable on domestic market and local consumers' prefer Gumz sheep meat than Rutana which show the superiority of one breed over another in terms of growth and carcass characteristics [42] [43] (**Table 10**).

]In both systems, sheep are one of the income sources for farmers which are similar to previous reports [4] . The money obtained from the sale of this animal is used to purchase different utilities (Table 11). This finding agrees with the earlier reports [25] indicating the importance of small ruminants as main source of household income for farmers and pastorals in various developing countries.

Table 10. Traded sheep by age and sex in 2015/16 as reported by respondents.

Age group	Sex	Smallholder (N = 180)		Large scale (N = 60)		Overall (N = 240)	
		N	%	N	%	N	%
<6 months	M	38	14.5	188	23.5	226	21.4
	F	31	11.8	166	20.9	197	18.6
6 - 12 months	M	52	19.7	196	24.6	248	23.4
	F	59	22.4	155	19.5	214	20.2
<12 months	M	45	17.1	61	7.7	106	10.0
	F	38	14.4	29	3.6	67	6.3
All ages	M	135	51.3	445	56.0	580	54.8
	F	128	48.7	350	44.0	478	45.2
Overall		263	100	795	100	1058	100

N = Number of households; M = Male; F = Female.

Table 11. Reported reasons for sale of sheep by smallholder respondents.

	Smallholder production systems			
Use of income	1	2	3	Index
Buy food	26.7	8.3	7.2	0.17
Buy seed	14.4	11.1	2.8	0.11
Buy breeding sheep	8.3	2.8	2.8	0.05
Buy cloth	9.4	17.8	12.8	0.14
Pay tax	11.7	19.4	6.7	0.13
Buy household supplies	9.4	16.4	7.2	0.11
Buy feed	7.8	7.8	10.6	0.08
Saving	0.0	1.1	9.4	0.02
Buy school supplies	3.9	8.9	24.4	0.09
Vet services	8.3	11	16.1	0.10

I = index; I = sum of [3 for rank 1 + 2 for rank 2 + 1 for rank 3] given for an individual uses divided by sum of [3 for rank 1 + 2 for rank 2 + 1 for rank 3] summed for all the uses.

4. CONCLUSION

The study revealed a number of relevant issues that help in developing effective breed conservation and improvement programmes of the breed. It identified two sheep production systems (smallholder and large-scale commercial) and two sheep breeds and their crossbreed. The two production systems varied in their sheep production, breeding and marketing strategies. The adapted local genetic resource of Gumz breed which accounted for the wide genetic diversity of sheep resources in Ethiopia and the thin-tailed local breed [12] is losing its genetic diversity and considered to be endangered. It is thus essential to design a conservation based breeding program for the Gumz breed. The design of breeding program needs to consider the sustainable utilization of the purebred Rutana sheep and the crossbreeding component. The program needs to be a comprehensive one involving the two sheep breeds and their crossbred in the area. Hence, we recommend designing of conservation-based breeding program to conserve the locally adapted Gumz breed as well as to improve sheep production and productivity in the area through rational utilization of all the two genotypes and their crossbred. For designing a sustainable breeding program for Gumz sheep, the traditional management system, production objectives, herding and breeding practices, marketing and production constraints specific to the two systems identified in this study shall be considered. Feed resource management interventions, efficient delivery of animal health services and profitable market outlets are required to increase the benefits of sheep rearing to the keepers of these valuable sheep resources.

ACKNOWLEDGEMENTS

We would like to thank Gondar Agricultural research center, Development Agents for their support during the field work and making the data available for the study. The financial support of the Ministry of Education and Bule Hora University is also highly acknowledged.

REFERENCES

1. CSA (Central Statistics Agency) (2015) Agricultural Sample Survey 2014/15. Volume II Report on Livestock and Livestock Characteristics (Private Peasant Holdings), Central Statistical Agency (CSA), Addis Ababa, Ethiopia, 188 p.
2. Markos, T. (2006) Productivity and Health of Indigenous Sheep Breeds and Crossbreds in the Central Ethiopian Highlands. PhD Dissertation, Department of Animal Breeding and Genetics, Faculty of Veterinary Medicine and Animal Sciences, Swedish University of Agricultural Science (SLU), Uppsala, Sweden, 74 p.
3. Tsegaye, T., Mengistu, U., Yoseph, M. and Merga, B. (2013) Pre-Weaning Growth Performance of Crossbred Lambs (Dorper × Indigenous Sheep Breeds) under Semi-Intensive Management in Eastern Ethiopia. Tropical Animal Health and Production, 46, 455-460.
4. Getachew, T., Haile, A., Tibbo, M., Sharma, A.K., Sölkner, J. and Wurzinger, M. (2010) Her Management and Breeding Practices of Sheep Owners in a Mixed Crop-Livestock and a Pastoral System of Ethiopia. African Journal of Agricultural Research, 5, 685-691
5. Hailemariam, F., Melesse, A. and Banerjee, S. (2013) Traditional Sheep Production and Breeding Practice in Gamogofa Zone, Southern Ethiopia. International Journal of Livestock Production Research, 1, 26-43.
6. Asegede, M., Bisrat, A., Hagos, Y. and Gugsa, G. (2015) Livestock Market Value Chain Assessment in Selected Sites of Tigray, North Ethiopia: Challenges and Opportunities for Enhancing Animal Product Export. Global Veterinaria, 14, 48-55.
7. Gizaw, S., Haile, A. and Dessie, T. (2010) Breeding Objectives and Breeding Plans for Washera Sheep under Subsistence and Market Oriented Production Systems. Ethiopian Journal of Animal Production, 10, 1-16.
8. IGAD (2011) The Contribution of Livestock to the Ethiopian Economy—Part II, IGAD Livestock Policy Initiative. Roy Behnke, Odessa Centre, Great Wolford, UK.
9. Legesse, G., Haile, A., Duncan, A.J., Dessie, T., Gizaw, S. and Rischkowsky, B. (2014) Sheep and Goat Value Chains in Ethiopia: A Synthesis of Opportunities and Constraints. ICARDA/ILRI Project Report, International Center for Agricultural Research in the Dry Areas/International Livestock Research Institute, Nairobi, Kenya.
10. Mulugeta, E., Gebremedhin, B., Hoekstra, D. and Mohammed, J. (2007) Analysis of the Ethio-Sudan Cross-Border Cattle Trade: IPMS (Improving Productivity and Market Success) of Ethiopian Farmers Project Working Paper 2. ILRI (International Livestock Research Institute), Nairobi, Kenya.
11. Solomon, A., Hegde, B.P. and Mengistie, T. (2011) Growth and Physical Body Characteristics of Gumuz Sheep under Traditional Management Systems in Amhara Regional State, Ethiopia. Livestock Research for Rural Development, 23, 1-5.

12. Gizaw, S., Abegaz, S., Rischkowsky, B., Haile, A., Mwai, A.O. and Dessie, T. (2013) Review of Sheep Research and Development Projects in Ethiopia. International Livestock Research Institute (ILRI), Nairobi, Kenya.
13. FAO (2012) Phenotypic Characterization of Animal Genetic Resources. FAO Animal Production and Health Guidelines No. 11, Rome.
14. Desalew, T., Tegegne, A., Nigatu, L. and Teka, W. (2010) Rangeland Condition and Feed Resources in Metema District, North Gondar Zone, Amhara Region, Ethiopia. IPMS (Improving Productivity and Market Success) of Ethiopian Farmers Project Working Paper 25, Nairobi, Kenya, ILRI, 74 p.
15. Tadesse, H. and Afework, B. (2008) Habitat Association of Insectivores and Rodents of Alatish National Park, Northwestern Ethiopia. Tropical Ecology, 49, 1-11.
16. Freund, R.J. and Wilson, W.J. (2003) Statistical Methods. Academic Press, New York.
17. Ayalew, W., Vandorland, A. and Rowlands, G.J., Eds. (2004) Designing, Execution and Analysis of the Livestock Breed Survey in Oromia Regional State, Ethiopia, OADB (Oromia Agricultural Development Bureau, Addis Ababa, Ethiopia, and ILRI (International Livestock Research Institute), Nairobi, Kenya, 260 p.
18. ILCA (1990) ILCA Annual Report 1989: 29-31. ILCA, Addis Ababa.
19. Kosgey, I.S. (2004) Breeding Objectives and Breeding Strategies for Small Ruminants in the Tropics. PhD Thesis, Wageningen University, The Netherlands, 272 p.
20. Kosgey, I.S., Rowlands, G.J., van Arendonk, J.A.M. and Baker, R.L. (2008) Small ruminant Production in Smallholder and Pastoral/Extensive Farming Systems in Kenya. Small Ruminant Research, 77, 11-24. https://doi.org/10.1016/j.smallrumres.2008.02.005
21. Mavule, B.S., Muchenje, V. and Kunene, N.W. (2013) Characterization of Zulu Sheep Production System: Implications for Conservation and Improvement, 8, 1226-1238.
22. Amare, S. (2006) Qualitative and Quantitative Aspects of Animal Feed in Different Agro-Ecological and Agriculture: Livestock in the Balance. FAO, Rome, 14.
23. Tegegne, F. and Asefa, G. (2010) Feed Resource Assessment in Amhara Regional State. Ethiopian Sanitary and Phytosanitary Standards and Livestock Meat Marketing Program (SPS-LMM), Texas A&M University System.
24. Hassen, A.S. and Tesfaye, Y. (2014) Sheep and Goat Production Objectives in Pastoral and Agro-Pastoral Production Systems in Chifra District of Afar, Ethiopia. Tropical Animal Health and Production, 46, 1467-1474. https://doi.org/10.1007/s11250-014-0668-4
25. Mengistie, T., Girma, A., Solomon, G., Sisay, L., Abebe, M. and Markos, T. (2010) Traditional Management Systems and Linear Body Measurements of Washera Sheep in the Western Highlands of the Amhara National Regional State, Ethiopia. Livestock Research for Rural Development, 22, 1-16.
26. Wurzinger, M., Ndumu, D., Maumung, R., Drucker, A.G., Okeyo, A.M., Semambo, D.K. and Sölkner, J. (2006) Stated Preferences through Use of Choice Experiments: Valuing (Re)Production versus Aesthetics in the Breeding Goals of Ugandan Ankole Cattle Breeds. Proceedings of the 8th World Congress on Genetics Applied to Livestock Production, Belo Horizonte, Brazil, 13-18 August 2006, 1-4.
27. Nigussie, H., Mekasha, Y., Kebede, K., Abegaz, S. and Pal, S.K. (2015) Indigenous Sheep Production System in Eastern Ethiopia: Implications for Genetic Improvement and Sustainable Use. American Scientific Research Journal for Engineering, Technology, and Sciences (ASRJETS), 11, 136-152.
28. Edea, Z., Haile, A., Tibbo, M., Sharma, A.K., Sölkner, J. and Wurzinger, M. (2012) Sheep Production Systems and Breeding Practices of Smallholders in Western and South-Western Ethiopia: Implications for Designing Community-Based Breeding Strategies. Livestock Research for Rural Development, 24, 7.
29. Mekoya, A., Yami, A. and Hailemariam, M. (2000) Management of Traditional Sheep Production in Lallomamma Mider District. North Shoa, Amhara, 143-153.
30. Fekerte, F. (2008) On-Farm Characterization of Blackhead Somali Sheep Breed and Its Production System in Shinile and Erer Districts of Shinile Zone. MSc Thesis, School of Graduate Studies of Haramaya University of Agriculture, Dire Dawa, Ethiopia, 115 p.
31. Gizaw, S., van Arendonk, J.A.M., Valle-Zárate, A., Haile, A., Rischkowsky, B., Dessie, T. and Mwai, A.O. (2014) Breeding Programmes for Smallholder Sheep Farming Systems: II.

Optimization of Cooperative Village Breeding Schemes. Journal of Animal Breeding and Genetics, 131, 350-357. https://doi.org/10.1111/jbg.12102

32. Seare, T., Gangwar, S.K. and Kebede, K. (2011) Performance and Physical Body Measurement of Abergell Sheep Breed under Traditional Management System of Tigray Regional State, Northern Ethiopia. International Journal of Science and Nature, 2, 225-230.
33. Mapiliyao, L. (2010) Sheep Production Practices, Flock Dynamics, Body Condition and Weight Variation in Two Ecologically Different Resource Poor Communal Farming Systems. MSc Thesis, University of Fort Hare, South Africa, 136 p.
34. Kynock, G. and Ulicki, T. (2001) Cross-Border Raiding and Community Conflict in the Lesotho South African Border Zone. Southern African Migration Project (SAMP), Migration Policy Series No. 21.
35. Beshaw, S. and Melaku, S. (2008) Bodyweight and Carcass Characteristics of Somali Goats Fed Hay Supplemented with Graded Levels of Peanut Cake and Wheat Bran Mixture. Tropical Animal Health and Production, 40, 553-560. https://doi.org/10.1007/s11250-008-9133-6
36. Hirpa, A. and Abebe, G. (2008) Economic Significance of Sheep and Goats. Sheep and Goat Production Handbook for Ethiopia, Branna Printing Enterprise, Addis Ababa, 2-24.
37. Hassen, A., Ebro, A., Kurtu, M. and Treydte, A.C. (2010) Livestock Feed Resources Utilization and Management as Influenced by Altitude in the Central Highlands of Ethiopia. Livestock Research for Rural Development, 22, 229-233.
38. Devendra, C. (2001) Small Ruminants: Imperatives for Productivity Enhancement Improved Livelihoods and Rural Growth—A Review. Asian-Australian Journal of Animal Sciences, 14, 1483-1496. https://doi.org/10.5713/ajas.2001.1483
39. Tesfay, Z., Anal, A.K. and Gebreyohanis, G. (2012) Assessment of the Sheep Production System of Northern Ethiopia in Relation to Sustainable Productivity and Sheep Meat Quality. International Journal of Advanced Biological Research, 2, 302-313.
40. Seleka, T.B. (2001) Determinants of Short-Run Supply of Small Ruminants in Botswana. Small Ruminant Research, 40, 203-214. https://doi.org/10.1016/S0921-4488(01)00182-1
41. Mirkena, T., Duguma, G., Haile, A., Tibbo, M., Okeyo, A.M., Wurzinger, M. and Solkner, J. (2010) Genetics of Adaptation in Domestic Farm Animals: A Review. Livestock Science, 132, 1-12. https://doi.org/10.1016/j.livsci.2010.05.003
42. Dossa, L.H., Sangare, M., Buerkert, A. and Schlecht, E. (2015) Production Objectives and Breeding Practices of Urban Goat and Sheep Keepers in West Africa: Regional Analysis and Implications for the Development of Supportive Breeding Programs. SpringerPlus, 4, 281-290. https://doi.org/10.1186/s40064-015-1075-7
43. Macfarlane, J.M. and Simm, G. (2007) The Contribution of Genetic Improvement for Lamb Meat Production. Paper presented at 3rd International Symposium about Goat and Sheep Meat Type, 3rd Sincorte, João Pessoa, Paraiba, Brazil.

CHAPTER 10

Genetics of a sex-linked recessive red eye color mutant of the tarnished plant bug, Lygus lineolaris

Margaret Louise Allen

Biological Control of Pests Research Unit, USDA Agricultural Research Service, Stoneville, USA

ABSTRACT

An inbred colony of the tarnished plant bug, Lygus lineolaris (Palisot de Beauvois) (Miridae: Hemiptera), was observed to contain specimens with abnormal traits including red eyes, deformed antennae, and deformed legs. These specimens were isolated and back crossed to create stable phenotypic strains. The only successful strain established was a red eyed strain named Cardinal. The trait was more prevalent and stable in males, suggesting that it could be sex linked. To test the hypothesis that the trait was based on a recessive sex linked allele, classical genetic crosses were performed. The hypothesis was confirmed, and the eye color phenotype was measured and characterized using color analysis software. The trait is similar to other red eyed phenotypes described in this species, but is clearly based on a different mutation since it is sex linked rather than autosomal. The results of crossing experiments also suggest that inbreeding in this species results in substantial fitness cost to laboratory insects.

Keywords: Genetics; Inbreeding; Fitness; Pigmentation; Ommochromes; Antennae; Spineless

1. INTRODUCTION

Variations in the eye color of insects can be found in nature and selected by inbreeding or induced by mutagenesis. Naturally occurring and induced mutations have served as a foundation for progress in both general genetics and in genetic manipulation of insects. A white eye color mutation was found to be linked to sex in Drosophila melanogaster, defining the phenomenon of sexlinkage, and establishing methods in classical genetics [1]. Transplantation studies using D. melanogaster eye disks from multiple eye color strains were vital to the establishment of biochemical pathway genetics [2]. Insect eye colors were later found to be more complex than could be explained by the one-gene-one-enzyme principle, and were shown to be related to transporter pathways [3]. More recently, eye color mutations have been utilized extensively to verify manipulation systems involving genetic transformation [4,5], especially when the eye-specific promoter 3xP3 is utilized [6].

Eye color mutations have been described in numerous insects other than D. melanogaster. Within the order Diptera, eye color mutations have been described in fruit flies, family Tephritidae [7-9], in blow flies, family Calliphoridae [10-14] and house flies [15-17], in the tsetse fly [18-22], and in many mosquitoes [23-33]. Eye color mutations have also been described in the order Lepidoptera [34-40], in the order Coleoptera [41-43], and in the order Hymenoptera [44-46]. In the order Hemiptera, eye color mutations have been described in kissing bugs (family Reduviidae) [47-51], in the tarnished plant bug, the subject of this paper, family Miridae [52,53], and recently in the brown planthopper, family Delphacidae [54].

The rapid pace of progress in genomics and bioinformatics presents remarkable opportunities for discovery and analysis of insect genes. Efforts to capitalize on this progress will be enhanced by increased focus on basic biological function phenomena in insects that have not yet been developed as model organisms. Naturally occurring strains with genetically defined visible markers will once again serve as valuable resources to scientific progress. A red-eyed strain, established through inbreeding of Lygus lineolaris, was reported previously [53]. A similar strain, established from a wild male specimen collected in the field in Arkansas was reported more recently [52]. Through genetic analysis, the eye color allele in these naturally occurring strains were characterized as autosomal recessive.

As part of an effort to perform in vivo genetic manipulations on L. lineolaris, a laboratory strain was established in quarantine and inbred for 5+ years. This strain was examined regularly for unusual phenotypes that might indicate mutation. Phenotypes associated with abnormal appendages were observed but could not be reproduced and maintained in the laboratory through normal breeding. However, an eye color mutation was observed and isolated. To verify that this mutant strain was not the same as the ones previously described, reciprocal crosses were undertaken to test the hypothesis that the mutation was sex linked. Additionally, digital image analysis was utilized to characterize the color of the phenotype. To differentiate the phenotype from those previously described, the strain was called Cardinal, with the eye color mutation cardinal. Strains generated by inbreeding could be highly useful in forthcoming genomics projects using this species of pest insect and other arthropods.

2. MATERIALS AND METHODS

2.1. Insects

A wild-type colony was maintained without introgression for 5+ years in the Stoneville Research Quarantine Facility (SRQF) in Mississippi. Insect stocks were kept in an environmental chamber (Percival Scientific, Inc. Perry, IA) set for 16:8 (L:D) h lighting regime with 25°C daytime and 19°C nighttime temperature and 55%RH. Insects were provided a standardized diet and oviposition system based on NI diet and 4% gelcarin oviposition substrate [55].

2.2. Classical Genetics

For first generation matings small groups of virgin insects, two males and four females, were housed in isolation. Second generation insects were grouped in sets of two males and seven females. Insects were sexed in the 5th instar [52]. Eggs were collected daily. Mature eggs were separated from oviposition substrate and kept on moist filter paper for observation until hatch, then transferred to 100 mm by 15 mm Petri dishes for rearing to the adult stage. Nymphs were provisioned with fresh red clover leaves as dietary supplement and refugia.

2.3. Imaging

Images of living specimens were collected using a stereoscopic zoom microscope (Nikon SMZ1500) and Nikon digital camera (DMX 1200). Images used to analyze the eye color were cropped to include only eyes and converted to .jpg files then analyzed using RGB software [56]. Each image was analyzed three times using different portions of the eye image for analysis. The means of these samples were then used to generate color match indicators for illustrations.

2.4. Statistical Analyses

Egg production data were analyzed using the mixed procedure SAS Enterprise Guide v. 4.2 (SAS 2006). Phenotype and sex ratio data were analyzed for goodness of fit to Mendelian ratios using the Chi-Square test [57].

3. RESULTS

After inbreeding a culture of *L.* lineolaris without introgression for overlapping generations of roughly 60 days for four years (approximately 24 generations), a red eyed individual was identified in a colony cage. For the next two years backcrossing and inspection of the parent colony for additional red eyed stock eventually produced a homozygous strain of red eyed specimens (Figure 1). When the first red-eyed individuals were identified they were invariably male, and were paired with wild type females from the inbred source colony. When back crosses of the offspring were obtained, male red eyed individuals were found but red-eyed females were rare. Thus the possibility that the eye color was a sex linked phenotype was recognized early on. It took several generations to establish a homozygous colony of red-eyed specimens (Cardinal), and the overall fitness of the strain seemed low, compared to the wild type (empirical observation). As Cardinal strain was becoming established, additional rare phenotypes began to appear regularly in both the wild and Cardinal strains. These phenotypes appeared as shortened and deformed appendages, initially short antennae (Figure 2). These individuals were also backcrossed and crossed with one another, but the phenotype was never reliably reproduced in offspring. The most common deformity other than short antennae found in the inbred colonies was truncation of legs (Figure 3). These deformities sometimes coincided with the cardinal mutation, but also appeared in wild types with no apparent linkage.

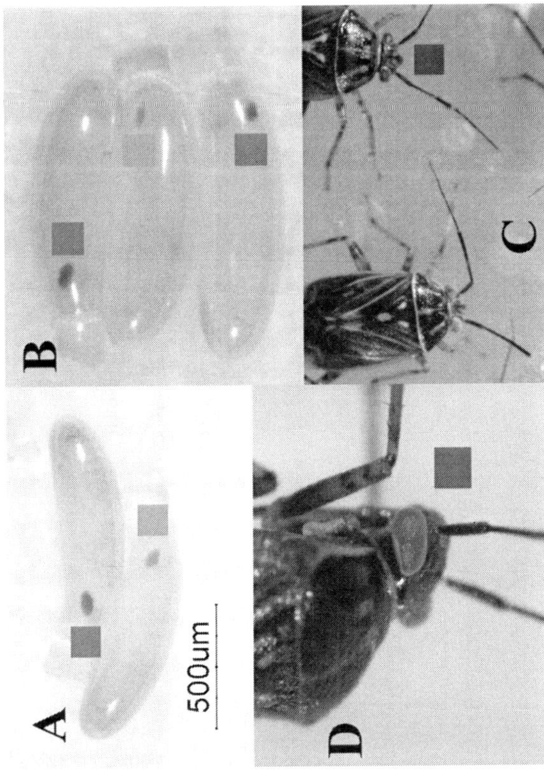

Figure 1. Eye colors of inbred strains of Lygus lineolaris (Palisot de Beauvois) (Hemiptera: Miridae) with square color match indicator boxes. (A) Eggs produced by homozygous insects: normal eye color above, cardinal eye color below; (B) Eggs produced by crossing phenotypes: upper, normal eye color; middle, cardinal eye color from crimson female parent; lower, heterozygous (normal) eye color from crimson female parent; (C) Young adult eye color comparison: left, cardinal; right, normal (wild type); (D) Eye color of an older male (30+ days post adult eclosion/ecdysis).

Because the phenotypes could not be established in permanent culture, no genetic data could be collected.

To test the hypothesis that fecundity was lower in the Cardinal strain, eggs were collected from small pools of individuals of wild type, Cardinal, and reciprocal crosses of the two strains. Egg production is shown in Table 1. While the differences in egg production were not significant in the first five or ten days of the test, by day 15, and continuing on day 20, the cumulative egg production data met the threshold ($P < 0.05$) to signify statistically significant difference (Table 1).

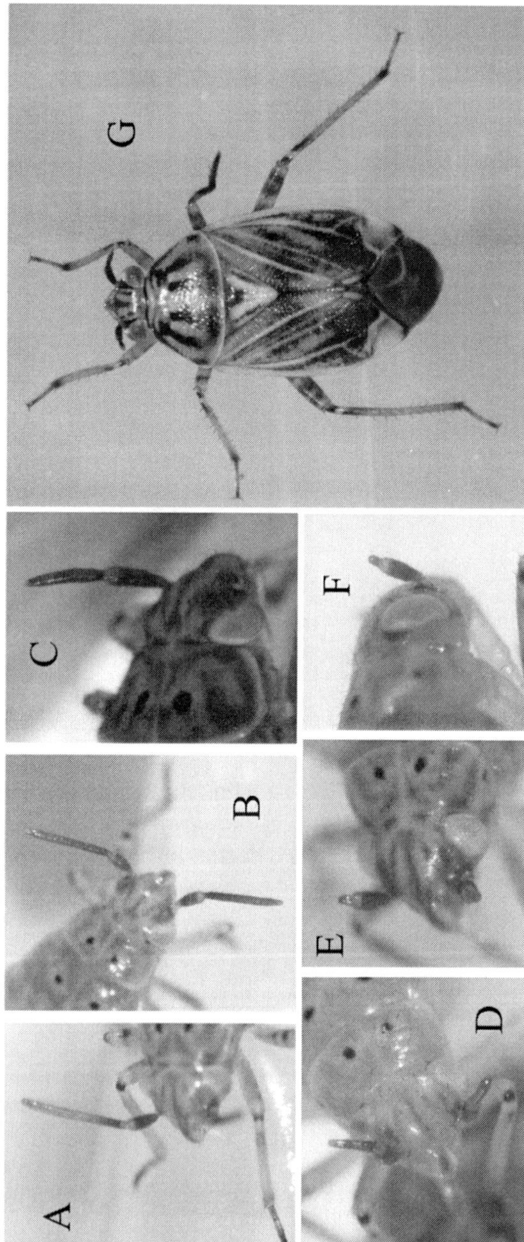

Figure 2. Inbred specimens of Lygus lineolaris that display deformed antennae. (A)-(F) nymphs; (G) adult. Note that the right middle leg of this specimen is also defective, tibia is shortened and tarsal segments are fused.

Table 1. Mean cumulative numbers of eggs (±SE) produced per female in crosses of inbred wild-type (W) and inbred cardinal-eyed (Ca) specimens of Lygus lineolaris.

Cross	Day							
♂ × ♀	5	n[a]	10	n	15	n	20	n
W × W	13.08 ± 2.39	12	25.58 ± 2.10	12	39.25 ± 2.29	12	51.14 ± 4.29	10
Ca × Ca	12.33 ± 2.51	12	20.42 ± 2.92	12	31.08 ± 3.29	12	41.25 ± 3.44	12
W × Ca	7.33 ± 4.51	12	18.42 ± 8.42	12	24.75 ± 10.00	12	32.00 ± 11.40	11
Ca × W	5.33 ± 1.92	12	8.00 ± 2.74	12	12.58 ± 2.43	12	16.42 ± 1.06	11
F	3.17		4.23		7.07		6.83	
$P > F$	0.11		0.06		0.02		0.02	

[a]The degrees of freedom for each comparison was 3.39.

Overall survival to the imaginal stage from fertile em bryos was only evaluated in the F_1 mating pools: from the male Ca parent 55.6% of wild type offspring survived and 64.9% of cardinal offspring survived; from the female Ca parent 69.5% of wild type offspring survived and 70.5% of cardinal offspring survived. No significant difference was detected ($P > 0.05$, student t-test, df = 2).

Differences of least squares means on day 20 showed no significance between homozygous Cardinal (Ca × Ca) egg production and homozygous wild type (W × W) egg production ($P = 0.262$, t = −1.24, df = 6), and also no significant difference between Ca x Ca and wild type male crossed with Cardinal female (W × Ca) ($P = 0.291$, t = 1.16, df = 6) nor between the W × W and W × Ca ($P = 0.054$, t = −2.40, df = 6). Additionally, the egg production of reciprocal crosses (W × Ca/Ca × W) did not differ from one another ($P = 0.099$, t = −1.95, df = 6). However, the Ca x W cross produced significantly lower quantities of eggs when compared to homozygous crosses W × W ($P = 0.0048$, t = −4.35, df = 6) and Ca × Ca (P = 0.021, t = 3.11, df = 6).

To test the hypothesis that the phenotype was sex linked, crosses of the Cardinal and wild type strains were evaluated, and yielded the expected ratios of offspring in the first generation for a recessive sex-linked mutation (Table 2A). All offspring of a male Ca parent were wild type, while all male offspring from a female Ca parent were red eyed and all female offspring were wild type. Second generation crosses of heterozygous females from both first generation crosses also yielded expected ratios of offspring (Table 2B). The crosses are illustrated diagrammatically in Figure 4.

Eye colors of the wild-type embryonic insects corresponded to RGB colors at R 193 ± 8, G 88 ± 7, B 70 ± 10 while the mutant embryonic eye colors corresponded to RGB colors R 240 ± 2, G 130 ± 7, B 68 ± 10. These colors persisted as developing embryos, nymphs, and through early adult stage. As the adults darken with age, particularly the males, the eyes darken to a more crimson color corresponding to R 161 ± 7, G 70 ± 18, B 44 ± 20. The fully mature wild-type adult eye corresponds to R 110 ± 7, G 58 ± 8, B 48 ± 6 (Figure 1).

4. DISCUSSION

Standardized laboratory rearing of L. lineolaris is primarily for the purpose of bioassays. The cultures of insects are expected to represent field performance of naturally occurring populations. Thus, colonies are frequently intermixed with new field collected insects in order to maintain colony health and heterozygosity [58]. The laboratory colony in the SRQF differed in that no introgression by field collected insects was allowed.

Table 2. First and second generation crosses of Cardinal eye color strain to parental wild type strain. (A) First generation heterozygous female parents are either combined or identified by the sex of the Cardinal-eyed parent. Second generation combined analysis included with pairwise comparisons of phenotype and male/female ratio. Critical value for all chi-square tests is 3.84, df = 1; (B) Second generation crosses, combined data (above) and data analyzed by parent carrying cardinal allele. Critical value for phenotype data is 7.82, df = 3.

A.

Cross	egg phenotype		χ^2	adult phenotype						adult survival					
				wild type		χ^2	cardinal		χ^2	Totals		χ^2	Totals		χ^2
♂ × ♀	wild type	cardinal		♂	♀		♂	♀		wild type	cardinal		♂	♀	
W × W	580	0	-	202	166	0.880	0	0	-	368	0	-	202	166	0.880
Ca × Ca	0	537	-	0	0	-	248	231	0.151	0	479	-	248	231	0.151
W × Ca	265	228	0.694	0	143	-	136	0	-	143	136	0.044	136	143	0.044
Ca × W	337	0	-	117	141	0.558	0	0	-	258	0	-	117	141	0.558
Ca × F_1	438	397	2.013	103	134	4.055[a]	131	125	0.141	237	256	0.732	234	259	1.268

B.

Cross	egg phenotype		χ^2	adult phenotype					adult survival					
				wild type		cardinal		χ^2	Totals		χ^2	Totals		χ^2
♂ × ♀	wild type	cardinal		♂	♀	♂	♀	(df = 3)	wild type	cardinal		♂	♀	
Ca × F_1	438	397	2.013	103	134	131	125	4.777	237	256	0.732	234	259	1.268
Ca × $F_{1♂}$	245	208	3.022	59	78	75	58	4.874	137	133	0.059	134	136	0.015
Ca × $F_{1♀}$	193	189	0.041	44	56	56	67	4.749	100	123	2.372	100	123	2.372

[a]Exceeds critical value for chi-square tests, 3.84, df = 1.

This culture strategy was specifically intended to limit heterozygosity and provide more inbreeding to support genetic analysis and genetic manipulation. Under laboratory inbreeding conditions, spontaneous mutations often are revealed. Novel eye color phenotypes are not uncommon in colonies of insects. Those arising in colonies of D. melanogaster kept by Thomas Morgan and his students formed a foundation for modern genetics.

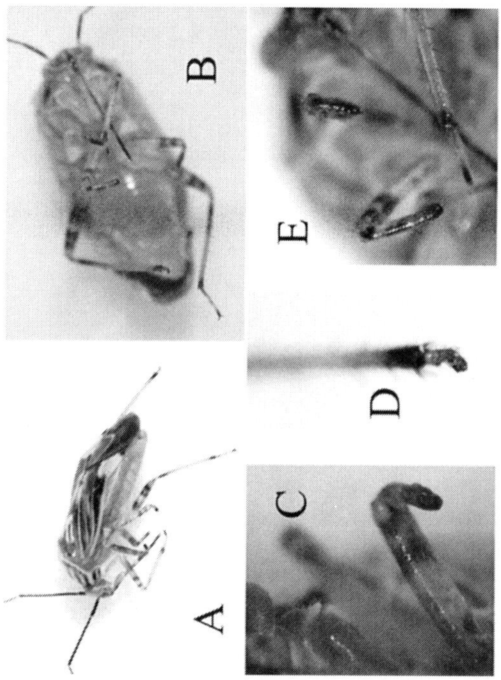

Figure 3. Individual specimen from Cardinal strain with severe leg deformities. Four of six legs have shortened and deformed termini. Tarsal segments appear to be fused. (A) Right lateral view; (B) Ventral view; (C) Left middle leg terminus; (D) Right hind leg; (E) Right foreleg and middle leg termini.

The truncated appendages observed repeatedly in the inbred colony bear resemblance to defects described in D. melanogaster and Tribolium castaneum related to the transcription factor spineless [59-61]. Further careful breeding of similar specimens will be required to determine whether these phenotypes are related. Unfortunately, the specimens we found were delicate and generally infertile, and were discarded after multiple attempts to establish a stable strain failed.

Naturally occurring orange or red eye color phenotypes of Hemiptera have been described [47-50,52-54]. Eye pigments in two species of Hemiptera in the family Reduviidae differ from one another, those from Triatoma infestans Klug being composed primarily of xanthommatin [47], while the eye pigments of Rhodnius prolixus are composed primarily of ommins [62]. The only enzyme identified from L. lineolaris from the ommochrome biosynthetic pathway is tryptophan oxygenase (TO) (http://www. ncbi. nlm.nih.gov/nuccore/307634529,http://www.ncbi.nlm.nih.gov/protein/3 07634530, submitted 10 August 2010, accessed 31 July 2012). This enzyme catalyzes the initial reaction in the conversion of tryptophan to ommochromes, the pigments in insect eyes. The white gene used to define sex linkage by Morgan (1910) was shown to be a membrane pigment transporter gene rather than a pigment synthesis enzyme gene (Sullivan and Sullivan 1975), and this may also be the case in L. lineolaris. The genetic identity of TO has been characterized in D. melanogaster [63], and the red flour beetle Tribolium castaneum [64], and putative homologues with a high similarity at the nucleotide level [65] can be identified in the human body louse Pediculus humanus corporis (XM_002423485.1, accessed from NCBI 16 July 2012) and the mountain pine beetle Dendroctonus ponderosae (BT128347.1, accessed from NCBI 16 July 2012) [66]. While TO may be mutated in either the autosomal or sex-linked phenotypes of red eyed L. lineolaris, any number of other biosynthesis or transport genes may also be responsible.

Both naturally occurring eye color strains of L. lineolaris previously identified were characterized as autosomal recessive alleles. The wild collected eye color variant [52] was reported to have little effect on behavior or physiology, although because of the low field prevalence of the phenotype it was implied that there could be a fitness disadvantage, possibly associated with mating disadvantage based on decreased visual ability. The red-eyed "R" strain was kept in culture for >5 years, but has since been discarded (G. Snodgrass, personal communication). The eye color mutation identified through inbreeding [53] was not tested for physiological characteristics, but was observed to display no obvious behavioral or developmental differences from wild type. The strain reported here and the progenitor wild type strain both exhibited apparent loss of fitness. The 20 day cumulative egg production reported for the wild caught red strain was over 120 eggs/female, while the specimens reported here produced less than half that in every cross (Table 1). The loss of fecundity is most distinct in the fecundity of the Ca × W cross. The reason for this loss of fecundity is not known. Interestingly, the total number of male vs. female wild type offspring produced by the Ca × F1 crosses exceeded the critical value for the expected m/f ratio (Table 2A), while the overall phenotype distribution fit the expected results (Table 2B). Inbreeding depression is an accepted phenomenon, relevant to evolution, conservation, and agriculture. Intentional inbreeding is also expected to play a vital and dominant role in future molecular genetics and functional genomic research [67].

If genetic manipulation technology improves to a point at which sex-specific applications such as the sterile insect technique (SIT) can be applied to insects in the order Hemiptera, identification of sex-specific genes and gene regulation will become critical to project success. Visible markers that can be induced by inbreeding or other methods of mutagenesis will facilitate progress towards these applications, and towards a better understanding of the genetics of these important pest insects.

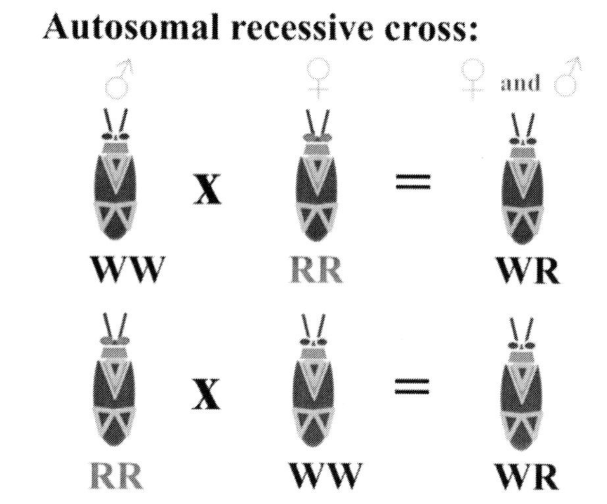

(a)

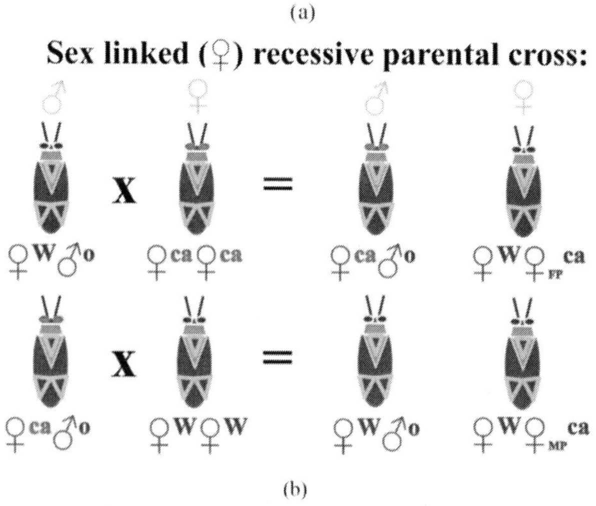

(b)

Sex linked (♀) recessive F1 cross:

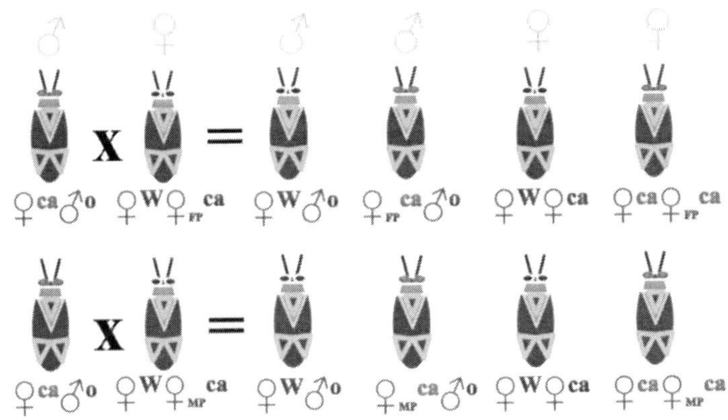

(c)

Figure 4. Diagrams showing expected results from crossing wild type insects with insects carrying an autosomal recessive red eye trait (R) compared with results from crossing wild type insects (WT) with insects carrying a sex linked recessive red eye trait such as cardinal (ca). MP = allele from male parent, FP = allele is from the female parent.

ACKNOWLEDGEMENTS

The author extends appreciation to the laboratory technicians who provided assistance and support in completing this work, Fannie M. Byrd, and especially to Catherine L. Smith who performed a vital role in setting up experiments and collecting data. Thanks go to Dr. Xixuan Jin, Dr. William B. Walker III, and Dr. William Rodney Cooper for reviewing a previous version of this manuscript and providing helpful suggestions. The United States Government has the right to retain a non-exclusive, royalty-free license in and to any copyright of this article. This article reports the results of research only. Mention of a commercial or proprietary product does not constitute an endorsement of the product by the United States Department of Agriculture. USDA is an equal opportunity provider and employer.

REFERENCES

1. Morgan, T.H. (1910) Sex limited inheritance in Drosophila. Science, 32, 120-122. doi:10.1126/science.32.812.120

2. Ephrussi, B. and Beadle, G.W. (1937) Development of eye colors in Drosophila: production and release of cn substance by the eyes of different eye color mutants. Genetics, 22, 479-783.
3. Sullivan, D.T. and Sullivan, M.C. (1975) Transport defects as the physiological basis for eye color mutants of Drosophila melanogaster. Biochemical Genetics, 13, 603- 613. doi:10.1007/BF00484918
4. Fridell, Y.W. and Searles, L.L. (1991) Vermilion as a small selectable marker gene for Drosophila transformation. Nucleic Acids Research, 19, 5082. doi:10.1093/nar/19.18.5082
5. Zwiebel, L.J., Saccone, G., Zacharopoulou, A., Besansky, N.J., Favia, G., Collins, F.H., Louis, C. and Kafatos, F.C. (1995) The white gene of Ceratitis capitata: A phenotypic marker for germline transformation. Science, 270, 2005- 2008. doi:10.1126/science.270.5244.2005
6. Horn, C., Jaunich, B. and Wimmer, E.A. (2000) Highly sensitive, fluorescent transformation marker for Drosophila transgenesis. Developmental Genes and Evolution, 210, 623-629. doi:10.1007/s004270000111
7. Yamada, S.M. and Selivon, D. (2001) Rose, an eye color mutation in a species of the Anastrepha fraterculus complex (Diptera: Tephritidae). Annals of the Entomological Society of America, 94, 592-595. doi:10.1603/0013-8746(2001)094 [0592: RAECMI] 2.0.CO;2
8. McCombs, S.D. and Saul, S.H. (1989) Genetic studies of the oriental fruit fly, Dacus dorsalis (Diptera: Tephritidae): the first eye color mutant. Annals of the Entomological Society of America, 82, 114-115.
9. Saul, S.H. (1982) Rosy-like mutant of the mediterranean fruit fly, Ceratitis capitata (Diptera: Tephritidae), and Its potential for use in a genetic sexing program. Annals of the Entomological Society of America, 75, 480-483.
10. Ullerich, F. (1996) Inheritance patterns of new genetic markers and occurrence of spontaneous mosaicism in the monogenic blowfly Chrysomya rufifacies (Diptera: Calliphoridae). Molecular and General Genetics, 253, 232- 241. doi:10.1007/s004380050317
11. Taylor, D.B. (1989) Genetics of screwworm: New genetic markers and preliminary linkage map. Journal of Heredity, 80, 425-432.
12. Gribakin, F.G. and Ukhanov, K. (1993) Light scattering in the eye of the blowfly chalky mutant: The effect on spectral sensitivity of photoreceptors R1-6. Vision Research, 33, 1185-1191. doi:10.1016/0042-6989(93)90207-D
13. Summers, K.M. and Howells, A.J. (1980) Functions of the white and topaz loci of Lucilia cuprina in the production of the eye pigment xanthommatin. Biochemical Genetics, 18, 643-653. doi:10.1007/BF00484582
14. Ullerich, F.H. and Langer, H. (1974) Lemon—Another eye colour mutant in the blowfly Calliphora erythrocephala. Molecular and General Genetics, 128, 273-276. doi:10.1007/BF00267116
15. Hiroyoshi, T. (1960) Some new mutants and linkage groups of the house fly. Journal of Economic Entomology, 53, 985-990.
16. Kung, K., Kao, C.Y. and Chei, K.T. (1963) Some biological and inherited characters of a mutant white-eyed strain of housefly and its use in bioassay. Acta Entomologica Sinica, 12, 262-267.
17. Wagoner, D.E. (1969) The ruby eye-color mutant in the house fly, Musca domestica L. A case of duplicate genes. Genetics, 62, 103-113.
18. Challoner, C.M. and Gooding, R.H. (1997) A white eye color mutant in the tsetse fly Glossina morsitans submotsitans Newstead (Diptera: Glossinidae). Genome, 40, 165- 169. doi:10.1139/g97-023
19. Distelmans, W., Makumyaviri, A.M., D'Haeseleer, F., Claes, Y., Le Ray, D. and Gooding, R.H. (1985) Influence of the salmon mutant of Glossina morsitans morsitans on the susceptibility to infection with Trypanosoma congolense. Acta Tropicalis, 42, 143-148.
20. Gooding, R.H. (1979) Genetics of Glossina morsitans morsitans (Diptera: Glossinidae). III. Salmon, a sex-linked, maternally influenced, semi-lethal eye color mutant. Canadian Entomologist, 111, 557-560. doi:10.4039/Ent111557-5

21. Gooding, R.H. and Rolseth, B.M. (1995) Genetics of Glossina palpalis palpalis: Designation of linkage groups and the mapping of eight biochemical and visible marker genes. Genome, 38, 833-837. doi:10.1139/g95-109
22. Rawlings, P. (1985) A sex-linked, recessive, non-deleterious eye-colour mutant in Glossina morsitans centralis. Entomologia Experimentalis et Applicata, 39, 211-212.doi:10.1111/j.1570-7458.1985.tb03565.x
23. Akhtar, K. and Sakai, R.K. (1985) Genetic analysis of three new eye colour mutations in the mosquito, Anopheles stephensi. Annals of Tropical Medicine and Parasitology, 79, 449-455.
24. Asman, S.M. (1977) Two sex-linked mutations in Culex tarsalis. Journal of Heredity, 68, 195-197.
25. Benedict, M.Q., Besansky, N.J., Chang, H., Mukabayire, O. and Collins, F.H. (1996) Mutations in the Anopheles gambiae pink-eye and white genes define distinct, tightly linked eye-color loci. Journal of Heredity, 87, 48-53. doi:10.1093/oxfordjournals.jhered.a022952
26. Bhalla, S.C., Santos, J.M. and De, F.B. (1975) Genetics of eye-gap and maroon-eye mutants in the mosquito. Journal of Heredity, 66, 349-352.
27. Ouda, N.A. and Wood, R.J. (1983) Inheritance of browneye and colourless-eye in the mosquito Aedes aegypti. Annals of Tropical Medicine and Parasitology, 77, 211- 218.
28. Rasgon, J.L. and Scott, T.W. (2004) Crimson: A novel sex-linked eye color mutant of Culex pipiens L. (Diptera: Culicidae). Journal of Medical Entomology, 41, 385-391. doi:10.1603/0022-2585-41.3.385
29. Rathor, H.R., Rashid, S. and Toqir, G. (1983) Genetic analysis of a new sex-linked mutant "chestnut eye" an allele of the white eye locus in the malaria vector Anopheles stephensi. Mosquito News, 43, 209-212.
30. Tadano, T. and Kanda, T. (1975) A new mutant red eye in Aedes riversi. Medical Entomology and Zoology, 27, 201- 202.
31. Tadano, T. (1977) Genetics of three new mutants, strawcolored larva, ruby eye and pigmented pupa, in Aedes (Finlaya) togoi (Diptera: Culicidae). Journal of Medical Entomology, 14, 33-37.
32. Wade, J.O. (1977) The genetics of white eye a sex linked mutant of Aedes cooki. Annals of Tropical Medicine and Parasitology, 71, 483-486.
33. Adak, T., Wattal, S., Kaur, S. and Sharma, V.P. (1999) Genetics of creamish white, an eye color mutant in Anopheles stephensi. Journal of Heredity, 90, 573-574. doi:10.1093/jhered/90.5.573
34. Bartlett, A.C. (1979) Genetics of the pink bollworm: sooty body and purple eye. Annals of the Entomological Society of America, 72, 256-259.
35. Beeman, R.W. (1983) Linkage analysis of genes affecting scale color, eye color, and resistance to malathion in the Indianmeal moth. Journal of Heredity, 74, 301-302.
36. Al-Hakkak, Z.S., Murad, A.M. and Hussain, A.F. (1985) Orange-eye: A sex-linked recessive mutation in Ephestia cautella (Walker) (Lepidoptera: Pyralidae). Journal of Stored Products Research, 21, 69-71. doi:10.1016/0022-474X(85)90023-2
37. Brown, T.M., Cho, S.Y., Evans, C.L., Park, S., Pimprale, S.S. and Bryson, P.K. (2001) A single gene (yes) controls pigmentation of eyes and scales in Heliothis virescens. Journal of Insect Science, 1, 1. http://www.insectscience.org/1.1/
38. Abraham, E.G., Sezutsu, H., Kanda, T., Sugasaki, T., Shimada, T. and Tamura, T. (2000) Identification and characterisation of a silkworm ABC transporter gene homologous to Drosophila white. Molecular and General Genetics, 264, 11-19. doi:10.1007/s004380000283
39. van den Abbeele, J. and D'Haeseleer, F. (1989) A new eye-colour mutant, brick, in the tsetse fly Glossina palpalis palpalis. Entomologia Experimentalis et Applicata, 52, 257-259. doi:10.1111/j.1570-7458.1989.tb01275.x
40. Bartlett, A.C. and Lewis, L.J. (1978) Genetics of the pink bollworm: 1 rust, orange, and garnet eye colors. Annals of the Entomological Society of America, 71, 813-816.
41. Dewees, A.A. and Bell, A.E. (1967) Close linkage of eye color genes in Tribolium castaneum. Genetics, 56, 633- 640.

42. Sokoloff, A., Lasley, E.L. and Shrode, R.R. (1960) Linkage studies in Tribolium castaneum Herbst. II. "Pygmy," "red" and "spotted" and their relationships to the sexlinked ma and pd genes. Canadian Journal of Genetics and Cytology, 2, 142-149.
43. Villavaso, E.J., Wilson, N.M. and Simmons, L.A. (1990) Lavender eye a recessive sex-linked mutant of boll weevil (Coleoptera: Curculionidae) and its map distance from the recessive sex-linked mutant dark scale. Annals of the Entomological Society of America, 83, 230-233.
44. Laidlaw, H.H., Banby, M.A. and Tucker, K.W. (1964) Five new eye-color mutants in the honey bee. Journal of Heredity, 55, 207-210.
45. Laidlaw, H.H. and Tucker, K.W. (1965) Umber eyecolor—A new mutant in honey bees. Journal of Heredity, 56, 271-272.
46. Lee, J.M., Hashino, Y., Hatakayama, M., Oishi, K. and Naito, T. (1998) Egg deposition behavior in the haplodiploid sawfly Athalia rosae ruficornis Jakovlev (Hymenoptera: Symphyta: Tenthredinidae). Journal of Insect Behavior, 11, 419-428. doi:10.1023/A:1020958831972
47. Moraes, A.S., Pimentel, E.R., Rodrigues, V.L.C.C. and Mello, M.L.S. (2005) Eye pigments of the blood-sucking insect, Triatoma infestans klug (Hemiptera, Reduviidae). Brazilian Journal of Biology, 65, 477-481. doi:10.1590/S1519-69842005000300013
48. Noé, C. and Silva, C. (1949) A new variety of Triatoma infestans. Boletin de Informaciones Parasitarias Chilenas, 4, 4-5.
49. Pires, H.H.R., Abrão, D.O., Machado, E.M.d.M., Schofield, C.J. and Diotaiuti, L. (2002) Eye colour as a genetic marker for fertility and fecundity of Triatoma infestans (Klug, 1834) Hemiptera, Reduviidae, Triatominae. Memórias do Instituto Oswaldo Cruz, 97, 675-678. doi:10.1590/S0074-02762002000500016
50. Rosabal, R. and Trejos, A. (1966) On a mutation of eye color in Rhodnius prolixus [Engl. summ.]. Revista de Biologia Tropical, 14, 99-103.
51. Wygodzinsky, P. and Briones, S. (1954) Nota sombre mutaciones del color del ojo en Triatoma (Klug). Anales del Instituto de Medicina Regional De la Universidad Nacional de Tucuman, 4, 115-118.
52. Snodgrass, G.L. (2002) Characteristics of a red-eye nutant of the tarnished plant bug (Heteroptera: Miridae). Annals of the Entomological Society of America, 95, 366-369. doi:10.1603/0013-8746(2002)095[0366:COAREM]2.0.CO;2
53. Slaymaker, P.H. and Tugwell, N.P. (1984) Inheritance of red eye color in Lygus lineolaris (Palisot de Beauvois) (Hemiptera: Miridae) an abnormal trait. Journal of the Kansas Entomological Society, 57, 343-344.
54. Seo, B.Y., Jung, J.K. and Kim, Y. (2011) An orange-eye mutant of the brown planthopper, Nilaparvata lugens (Hemiptera: Delphacidae). Journal of Asia-Pacific Entomology, 14, 469-472. doi:10.1016/j.aspen.2011.06.005
55. Cohen, A.C. (2000) New oligidic production diet for Lygus hesperus Knight and L. lineolaris (Palisot de Beauvois). Journal of Entomological Science, 35, 301- 310.
56. Byers, J.A. (2006) Analysis of insect and plant colors in digital images using Java software on the Internet. Annals of the Entomological Society of America, 99, 865-874. doi:10.1603/0013-8746(2006)99[865:AOIAPC]2.0.CO;2
57. Zar, J.H. (1996) Biostatistical analysis. Prentice-Hall, Inc., New York.
58. Abel, C.A., Snodgrass, G.L., Jackson, R. and Allen, C. (2010) Oviposition and development of the tarnished plant bug (Heteroptera: Miridae) on field maize. Environmental Entomology, 39, 1085-1091. doi:10.1603/EN10010
59. Emmons, R.B., Duncan, D. and Duncan, I. (2007) Regulation of the Drosophila distal antennal determinant spineless. Developmental Biology, 302, 412-426. doi: 10.1016/j.ydbio.2006.09.044
60. Emmons, R.B., Duncan, D., Estes, P.A., Kiefel, P., Mosher, J.T., Sonnenfeld, M., Ward, M.P., Duncan, I. and Crews, S.T. (1999) The spineless-aristapedia and tango bHLH-PAS proteins interact to control antennal and tarsal development in Drosophila. Development, 126, 3937- 3945.

61. Shippy, T.D., Yeager, S.J. and Denell, R.E. (2009) The Tribolium spineless ortholog specifies both larval and adult antennal identity. Development Genes and Evolution, 219, 45-51. doi:10.1007/s00427-008-0261-9
62. Linzen, B. (1974) The tryptophan-ommochrome pathway in insects. In: Beament, J.W. and Treherne, J.E., Eds., Advances in Insect Physiology 10, Academic Press, Waltham, pp. 117-246.
63. Searles, L.L. and Voelker, R.A. (1986) Molecular characterization of the Drosophila vermilion locus and its suppressible alleles. Proceedings of the National Academy of Sciences USA, 83, 404-408. doi:10.1073/pnas.83.2.404
64. Lorenzen, M.D., Brown, S.J., Denell, R.E. and Beeman, R.W. (2002) Cloning and characterization of the Tribolium castaneum eye-color genes encoding tryptophan oxygenase and kynurenine 3-monooxygenase. Genetics, 160, 225-234.
65. Altschul, S.F., Madden, T.L., Schaffer, A.A., Zhang, J., Zhang, Z., Miller, W. and Lipman, D.J. (1997) Gapped BLAST and PSI-BLAST: A new generation of protein database search programs. Nucleic Acids Research, 25, 3389-3402. doi:10.1093/nar/25.17.3389
66. Keeling, C.I., Henderson, H., Li, M., Yuen, M., Clark, E.L., Fraser, J.D., Huber, D.P., Liao, N.Y., Docking, T.R., Birol, I., Chan, S.K., Taylor, G.A., Palmquist, D., Jones, S.J. and Bohlmann, J. (2012) Transcriptome and fulllength cDNA resources for the mountain pine beetle, Dendroctonus ponderosae Hopkins, a major insect pest of pine forests. Insect Biochemistry and Molecular Biology, 42, 525-536. doi:10.1016/j.ibmb.2012.03.010
67. Ayroles, J.F., Hughes, K.A., Rowe, K.C., Reedy, M.M., Rodriguez-Zas, S.L., Drnevich, J.M., Caceres, C.E. and Paige, K.N. (2009) A genomewide assessment of inbreeding depression: Gene number, function, and mode of action. Conservation Biology, 23, 920-930. doi:10.1111/j.1523-1739.2009.01186.x

CHAPTER 11

Genetics and genomics of disease resistance in salmonid species

José M.Yáñez[1,2] *, Ross D. Houston[3] and Scott Newman[4]

[1]*Faculty of Veterinary and Animal Sciences, University of Chile, Santiago, Chile*

[2]*Aquainnovo, Puerto Montt, Chile*

[3]*The Roslin Institute and Royal (Dick) School of Veterinary Studies, University of Edinburgh, Midlothian, UK*

[4]*Genus plc, Hendersonville, TN, USA*

Infectious and parasitic diseases generate large economic losses in salmon farming. A feasible and sustainable alternative to prevent disease outbreaks may be represented by genetic improvement for disease resistance. To include disease resistance into the breeding goal, prior knowledge of the levels of genetic variation for these traits is required. Furthermore, the information from the genetic architecture and molecular factors involved in resistance against diseases may be used to accelerate the genetic progress for these traits. In this regard, marker assisted selection and genomic selection are approaches which incorporate molecular information to increase the accuracy when predicting the genetic merit of selection candidates. In this article we review and discuss key aspects related to disease resistance in salmonid species, from both a genetic and genomic perspective, with emphasis in the applicability of disease resistance traits into breeding programs in salmonids.

Keywords: salmon, disease resistance, breeding programs, QTL, genomic selection

1. INTRODUCTION

Farming of salmonid species is one of the largest aquaculture industries, with a worldwide production of approximately 1.9 million tons of high value product in 2010 (Food and Agriculture Organization of the United Nations [FAO], 2012). As in other animal production systems, the success and sustainability of salmonid aquaculture largely depends on the control of diseases. A clear example of the negative impact of infectious diseases in salmon farming is the unprecedented economic loss caused by outbreaks of the viral disease infectious salmon anemia (ISA) between 2007 and 2009 in Chile (Asche et al., 2010).

Genetic improvement programs are focused on increasing economic return of aquaculture systems via selective breeding (Gjedrem, 2012). In this regard, all heritable and economically relevant traits should be included in the breeding objective. Thus, in salmonid species, traits such as growth rate, flesh color, and resistance to viral, bacterial, and parasitic diseases should be included (Gjedrem, 2000, 2012). Selective breeding can utilize trait information recorded on selection candidates themselves or, particularly in the case of disease or invasive traits, on relatives. Until now, salmon breeding programs have typically included disease resistance based only on information from relatives, which affects the degree of genetic progress achievable on each generation. This is because of the lower accuracy of estimated breeding values (EBVs) when using only sib information compared to the accuracy obtained when using information of the selection candidates themselves (Falconer and Mackay, 1996).

Recent advances in molecular biology techniques, such as next generation sequencing and high throughput genotyping methods, have helped identify genetic variants influencing phenotypic variation for different traits in a wide range of organisms (Goddard and Hayes, 2009). Molecular markers can be used for a variety of applications in livestock and aquaculture species, such as strain and hybrid identification, genetic variability and genetic diversity evaluation, parentage analyses, quantitative trait loci (QTL) mapping, marker assisted selection (MAS), and genomic selection (GS; Liu and Cordes, 2004; Goddard and Hayes, 2009). Information from a few molecular markers linked to QTL (i.e., genomic regions harboring genes with a significant effect on the trait) might be implemented in breeding schemes through MAS, if they explain a high proportion of genetic variation in the trait. Additionally, the information of 1000s of markers might be simultaneously incorporated into genetic evaluation to estimate genomic breeding values (GEBVs; Meuwissen et al., 2001). These marker-based methods may be particularly useful for the improvement of traits that are complicated or impossible to measure directly on selection candidates, as is the case of resistance to disease (Sonesson and Meuwissen, 2009;

Villanueva et al., 2011; Taylor, 2014). A typical first step to implement MAS or GS is to quantify the level of genetic variation in the trait by dissecting its genetic architecture. In salmonids, there is limited information on the genetic architecture of disease resistance. Nevertheless, it is expected that more knowledge on the QTL or genes affecting disease resistance traits will be revealed in the near future, facilitated by the increasing availability of genomic resources and better understanding of the biology of immune response in these species.

This paper reviews aspects of conventional breeding to improve disease resistance in salmonids and the application of molecular tools for the identification of genetic factors involved in these traits. Additionally, the incorporation of molecular information into breeding schemes to improve disease resistance is discussed.

2. IMPORTANCE OF DISEASE CONTROL IN SALMON FARMING

The health status of farmed fish is one of the main factors affecting the economic return in the salmon industry. Despite scientific, professional, and technical strategies aimed at improving health management, many novel pathological conditions have emerged in salmonid fish species worldwide in recent decades. A detailed description of each disease affecting culture of salmonid species would greatly exceed the purpose of this review. However, some particular examples are discussed to demonstrate the large economic impact diseases can cause in salmon production.

One of the most striking cases affecting salmon farming was the economic crisis triggered by ISA virus outbreaks since mid-2007 in Chile. The production of Chilean Atlantic salmon suffered a dramatic decrease due to increasingly frequent outbreaks between 2007 and 2009. In fact, total production of Atlantic salmon between 2005 and 2010 decreased by more than 60% in volume (Asche et al., 2010). Currently, ISA virus outbreaks appear to be controlled to a low number of events per year in Europe and North and South America. However, the prevalence and emergence of other viral diseases is still of concern. In Northern European countries including Norway and Scotland, ISA outbreaks have been rare in recent years. Nevertheless, infectious pancreatic necrosis (IPN), caused by an aquatic birnavirus, has caused large levels of mortality in Europe, particularly during the window of susceptibility following transfer from freshwater to seawater (Roberts and Pearson, 2005). Other viral diseases have also emerged and pose serious threats to salmon aquaculture, such as skeletal muscle inflammation (HSMI) – a piscine reovirus – and pancreas disease

(PD) – an alphavirus, which has shown an increase in recent years (Biering et al., 2012). These viruses cause direct economic losses through mortality and indirect losses through reduced growth rate and treatment costs.

Among bacterial diseases with a negative impact on salmon farming, salmon rickettsial syndrome (SRS) caused by the Gram-negative bacterium *Piscirickettsia salmonis*, is one of the main sanitary challenges in Chilean salmon industry. This disease affects different salmonid species, including Atlantic salmon (*Salmo salar*), coho salmon (*Oncorhynchus kisutch*), and rainbow trout (*O. mykiss*; Fryer and Hedrick, 2003) and can generate economic losses equivalent to 25% of total profit in salmon exports in Chile (Rozas and Enríquez, 2014). Other bacterial diseases, such as those caused by *Aeromonas salmonicida*, *Vibrio anguillarum*, and *Vibrio/Aliivibrio salminonicida*, are recognized to be efficiently controlled by vaccination and do not currently represent a major economic threat for salmon (Biering et al., 2012).

In terms of parasitic diseases, two different species of sea lice, *Lepeophtheirus salmonis* and *Caligus rogercresseyi*, are most detrimental parasites for salmon farming at a worldwide level. In this regard, it has been estimated that on average, the economic impact of sea lice infestation is about 6% of the total value produced by the world salmon industry (Costello, 2009). An emerging threat to salmon production worldwide is Amoebic Gill Disease (AGD), which has been the major disease of farmed salmonid production in Tasmania for several decades (Mitchell and Rodger, 2011) and has appeared relatively recently in most major salmon-producing countries (Ruane and Jones, 2013). The free-living amoebic protozoan *Neoparamoeba pemaquidensis* is the primary causative agent for the disease that can cause serious morbidity and reduced growth, in addition to increasing susceptibility to other pathogens (Mitchell and Rodger, 2011).

The measures used for prevention and treatment (vaccinations, antibiotics, and antiparasitic drugs, biosecurity measures) of some of the diseases presented above have typically been only partially effective in field conditions (Bravo et al., 2013; Jones et al., 2013; Rozas and Enríquez, 2014). Where effective vaccines do exist, administration typically requires individual handling and treatment of all production fish, which can be expensive and impractical in a large-scale production environment. Due to the fact that improvement in economic efficiency of salmon farming is dependent on disease prevention and control, (Asche and Roll, 2013) it is imperative to develop alternative effective and sustainable strategies. Genetic improvement of disease resistance represents a feasible solution to increase the sanitary status in animal production (Stear et al., 2001; Bishop, 2010). In this regard, there is increasing scientific literature aiming at both quantifying levels of host genetic variation for resistance against different diseases and identifying the specific genetic factors that influence these traits in salmonid species, as discussed below.

3. CONVENTIONAL BREEDING FOR DISEASE RESISTANCE IN SALMONIDS

Resistance to diseases can be defined as the ability of the host to limit infection by reducing pathogen replication (Råberg et al., 2007; Doeschl-Wilson et al., 2012). Selecting animals with increased resistance to specific diseases is a feasible method to improve productivity and animal welfare and offers advantages over other control methods against infection, such as the cumulative and permanent benefits of the improved resistance (Stear et al., 2001; Bishop, 2010). Disease resistance has been a target trait for the salmon breeding industry for at least 20 years, with a Norwegian salmon breeding program including resistance to bacterial and viral diseases into its breeding goal since 1993 (Gjøen and Bentsen, 1997). However, the study of disease resistance and its incorporation into breeding programs can be hindered by the difficulty in determining and measuring accurate and appropriate phenotypes (Bishop and Woolliams, 2014). This in turn influences the accuracy of disease resistance EBVs that can be achieved. Another limiting step is that disease information is typically only available from relatives of the selection candidates and not directly from the candidates themselves. In the following, we review the main aspects of breeding for resistance to infectious diseases in salmonids and discuss current status and future directions of research in this area.

4. CHALLENGE AGAINST PATHOGENS

Host resistance to viral and bacterial pathogens can often be measured, in practical terms, as survival (and/or mortality) of individuals during an outbreak (Ødegård et al., 2011). Data and samples from field outbreaks can be used opportunistically to make inference about genetic resistance to infectious diseases. For this purpose, it is necessary that the pedigree of the population be accurately determined, often using genetic markers or electronic tagging of the fish (Guy et al., 2006). However, using the information from field outbreaks has some disadvantages, such as difficulty to identify the exact cause of death because the factors that influence survival under these conditions are likely to be diverse. Furthermore, the availability of information depends on the occurrence of high-mortality outbreaks, which are usually prevented or controlled to avoid serious economic loss. Moreover, the inference of pedigree using molecular markers can be expensive and laborious. Therefore, survival data are often obtained from experimental challenges, which can readily be standardized to control other variables and potentially allow a clearer interpretation of the results. In this case, it is necessary that a high genetic correlation between the trait measured in experimental and field conditions exists. High genetic correlations ($r_g \geq 0.95$) between field trials and experimental challenges to furunculosis in Atlantic salmon have been reported (Gjøen et al., 1997; Ødegård et al., 2006), suggesting that results from experimental challenges are likely to be directly applicable to commercial production systems. Therefore, challenge tests will often be more accurate and reliable than field outbreaks, due to decreased environmental variability and higher practical feasibility. In fact, challenge testing is currently used to select for resistance to viral, bacterial, and parasitic diseases in breeding programs for Atlantic salmon and rainbow trout (Gjøen and Bentsen, 1997; Leeds et al., 2010; Yáñez and Martínez, 2010; Ødegård et al., 2011; Gjedrem, 2012; Wiens et al., 2013a).

5. IMMUNOLOGICAL AND PHYSIOLOGICAL VARIABLES AS INDIRECT MEASURES OF RESISTANCE

Direct genetic selection for improved disease resistance based on challenge testing can be costly and time consuming, and has negative animal welfare implications. Furthermore, selection decisions using this strategy can only be carried out using information from relatives and not the candidates themselves. Indirect selection based on the measurement of other characteristics that are genetically correlated with disease resistance, would simplify the data collection and allow the incorporation of individual information. Some studies have aimed at determining the genetic variation of physiological and immunological variables, and the correlation between them and survival in challenge tests in salmon. Examples of variables that have been studied to date are hemolytic activity of serum and lysozyme activity (Røed et al., 1993; Lund et al., 1995), plasma levels of cortisol (Fevolden et al., 1993; Weber et al., 2008), and levels of IgM and antibody titer (Lund et al., 1995), serum α2-antiplasmin (Salte et al., 1993), bactericidal and complement activity (Hollebecq et al., 1995). However, even when some studies show significant correlations between resistance and immune parameters, the proportion of the total variation in survival that could be explained by immune variables has been considered too low to be useful as a selection criterion. Hence, the prediction of breeding values for survival based on these variables may not be practically useful (Gjøen and Bentsen, 1997). This may be due in part to the complexity of the mechanisms involved in the immune response and the large number of factors that may be involved in disease resistance, which results in a great difficulty when trying to use the information from a single parameter for the genetic evaluation of disease resistance.

6. GENETIC VARIATION IN RESISTANCE TO INFECTIOUS DISEASES

A requirement to improve a trait by means of artificial selection is that sufficient genetic variation for this trait exists in the population. Heritability is the proportion of the total phenotypic variance that is attributable to additive genetic variation (Falconer and Mackay, 1996). For disease resistance traits, heritability estimates can vary due to differences in trait definitions and the statistical models used in the analysis (Yáñez and Martínez, 2010; Ødegård et al., 2011). For example, some studies consider the binary trait of survival or mortality as a measure of resistance, whereas others consider the survival time following challenge. These require different statistical approaches to analysis, and are likely to inform on different components of the host response to infection. Nonetheless, there have been several studies aimed at determining levels of additive genetic variation for resistance to different diseases affecting salmonid species (see Table 1). The results from these studies show the feasibility of improving disease resistance through genetic improvement and the potential of this approach for helping in the control of disease problems in salmonids.

7. GENETIC CORRELATIONS BETWEEN DISEASE RESISTANCE AND OTHER TRAITS

The potential to simultaneously improve resistance to different diseases and other economically important traits is partly dependent on the genetic correlations between the traits. Few studies to date have aimed to determine the genetic correlations for disease resistance traits in salmon. Some studies in Atlantic salmon have indicated positive genetic correlations between resistance to bacterial diseases such as furunculosis, BKD, and cold water Vibriosis (Gjedrem and Gjøen, 1995; Gjøen et al., 1997). Weak negative genetic correlations between resistance against ISA and bacterial diseases such as furunculosis, vibriosis, and cold-water vibriosis have been reported (Gjøen et al., 1997).

Table 1. Heritabilty values (*h* (²and their SE for resistance to different infectious and parasitic diseases in salmonid species.

Salmonid species	Disease agent	$h^2 \pm$ SE	Reference
Salmo salar	*Neoparamoeba spp*	0.40 ± 0.08 – 0.49 ± 0.09	Taylor et al. (2009)
	Renibacterium salmoninarum	0.2 ± 0.1	Gjedrem and Gjøen (1995)
	Aeromonas salmonicida	0.48 ± 0.17	Gjedrem et al. (1991)
	A. salmonicida	0.34 ± 0.13	Gjøen et al. (1997)
	A. salmonicida	0.38 ± 0.09	Ødegård et al. (2006)
	A. salmonicida	0.43 ± 0.02	Ødegård et al. (2007b)
	A. salmonicida	0.59 ± 0.06	Olesen et al. (2007)
	A. salmonicida	0.62	Kjøglum et al. (2008)
	A. salmonicida	0.47 ± 0.05	Gjerde et al. (2009)
	Gyrodactylus salaris	0.32 ± 0.10	Salte et al. (2009)
	IPNV	0.43	Guy et al. (2006)
	IPNV	0.5	Wetten et al. (2007)
	IPNV	0.55	Kjøglum et al. (2008)
	ISAV	0.13 ± 0.03	Gjøen et al. (1997)
	ISAV	0.16 ± 0.01	Ødegård et al. (2007a)
	ISAV	0.32 ± 0.02	Ødegård et al. (2007b)
	ISAV	0.24 ± 0.03	Olesen et al. (2007)
	ISAV	0.37	Kjøglum et al. (2008)
	ISAV	0.40 ± 0.04	Gjerde et al. (2009)
	Caligus elongatus	0.22	Mustafa and MacKinnon (1999)
	Lepeophtheirus salmonis	0.14 ± 0.02	Kolstad et al. (2005)
	Lepeophtheirus salmonis	0.07 ± 0.02	Glover et al. (2005)
	Vibrio anguillarum	0.38 ± 1.07	Gjøen et al. (1997)
	V. salmonicida	0.13 ± 0.08	Gjedrem and Gjøen (1995)
	Caligus royercresseyi	0.10 ± 0.03	Yáñez et al. (2014a)
	Caligus royercresseyi	0.12 ± 0.07 – 0.34 ± 0.07	Lhorente et al. (2012)
	SPDV	0.21 ± 0.005	Norris et al. (2008)
	Piscirickettsia salmonis	0.11 ± 0.02 - 0.41	Yáñez et al. (2013)
	P. salmonis	0.18 ± 0.03	Yáñez et al. (2014a)
Salvelinus fontinalis	*Aeromonas salmonicida*	0.51 ± 0.03	Perry et al. (2004)
Oncorhynchus tshawytscha	*R. salmoninarum*	0 ± 0.05	Beacham and Evelyn (1992)
	A. salmonicida	0.21 ± 0.14	Beacham and Evelyn (1992)
	V. anguillarum	0.14 ± 0.11	Beacham and Evelyn (1992)
O. kisutch	*R. salmoninarum*	0.53 ± 0.16	Withler and Evelyn (1990)
O. mykiss	*Yersinia ruckeri*	0.21 ± 0.05	Henryon et al. (2005)
	Flavobacterium psychrophilum	0.35 ± 0.09	Silverstein et al. (2009)
	F. psychrophilum	0.23 ± 0.03	Leeds et al. (2010)
	F. psychrophilum	0.53 ± 0.09	Vallejo et al. (2010)
	F. psychrophilum	0.07 ± 0.02	Henryon et al. (2005)
	VHS	0.63 ± 0.26	Dorson et al. (1995)
	VHS	0.11 ± 0.1	Henryon et al. (2005)

However, positive genetic correlations between resistance against ISA virus and furunculosis have also been reported (Ødegård et al., 2007b). In rainbow trout, weak genetic correlations between viral hemorrhagic septicemia (VHS) and bacterial diseases such as enteric redmouth disease (ERM) and rainbow trout fry syndrome have been found (Henryon et al., 2005). Kjøglum et al. (2008) reported only weak genetic correlations when they estimated genetic correlations between resistance to IPN, ISA, and furunculosis. Verrier et al. (2013b) also failed to detect any genetic correlation between host resistance to two rhabdoviral pathogens (VHS and Infectious Hematopoietic Virus). Additionally, weak genetic correlations have also been calculated between resistance against SRS and *C. rogercresseyi* in Atlantic salmon (Yáñez et al., 2014a). In general, these results suggest no clear-cut relationship between genetic resistance to one pathogen and genetic resistance to another pathogen.

Moreover, it is important to know the genetic correlations between disease resistance and other economically important traits in salmon production, especially for selection index development. Previous reports of correlations between disease resistance and production traits have ranged from zero (Silverstein et al., 2009), low to moderately negative (Henryon et al., 2002), inconsistent (Beacham and Evelyn, 1992; Henryon et al., 2002), and low to moderately positive (Gjedrem et al., 1991; Perry et al., 2004; Yáñez et al., 2014a). Thus, it is important to evaluate these parameters for each particular breeding population to maximize the impact of selecting for improved disease resistance. Additionally, the relationship between ploidy levels and disease resistance is of interest because all female triploid rainbow trout are advantageous for production. Weber et al. (2013) showed that triploid fish are generally slightly more susceptible to Bacterial Coldwater Disease than diploids, but that selection for improved resistance in diploids is also effective for triploid production.

8. RESISTANCE VERSUS TOLERANCE

If resistance is defined as an individual's ability to block the reproduction of a pathogen, then disease tolerance can be defined as the ability to limit the impact of infection on a host (Råberg et al., 2007; Doeschl-Wilson et al., 2012). Information relating to the genetic basis of disease tolerance has been sparse in animal genetics studies to date (Doeschl-Wilson et al., 2012). However, it is important to disentangle resistance from tolerance because the appropriate trait to select for may differ depending on the disease, host species, and environment. It is also possible that the two traits will be antagonistic, which could result in inadvertent undesirable outcomes of selection for disease resistance (Doeschl-Wilson et al., 2012). The implications of selection for resistance or tolerance on the host–pathogen interaction and pathogen evolution have also been considered. For example, selection for resistance *per se* may lead to selection pressure for higher virulence in the pathogen, whereas selection for tolerance could result in co-existence of pathogen and host with minimal impact on the performance of the host population. Therefore, in future disease studies and in selective breeding programs, it will be important to carefully consider the optimal disease trait to target for maximum benefit at the population level, but also the analytic issues of resistance versus tolerance.

9. STRATEGIES FOR GENETIC DISSECTION OF DISEASE RESISTANCE TRAITS

Major advances in nucleotide sequencing, ever-improving bioinformatics pipelines, and high-throughput genotyping tools have helped to identify genes associated with complex traits in various species of vertebrates. In salmonids, this is reflected in the substantial increase in genomic resources for these species during recent years and, for example, in the formation of an international collaboration to sequence the Atlantic salmon genome (Davidson et al., 2010). The salmon genome has undergone a relatively recent duplication and has a very high content of long repeat elements. This has hampered the sequencing and assembly of a reference genome for salmonids. However, a high-quality reference sequence has been published for rainbow trout (Berthelot et al., 2014) and is available for Atlantic salmon (Davidson et al., 2010). Further, high density SNP genotyping arrays were recently developed for Atlantic salmon (Houston et al., 2014; Yáñez et al., 2014b), and a lower density platform (Lien et al., 2011) has previously been used for QTL mapping and population genetics. Currently, these genomic resources are increasingly being used for the identification of the genetic factors involved in the resistance to different diseases in salmonids. The strategies used in the study of the genetic architecture of disease resistance can be classified as: (i) candidate gene approaches, (ii) QTL mapping, and (iii) gene expression studies.

10. CANDIDATE GENE APPROACH

Candidate gene theory states that a significant proportion of phenotypic variance of one trait in a population is determined by the presence of polymorphisms within genes known to be involved in the physiological regulation of that trait (Rothschild and Soller, 1997). This approach requires previous knowledge on the biology of the species, biochemical pathways, and especially gene sequences, to study the variation within specific candidate genes. In aquaculture species, the availability of annotated gene sequences of known function is typically low, but is likely to increase in the short term with the use of high-throughput sequencing and ongoing genome sequencing projects.

In vertebrates, the major histocompatibility complex (MHC) has attracted much attention in studies of association between genetic variants and disease resistance. However, other genes are likely to play an important role in the mechanisms of disease resistance in production animals, model organisms, and humans (Hill, 1999; Qureshi et al., 1999). To our knowledge, there are no studies aimed at establishing association between candidate genes, other than the MHC, and resistance to infectious diseases in salmonid species.

11. MAJOR HISTOCOMPATIBILITY COMPLEX

The MHC is a multigene family that acts at the interface between the immune system and infectious pathogens. The MHC gene family comprises two subfamilies: class I and II. Both classes are membrane glycoproteins involved in the processing and removal of pathogens (Thorgaard et al., 2002). MHC genes have been identified, cloned, and characterized in Atlantic salmon, rainbow trout, and other salmonids (Grimholt et al., 1993; Hordvik et al., 1993; Hansen et al., 1996; Shum et al., 1999, 2002). Furthermore, it has been shown that these genes are highly polymorphic in these species (Grimholt et al., 1994, 2002; Miller and Withler, 1996; Hansen et al., 1999; Garrigan and Hedrick, 2001; Aoyagi et al., 2002). As in other vertebrates there are two types of class I genes; the UAA which are highly divergent, non-polymorphic and expressed at low levels, and the UBA which are polymorphic, expressed at high levels in spleen, and with structural features similar to those of class Ia molecules (classical) which present antigen to T lymphocytes (Shum et al., 1999). The class II genes are divided into Class II A (DAA) and II B (DAB), depending on whether encoding α or β chain of the molecule, respectively (Grimholt et al., 2000; Stet et al., 2002). Both loci (DAA and DAB) co-segregate as haplotypes, suggesting a close physical linkage between them in Atlantic salmon (Stet et al., 2002).

The association between a polymorphism linked to MHC class II genes and resistance to virus infectious hematopoietic necrosis (IHN) in backcrosses of rainbow trout and cutthroat trout (*O. clarki*) has been found, but was relatively weak and dependent on the family analyzed (Palti et al., 2001). Suggestive associations between rainbow trout MHC class IB alleles and bacterial cold water disease have also been shown (Johnson et al., 2008). In Atlantic salmon, the association between MH class IIB alleles and resistance against *A. salmonicida* has been reported (Langefors et al., 2001; Lohm et al., 2002). In the same species, MHC class I and class II variants have been associated with susceptibility to IHN (Miller et al., 2004) and resistance to furunculosis and ISA (Grimholt et al., 2003; Kjøglum et al., 2006). Although associations between MH gene variants and disease resistance have been established, the MHC is most likely not the only factor influencing genetic variation in disease resistance. For instance, in Atlantic salmon, a non-MHC effect for resistance to IPN, furunculosis and ISA has been detected (Kjøglum et al., 2005). Because disease resistance traits will typically be polygenic in nature, it is important to consider variants in a genome-wide context and possible interactions that can occur between genes (epistasis), which hinders the candidate gene approach as a comprehensive strategy for incorporating molecular information into the genetic evaluation of these traits.

12. QTL MAPPING

The development of DNA markers has had a major impact on studies of genetic variation in animals and fish. The categories of DNA markers widely used historically include Restriction Fragment Length Polymorphisms (RFLP), Random Amplified Polymorphic DNA (RAPD), Amplified Fragment Length Polymorphism (AFLP), microsatellites, and single nucleotide polymorphisms (SNPs). More recently, SNP markers have become the predominant marker due to their abundance, ease of discovery and low cost of genotyping per locus (Houston et al., 2014). However, each of these marker types vary from each other in their mode of inheritance (i.e., dominant or codominant), identification and detection methods, number of spanning loci and polymorphic information content (Liu and Cordes, 2004).

Microsatellites are tandem repeat nucleotide sequences that generally span between one to six base pairs. Different alleles are generated due to the variation in the number of repeats. Their main useful features for genetic studies include high variability, codominant inheritance, abundance, and wide distribution across the genome. One of the major advantages of microsatellites is their high degree of polymorphism, with 10s of alleles often observed at a single locus in an outbred population. Additionally, their genotyping is typically based on a simple DNA amplification by means of polymerase chain reaction (PCR). Thus, genotyping can be relatively rapid, cheap, and the amount of DNA required is minimal (nanograms). However, the scoring of microsatellites often requires optimization and manual input, which reduces their scalability for large genetic studies. There are a considerable number of microsatellites identified for different salmonid species, available for use in genetic studies (e.g., Cairney et al., 2000; Gilbey et al., 2004; Phillips et al., 2009).

Historically, AFLPs (which typically score SNPs) had the advantage of being generated more easily and cheaply than SNPs and microsatellites, because previous knowledge on genomic sequence is not needed for their generation. However, their mode of inheritance, similar to RAPDs, is dominant, i.e., it is not possible to distinguish between heterozygous and one category of homozygous genotypes without the use of special equipment and software (Piepho and Koch, 2000). This reduces the amount of information provided by these kinds of markers.

Single nucleotide polymorphismss have several key advantages for genetic studies, which have led to a rapid increase in their popularity over recent years. Some of these include high abundance, amenability to automated scoring in large numbers, simultaneous assays of thousands of markers (e.g., using SNP arrays) and presence in both coding and non-coding regions. Therefore, SNPs have been applied recently for construction of dense genetic maps, which can be used for fine mapping of QTLs and facilitate the identification of causative genes involved in the genetic variation of specific characters (e.g., Lien et al., 2011; Gonen et al., 2014). Recently, there has been a rapid increase in available SNPs for salmonid species, mainly for Atlantic salmon and rainbow trout (Hayes et al., 2007; Sanchez et al., 2009; Everett et al., 2011, 2012; Hohenlohe et al., 2011; Lien et al., 2011; Houston et al., 2012, 2014; Salem et al., 2012). Additionally, SNP arrays are available for simultaneous genotyping of 10s to 100s of 1000s of markers in rainbow trout (Palti et al., 2014b) and Atlantic salmon (Houston et al., 2014; Yáñez et al., 2014b). These SNP resources are likely to be increasingly applied to high-resolution mapping of disease resistance genes in salmonid species.

In recent years, the same sequencing technology that has led to the increased ease of high-density SNP discovery described above has enabled direct genotyping of individual fish based on sequence information alone. Such techniques are collectively termed 'genotyping by sequencing' (GBS). Although GBS techniques encompass a diverse range of laboratory and bioinformatic pipelines, they are all based on the principle of using nucleotide barcodes ligated to the fragmented genomic DNA of individual fish and high-throughput sequencing in multiplexed pools (Davey et al., 2013). Partly due to the lack of established genomics resources for many farmed fish species, the aquaculture genetics community has been early adopters and developers of these techniques. In particular, RAD sequencing has been utilized in both Atlantic salmon (e.g., Houston et al., 2012) and rainbow trout (Palti et al., 2014a) as a conduit to incorporating genomic information into aquaculture breeding programs. While the application of GBS has huge potential for applied research in salmonids, there are also some challenges in managing and interpreting these datasets. For example, discovery of 'false' SNPs using sequencing can occur, particularly with the recent whole genome duplication of the salmonid species. Therefore, species-tailored bioinformatics pipelines are typically required to minimize these issues.

13. DETECTION OF QTL AFFECTING DISEASE RESISTANCE IN SALMONID SPECIES

Salmon may be more likely than terrestrial farmed species to have QTL of major effect because they have had fewer generations of selection in the farmed environment and therefore the standing genetic variation for traits of economic importance may still be very large. The development of a genetic map based on linkage between genetic markers is the first step towards the identification of QTL. To date, several linkage maps have been constructed using different marker types for rainbow trout (e.g., Young et al., 1998; Sakamoto et al., 2000; Nichols et al., 2003; Guyomard et al., 2006; Rexroad et al., 2008; Palti et al., 2011, 2012; Guyomard et al., 2012); Atlantic salmon (Gilbey et al., 2004; Moen et al., 2004b; Lien et al., 2011; Gonen et al., 2014); coho salmon (McClelland and Naish, 2008); brown trout (*Salmo trutta*; Gharbi et al., 2006); Arctic Trout (*Salvelinus alpinus*; Woram et al., 2004); and sockeye salmon (*Salmo nerka*; Everett et al., 2012).

Using microsatellite markers, two QTL with major effects on resistance to IPN in rainbow trout has been detected. These loci explained a large proportion (27 and 34%) of the phenotypic variation in a family from a backcross between a strain susceptible to IPN (YK-RT101) and a resistant one (YN-RT201; Ozaki et al., 2001). Using AFLP and microsatellite markers, QTLs for IHN resistance have been identified in three different linkage groups of the same species (Rodriguez et al., 2004). For the same disease, RFLP markers were associated with resistant and susceptible families in backcrosses of rainbow trout and cutthroat trout (Palti et al., 1999). Extensive research into the genetic architecture of resistance to Bacterial Coldwater Disease has also been undertaken, with evidence for QTL of major effect (Wiens et al., 2013b; Vallejo et al., 2014a,b). Additionally, major QTL have been detected for resistance to whirling disease caused by the myxosporean parasite *Myxobolus cerebralis* (Baerwald et al., 2010) and VHS (Verrier et al., 2013a).

In Atlantic salmon, using AFLP markers, two QTL associated with resistance to ISA have been detected in two full sib families (Moen et al., 2004a). One of these QTLs has been validated using microsatellite markers from a higher number of genotyped fish. This QTL explained 6% of the phenotypic variation for resistance to ISA and has been mapped to linkage group VIII of the Atlantic salmon genome (following SALMAP notation; Moen et al., 2007). In the same species, a major QTL for resistance to IPN has been identified using data collected from a 'field' seawater outbreak of the disease (Houston et al., 2008a). The QTL detection strategy involved utilizing the low male recombination rate observed in salmonids by using just two to three microsatellites per chromosome and male segregation to determine linkage groups with a significant effect, and secondly, a higher number of markers per linkage group. Female segregation was used to confirm the previously detected QTL and position it within the linkage group (Houston et al., 2008a). The major QTL, mapped to linkage group 21, was subsequently confirmed by analyzing nine additional families and a higher saturation of markers (Houston et al., 2008b). The same QTL was then confirmed and fine mapped in an independent population, and haplotypes of markers that could predict the genotype at the QTL were identified based on the linked microsatellites (Moen et al., 2009). Using a restriction-site associated DNA sequencing approach (RAD sequencing), several additional SNP markers linked to the QTL were identified and two SNP markers showed a significant population-level association with resistance in two year classes from an Atlantic salmon breeding population (Houston et al., 2012). The high proportion of the total variance for IPN resistance that this QTL explains has allowed the incorporation of markers linked to it into MAS schemes for the genetic improvement of this trait in Atlantic salmon in both Norway (Moen et al., 2009) and Scotland (Houston et al., 2010).

In general, confidence intervals for mapped QTLs mapped are large. This issue has two consequences: first, widespread confidence intervals might contain a large number of genes (1000s) and, therefore, identification of the causative polymorphism is challenging. Second, the use of markers linked to QTLs in MAS programs is complicated, since the linkage phase between the marker and the QTL throughout the population may be different from family to family. An alternative for QTL fine mapping is to use information from linkage analysis in conjunction with information from linkage disequilibrium (LD) across the population (Meuwissen et al., 2002). Through simulation studies, power and accuracy of this combined approach has been successfully tested for QTL fine mapping, accounting for the structure of commercial salmon populations (Hayes et al., 2006). Additionally, the availability of both high-density SNPs marker panels and high-resolution genetic maps will contribute to detect association between markers and QTLs with higher precision by means of using across-population LD mapping (Goddard and Hayes, 2009). These strategies along with a reference sequence and a consolidated physical map of the genome, will facilitate the identification of causative mutations affecting disease resistance traits through positional studies in salmonid species.

14. GLOBAL GENE EXPRESSION

Functional genomics, defined as the application of experimental methods of genomic or systemic coverage to assess gene function using data from structural genomics (mapping and sequencing), has been recognized as an area of primary interest in disease studies (Hiendleder et al., 2005). These methodologies broaden the spectrum of biological research to study, simultaneously, the expression of thousands of genes at the transcriptional level. Currently, genomic resources and new sequencing technologies have helped to assess differential gene expression levels in the response against diseases in salmonid species. These data can help to pinpoint functional genetic variation underlying disease resistance.

An early example, using suppression subtractive hybridization (SSH) and liver samples from individuals injected with a *V. Anguillarum* bacterium and normal individuals, more than 25 genes important in the immune response in rainbow trout were identified, including sequences of proteins of acute phase of inflammation, complement, and coagulation system (Bayne et al., 2001). Using the same technique, genes involved in signal transduction and innate immunity (among others) have been identified as relevant factors in response to a challenge against *A. salmonicida* in Atlantic salmon (Tsoi et al., 2004).

The availability of ESTs (expression sequence tags) and cDNA libraries have allowed the development of DNA microarrays which can be used to study the differential expression patterns of a large number of genes simultaneously in salmonids (Rise et al., 2004b; Ewart et al., 2005; von Schalburg et al., 2005). Using a microarray of human cDNAs, differentially expressed transcripts against a challenge with *A. Salmonicida* have been identified in Atlantic salmon. However, due to species divergence, only 6% of the sequences of the microarray showed detectable hybridization against salmon liver cDNA (Tsoi et al., 2003). Rise et al. (2004a) conducted the first study using a microarray constructed from salmon cDNA libraries. Following this, differential gene expression has been assessed in between macrophages infected and uninfected with *P. salmonis* and hematopoietic kidney from Atlantic salmon individuals challenged or not against the same pathogen. Differentially expressed genes were proposed to be relevant in immune response and as potential biomarkers of infection with *P. salmonis* (Rise et al., 2004a). Additionally, using an Atlantic salmon cDNA microarray including more than 4,000 genes extracted from liver, spleen and hematopoietic kidney, several differentially expressed genes in response to infection by *A. salmonicida* have been found (Ewart et al., 2005). Furthermore, the transcriptomic response against vaccination with an *A. salmonicida* bacterium has been assessed in Atlantic salmon, revealing temporal and tissue differences in terms of expression levels, which may be relevant in establishing protection (Martin et al., 2006). In addition, gene expression profiles in response to a DNA vaccine for the IHN virus has been studied in rainbow trout, identifying 910 genes modulated in the injection site, and also determining the over-expression of genes of the type I Interferon system (IFN-1) in other tissues, suggesting that this system forms the basis of early antiviral immunity (Purcell et al., 2006). In the studies presented above, certain transcripts that showed variation in their expression levels during infection have low levels of homology with well-characterized genes available in public databases and, thus, do not have a known function (Rise et al., 2004a; Ewart et al., 2005; Martin et al., 2006; Purcell et al., 2006). Therefore, studies based on sequences of the encoded proteins aimed at decrypting their role in the immune response against infection are still needed.

To date, few studies have focused on the analysis of differential expression between resistant and susceptible fish for a particular disease. For example, Sutherland et al. (2014) analyzed gene expression profile differences between chum and pink salmon during infections with sea lice (*L. salmonis*) to gain insight into the functional mechanisms underlying the divergent resistance to lice observed in these two species. Zhang et al. (2011) took a similar approach to examine the differences between resistant and susceptible salmon families to the pathogen *A. salmonicida* to highlight the importance of several innate immune response genes. Cofre et al. (2014) also examined differential expression of several candidate genes in families with divergent resistance to IPNV. Finally, Langevin et al. (2012) compared the transcriptional response of resistant and susceptible clonal lines for Bacterial Coldwater Disease. The information provided by this type of analysis might be useful in the discovery of new sets of genes, with or without an assigned function, which may be associated with disease resistance (Walsh and Henderson, 2004). Another possibility is to consider the differential expression levels as a quantitative trait. This may allow identification of QTLs associated with differences in gene expression patterns between resistant and susceptible individuals (eQTL; Pomp et al., 2004; de Koning et al., 2005). However, it remains unclear how the information given by expression analysis can be used in breeding programs (Walsh and Henderson, 2004).

15. MOLECULAR MARKER-ASSISTED SELECTION (MAS) AND GENOMIC SELECTION (GS)

While selective breeding based on performance information from the selection candidate and its relatives is a highly successful means of genetic improvement in the trait of interest, utilizing genetic markers can provide an improvement in both genetic gain and selection accuracy (Goddard and Hayes, 2009). Historically, the use of genetic markers required mapping of QTL and therefore a key parameter for subsequent application in MAS programs is the level of LD between markers and causative mutations at a population level (Goddard and Meuwissen, 2005).

One factor reducing the level of LD each generation is recombination. When the LD between the QTL and the marker only exists within families and not across families, recombination can break the association between marker alleles and the QTL between families. Therefore, the linkage phase between the marker and the QTL should be determined in each generation and separately for each family if it is to be utilized in selection (Wientjes et al., 2013). To determine if the marker and QTL are in LD within each family phenotypic records and genotypes are needed on each generation. This makes unattractive the implementation of MAS exploiting only within-family LD (i.e., linkage) between the QTL and markers (Dekkers and Van der Werf, 2007). In the case of disease resistance traits, this means that all families comprising the breeding nucleus would need to be challenged each generation.

Marker assisted selection schemes are likely to be optimal when the markers explain a large proportion of the total variance of the trait, as in the case of the QTL for resistance to IPN in Atlantic salmon (Houston et al., 2008a, 2012; Moen et al., 2009). The information from markers linked to this QTL is being applied in MAS programs in different breeding populations of Atlantic salmon. Although SNP variants associated with IPN resistance in two different year classes of a breeding nucleus have been discovered (Houston et al., 2012), there is still a need to validate if these alleles are associated with resistance in independent Atlantic salmon populations. This will depend on the LD relationship between the markers and the causative mutation in the population(s) of interest. As such, testing the association between markers and IPNV mortality in these populations should be considered a prerequisite for effective commercial application.

Another alternative to exploit the LD at the population level is using information from high density panels of markers to predict GEBVs of selection candidates (Meuwissen et al., 2001). This approach effectively takes into account all markers when estimating the breeding value of the candidate, without the need to surpass a significance threshold for association with a particular trait of interest. The effectiveness of this strategy will also depend on the magnitude of the effects associated with the markers. When the effects of the markers across the entire genome are estimated, these effects can be used to select individuals lacking phenotypes (Meuwissen et al., 2001). As such, in the case of disease resistance, there would be fewer requirements for challenge testing of siblings for diseases of economic importance. Recombination results in decay of LD on each generation and the magnitude of this reduction depend on several population features (Porto-Neto et al., 2014). This means that in practice, it is necessary to corroborate the accuracy in estimating the GEBVs and response selection in each generation.

Best linear unbiased prediction (BLUP) combines information from pedigree and phenotypes to predict breeding values (EBVs) of individuals. Molecular markers provide a new source of information, which will give higher accuracy when predicting EBVs. Thus, selection response will be potentially higher in traits in which the accuracy is low, i.e., traits with low heritability or traits that cannot be measured in the selection candidate, such as disease resistance (Sonesson and Meuwissen, 2009; Villanueva et al., 2011; Taylor, 2014). The relative increase in accuracy depends on the amount of variation explained by the markers. It is also possible to combine breeding values for disease resistance and other performance traits into a selection index with specified weightings based on their economic importance for multi-trait genetic improvement.

In livestock species the effects of QTL have been shown to exhibit a moderate leptokurtic gamma distribution, suggesting a small number of loci of large effect and high number of loci of small effect (Hayes and Goddard, 2001), which is likely to be expected in aquaculture species for most economically relevant traits. Therefore, it is expected that more than one marker is necessary to efficiently assist breeding programs using molecular information. The availability of high throughput SNP genotyping platforms will allow the use of these markers in selective breeding for aquaculture species by means of the implementation of GS schemes (Goddard and Hayes, 2009; Sonesson and Meuwissen, 2009; Taylor, 2014). There are currently at least three independent initiatives to generate high density SNP arrays for Atlantic salmon, run by groups from three of the leading producers of this species: Chile, Scotland and Norway. The first of these to be published is that of Houston et al. (2014) whereby approximately 132 K SNP markers were identified and verified in several populations of farmed and wild Atlantic salmon. The Chilean group validated almost 160 K SNP markers from a 200 K SNP platform for Atlantic salmon, genotyping fish representing wild and farmed populations from Europe, North America, and Chile (Yáñez et al., 2014b). Additionally, a 57 K SNP chip is available for rainbow trout (Palti et al., 2014b). Therefore, it can be expected that programs using a high number of markers in population LD will be implemented in the near future. However, it is necessary to determine the economic benefit of using high-density panels, versus the use of low-density panels supported by imputation methodologies and establish a strategy for GS taking into account technical and economic feasibility.

16. CONCLUSION

There is increasing information related to the determination of the genetic basis for disease resistance in salmonid species. The rapid development of genomic resources in these species will provide new tools for genetic dissection of these traits. The use of these tools will be of great help in identifying loci involved in the genetic variation of disease resistance. This information will increase our understanding of the underlying biology of resistance to disease. Further, it will be crucial for the implementation of MAS or GS programs that include disease resistance within the breeding objective. These methods will increase the accuracy of selection candidates, thereby improving the selection response. However, the economic feasibility and profitability of the implementation of these new strategies and its comparison with conventional selection schemes must be studied for each particular breeding program in salmonids. It will also be necessary to assess the long-term impact of these strategies on the control of each specific disease, in an epidemiological context.

ACKNOWLEDGMENTS

This work was partially funded by grants from CORFO (11IEI-12843 and 12PIE-17669), Government of Chile and from Programa U-Inicia, Vicerrectoría de Investigación y Desarrollo, Universidad de Chile. José M. Yáñez would like to thank María E. López and Liane Bassini for their help in revising the manuscript.

REFERENCES

1. Aoyagi, K., Dijkstra, J. M., Xia, C., Denda, I., Ototake, M., Hashimoto, K.,et al. (2002). Classical MHC class I genes composed of highly divergent sequence lineages share a single locus in rainbow trout (*Oncorhynchus mykiss*). *J. Immunol.* 168, 260–273. doi: 10.4049/jimmunol.168.1.260
2. Asche, F., Hansen, H., Tveteras, R., and Tveteras, S. (2010). The salmon disease crisis in Chile. *Marine Resour. Econ.* 24, 405–411. doi: 10.5950/0738-1360-24.4.405
3. Asche, F., and Roll, K. H. (2013). Determinants of inefficiency in Norwegian salmon aquaculture. *Aquac. Econ. Manag.* 17, 300–321. doi: 10.1080/13657305.2013.812154
4. Baerwald, M. R., Petersen, J. L., Hedrick, R. P., Schisler, G. J., and May, B. (2010). A major effect quantitative trait locus for whirling disease resistance identified in rainbow trout (*Oncorhynchus mykiss*). *Heredity (Edinb.)* 106, 920–926. doi: 10.1038/hdy.2010.137
5. Bayne, C. J., Gerwick, L., Fujiki, K., Nakao, M., and Yano, T. (2001). Immune relevant (including acute phase) genes identified in the livers of rainbow trout, *Oncorhynchus mykiss*, by means of suppression subtractive hybridization. *Dev. Comp. Immunol.* 25, 205–217. doi: 10.1016/S0145-305X(00)00057-4
6. Beacham, T. D., and Evelyn, T. P. T. (1992). Genetic variation in disease resistance and growth and growth of chinook, coho, and chum salmon with respect to vibriosis, furunculosis, and bacterial kidney disease. *Trans. Am. Fish Soc.* 121, 456–485. doi: 10.1577/1548-8659(1992)121<0456:GVIDRA>2.3.CO;2

7. Berthelot, C., Brunet, F., Chalopin, D., Juanchich, A., Bernard, M., Noël, B.,et al. (2014). The rainbow trout genome provides novel insights into evolution after whole-genome duplication in vertebrates. *Nat. Commun.* 5, 3657. doi: 10.1038/ncomms4657
8. Biering, E., Madhun, A., Isachsen, C. H., Omdal, L. M., Einen, A. C. B., Garseth, Å. H.,et al. (2012). *Annual Report on Health Monitoring of Wild Anadromous Salmonids in Norway. Annual Report 2012, No 6-2013.* Bergen: Institute of Marine Research.
9. Bishop, S. C. (2010). "Disease resistance: genetics," in *Encyclopedia of Animal Science*, eds W. G. Pond and A. W. Bell (New York, NY: Marcel Dekker, Inc.) 288–290.
10. Bishop, S. C., and Woolliams, J. A. (2014). Genomics and disease resistance studies in livestock. *Livest. Sci.* 166, 190–198. doi: 10.1016/j.livsci.2014.04.034
11. Bravo, S., Nunez, M., and Silva, M. T. (2013). Efficacy of the treatments used for the control of *Caligus rogercresseyi* infecting Atlantic salmon, *Salmo salar* L., in a new fish-farming location in Region XI, Chile.w *J. Fish Dis.* 36, 221–228. doi: 10.1111/jfd.12023
12. Cairney, M., Taggart, J. B., and Høyheim, B. (2000). Characterization of microsatellite and minisatellite loci in Atlantic salmon (*Salmo salar* L.) and cross-species amplification in other salmonids. *Mol. Ecol.* 9, 2175–2178. doi: 10.1046/j.1365-294X.2000.105312.x
13. Cofre, C., Gonzalez, R., Moya, J., and Vidal, R. (2014). Phenotype gene expression differences between resistant and susceptible salmon families to IPNV. *Fish Physiol. Biochem.* 40, 887–896. doi: 10.1007/s10695-013-9894-3
14. Costello, M. J. (2009). The global economic cost of sea lice to the salmonid farming industry. *J. Fish Dis.* 32, 115–118. doi: 10.1111/j.1365-2761.2008.01011.x
15. Davey, J. W., Cezard, T., Fuentes-Utrilla, P., Eland, C., Gharbi, K., and Blaxter, M. L. (2013). Special features of RAD Sequencing data: implications for genotyping. *Mol. Ecol.* 22, 3151–3164. doi: 10.1111/mec.12084
16. Davidson, W., Koop, B., Jones, S., Iturra, P., Vidal, R., Maass, A.,et al. (2010). Sequencing the genome of the Atlantic salmon (*Salmo salar*). *Genome Biol.* 11, 403. doi: 10.1186/gb-2010-11-9-403
17. de Koning, D. J., Carlborg, O., and Haley, C. S. (2005). The genetic dissection of immune response using gene-expression studies and genome mapping. *Vet. Immunol. Immunopathol.* 105, 343–352. doi: 10.1016/j.vetimm.2005.02.007
18. Dekkers, J., and Van der Werf, J. (2007). "Strategies, limitations and opportunities for marker-assisted selection in livestock," in *Marker-Assisted Selection – Current Status and Future Perspectives in Crops, Livestock, Forestry and Fish*, eds E. P. Guimarães, J. Ruane, B. D. Scherf, A. Sonnino, and J. D. Dargie (Rome: FAO), 167–184.
19. Doeschl-Wilson, A. B., Bishop, S. C., Kyriazakis, I., and Villanueva, B. (2012). Novel methods for quantifying individual host response to infectious pathogens for genetic analyses. *Front. Genet.* 3:266. doi: 10.3389/fgene.2012.00266
20. Dorson, M., Quillet, E., Hollebecq, M., Torhy, C., and Chevassus, B. (1995). Selection of rainbow trout resistant to viral haemorrhagic septicaemia virus and transmission of resistance by gynogenesis. *Vet. Res.* 26, 361–368.
21. Everett, M., Grau, E., and Seeb, J. (2011). Short reads and nonmodel species: exploring the complexities of next-generation sequence assembly and SNP discovery in the absence of a reference genome. *Mol. Ecol. Resour.* 11, 93–108. doi: 10.1111/j.1755-0998.2010.02969.x
22. Everett, M. V., Miller, M. R., and Seeb, J. E. (2012). Meiotic maps of sockeye salmon derived from massively parallel DNA sequencing. *BMC Genomics* 13:521. doi: 10.1186/1471-2164-13-521
23. Ewart, K., Belanger, J., Williams, J., Karakach, T., Penny, S., Tsoi, S.,et al. (2005). Identification of genes differentially expressed in Atlantic salmon (*S. salar*) in response to infection by *Aeromonas salmonicida* using cDNA microarray technology. *Dev. Comp. Immunol.* 29, 333–347. doi: 10.1016/j.dci.2004.08.004
24. Falconer, D. S., and Mackay, T. F. C. (1996). *Introduction to Quantitative Genetics*, 4 Edn. Essex: Longman.
25. Fevolden, S. E., Nordmo, R., Refstie, T., and Røed, K. H. (1993). Disease resistance in Atlantic salmon (*Salmo salar*) selected for high or low responses to stress. *Aquaculture* 109, 215–224. doi: 10.1016/0044-8486(93)90164-T

26. Food and Agriculture Organization of the United Nations [FAO]. (2012). *The State of World Fisheries and Aquaculture.* Italy: FAO.
27. Fryer, J., and Hedrick, R. (2003). *Piscirickettsia salmonis*: a Gram-negative intracellular bacterial pathogen of fish. *J. Fish Dis.* 26, 251–262. doi: 10.1046/j.1365-2761.2003.00460.x
28. Garrigan, D., and Hedrick, P. W. (2001). Class I MHC polymorphism and evolution in endangered California Chinook and other Pacific salmon. *Immunogenetics* 53, 483–489. doi: 10.1007/s002510100352
29. Gharbi, K., Gautier, A., Danzmann, R. G., Gharbi, S., Sakamoto, T., Hoyheim, B.,et al. (2006). A linkage map for brown trout (*Salmo trutta*): chromosome homeologies and comparative genome organization with other salmonid fish. *Genetics* 172, 2405–2419. doi: 10.1534/genetics.105.048330
30. Gilbey, J., Verspoor, E., Mclay, A., and Houlihan, D. (2004). A microsatellite linkage map for Atlantic salmon (*Salmo salar*). *Anim. Genet.* 35, 98–105. doi: 10.1111/j.1365-2052.2004.01091.x
31. Gjedrem, T. (2000). Genetic improvement of cold-water fish species. *Aquac. Res.* 31, 25–33. doi: 10.1046/j.1365-2109.2000.00389.x
32. Gjedrem, T. (2012). Genetic improvement for the development of efficient global aquaculture: a personal opinion review. *Aquaculture* 344, 12–22. doi: 10. 1016/ j.aquaculture.2012.03.003
33. Gjedrem, T., and Gjøen, H. (1995). Genetic variation in susceptibility of Atlantic salmon, *Salmo salar* L., to furunculosis, BKD and cold water vibriosis. *Aquac. Res.* 26, 129–134. doi: 10.1111/j.1365-2109.1995.tb00892.x
34. Gjedrem, T., Salte, R., and Gjøen, H. M. (1991). Genetic variation in susceptibility of Atlantic salmon to furunculosis. *Aquaculture* 97, 1–6. doi: 10.1016/0044-8486(91)90274-B
35. Gjerde, B., Evensen,Ø., Bentsen, H. B., and Storset, A. (2009). Genetic (co) variation of vaccine injuries and innate resistance to furunculosis (*Aeromonas salmonicida*) and infectious salmon anaemia (ISA) in Atlantic salmon (*Salmo salar*). *Aquaculture* 287, 52–58. doi: 10.1016/j.aquaculture.2008.10.028
36. Gjøen, H., and Bentsen, H. (1997). Past, present, and future of genetic improvement in salmon aquaculture. *ICES J. Marine Sci.* 54, 1009–1014. doi: 10.1016/S1054-3139(97)80005-7
37. Gjøen, H. M., Refstie, T., Ulla, O., and Gjerde, B. (1997). Genetic correlations between survival of Atlantic salmon in challenge and field tests. *Aquaculture* 158, 277–288. doi: 10.1016/S0044-8486(97)00203-2
38. Glover, K., Aasmundstad, T., Nilsen, F., Storset, A., and Skaala, Ø. (2005). Variation of Atlantic salmon families (*Salmo salar* L.) in susceptibility to the sea lice *Lepeophtheirus salmonis* and *Caligus elongatus. Aquaculture* 245, 19–30. doi: 10.1016/j.aquaculture.2004.11.047
39. Goddard, M. E., and Hayes, B. J. (2009). Mapping genes for complex traits in domestic animals and their use in breeding programmes. *Nat. Rev. Genet.* 10, 381–391. doi: 10.1038/nrg2575
40. Goddard, M. E., and Meuwissen, T. H. E. (2005). The use of linkage disequilibrium to map quantitative trait loci. *Aust. J. Exp. Agric.* 45, 837–845.
41. Gonen, S., Lowe, N. R., Cezard, T., Gharbi, K., Bishop, S. C., and Houston, R. D. (2014). Linkage maps of the Atlantic salmon (*Salmo salar*) genome derived from RAD sequencing. *BMC Genomics* 15:166. doi: 10.1186/1471-2164-15-166
42. Grimholt, U., Drablos, F., Jorgensen, S. M., Hoyheim, B., and Stet, R. J. (2002). The major histocompatibility class I locus in Atlantic salmon (*Salmo salar* L.): polymorphism, linkage analysis and protein modelling. *Immunogenetics* 54, 570–581. doi: 10.1007/s00251-002-0499-8
43. Grimholt, U., Getahun, A., Hermsen, T., and Stet, R. J. (2000). The major histocompatibility class II alpha chain in salmonid fishes. *Dev. Comp. Immunol.* 24, 751–763. doi: 10.1016/S0145-305X(00)00034-3

44. Grimholt, U., Hordvik, I., Fosse, V. M., Olsaker, I., Endresen, C., and Lie, Ø. (1993). Molecular cloning of major histocompatibility complex class I cDNAs from Atlantic salmon (*Salmo salar*). *Immunogenetics* 37, 469–473. doi: 10.1007/BF00222473
45. Grimholt, U., Larsen, S., Nordmo, R., Midtlyng, P., Kjoeglum, S., Storset, A.,et al. (2003). MHC polymorphism and disease resistance in Atlantic salmon (*Salmo salar*), facing pathogens with single expressed major histocompatibility class I and class II loci. *Immunogenetics* 55, 210–219. doi: 10.1007/s00251-003-0567-8
46. Grimholt, U., Olsaker, I., Lindstrøm, C., and Lie, Ø. (1994). A study of variability in the MHC class II β1 and class I α2 domain exons of Atlantic salmon, *Salmo salar* L. *Anim. Genet.* 25, 147–153. doi: 10.1111/j.1365-2052.1994.tb00103.x
47. Guy, D. R., Bishop, S. C., Brotherstone, S., Hamilton, A., Roberts, R. J., Mcandrew, B.,et al. (2006). Analysis of the incidence of infectious pancreatic necrosis mortality in pedigreed Atlantic salmon, *Salmo salar* L., populations. *J. Fish Dis.* 29, 637–647. doi: 10.1111/j.1365-2761.2006.00758.x
48. Guyomard, R., Boussaha, M., Krieg, F., Hervet, C., and Quillet, E. (2012). A synthetic rainbow trout linkage map provides new insights into the salmonid whole genome duplication and the conservation of synteny among teleosts. *BMC Genet.* 13:15. doi: 10.1186/1471-2156-13-15
49. Guyomard, R., Mauger, S., Tabet-Canale, K., Martineau, S., Genet, C., Krieg, F.,et al. (2006). A type I and type II microsatellite linkage map of rainbow trout (*Oncorhynchus mykiss*) with presumptive coverage of all chromosome arms. *BMC Genomics* 7:302. doi: 10.1186/1471-2164-7-302
50. Hansen, J. D., Strassburger, P., and Du Pasquier, L. (1996). Conservation of an alpha 2 domain within the teleostean world, mhc class i from the rainbow trout *Oncorhynchus mykiss*. *Dev. Comp. Immunol.* 20, 417–425. doi: 10.1016/S0145-305X(96)00030-4
51. Hansen, J. D., Strassburger, P., Thorgaard, G. H., Young, W. P., and Du Pasquier, L. (1999). Expression, linkage, and polymorphism of MHC-related genes in rainbow trout, *Oncorhynchus mykiss*. *J. Immunol.* 163, 774–786.
52. Hayes, B., and Goddard, M. E. (2001). The distribution of the effects of genes affecting quantitative traits in livestock. *Genet. Select. Evol.* 33, 209–230. doi: 10.1186/1297-9686-33-3-209
53. Hayes, B., Laerdahl, J., Lien, S., Moen, T., Berg, P., Hindar, K.,et al. (2007). An extensive resource of single nucleotide polymorphism markers associated with Atlantic salmon (*Salmo salar*) expressed sequences. *Aquaculture* 265, 82–90. doi: 10.1016/j.aquaculture.2007.01.037
54. Hayes, B. J., Gjuvsland, A., and Omholt, S. (2006). Power of QTL mapping experiments in commercial Atlantic salmon populations, exploiting linkage and linkage disequilibrium and effect of limited recombination in males. *Heredity (Edinb.)* 97, 19–26. doi: 10.1038/sj.hdy.6800827
55. Henryon, M., Berg, P., Olesen, N. J., Kjær, T. E., Slierendrecht, W. J., Jokumsen, A.,et al. (2005). Selective breeding provides an approach to increase resistance of rainbow trout (*Oncorhynchus mykiss*) to the diseases, enteric redmouth disease, rainbow trout fry syndrome, and viral haemorrhagic septicaemia. *Aquaculture* 250, 621–636. doi: 10.1016/j.aquaculture.2004.12.022
56. Henryon, M., Jokumsen, A., Berg, P., Lund, I., Pedersen, P. B., Olesen, N. J.,et al. (2002). Genetic variation for growth rate, feed conversion efficiency, and disease resistance exists within a farmed population of rainbow trout. *Aquaculture* 209, 59–76. doi: 10.1016/S0044-8486(01)00729-3
57. Hiendleder, S., Bauersachs, S., Boulesteix, A., Blum, H., Arnold, G., Frohlich, T.,et al. (2005). Functional genomics: tools for improving farm animal health and welfare. *Rev. Sci. Tech.* 24, 355. doi: 10.1186/1476-4598-13-88
58. Hill, A. V. (1999). Genetics and genomics of infectious disease susceptibility. *Br. Med. Bull.* 55, 401–213. doi: 10.1258/0007142991902457
59. Hohenlohe, P. A., Amish, S. J., Catchen, J. M., Allendorf, F. W., and Luikart, G. (2011). Next-generation RAD sequencing identifies thousands of SNPs for assessing hybridization

between rainbow and westslope cutthroat trout. *Mol. Ecol. Resour.* 11(Suppl. 1), 117–122. doi: 10.1111/j.1755-0998.2010.02967.x

60. Hollebecq, M.-G., Faivre, B., Bourmaud, C., and Michel, C. (1995). Spontaneous bactericidal and complement activities in serum of rainbow trout (*Oncorhynchus mykiss*) genetically selected for resistance or susceptibility to furunculosis. *Fish Shellfish Immunol.* 5, 407–426. doi: 10.1006/fsim.1995.0040
61. Hordvik, I., Grimholt, U., Fosse, V. M., Lie,Ø., and Endresen, C. (1993). Cloning and sequence analysis of cDNAs encoding the MHC class II β chain in Atlantic salmon (*Salmo salar*). *Immunogenetics* 37, 437–441. doi: 10.1007/BF00222467
62. Houston, R. D., Davey, J. W., Bishop, S. C., Lowe, N. R., Mota-Velasco, J. C., Hamilton, A.,et al. (2012). Characterisation of QTL-linked and genome-wide restriction site-associated DNA (RAD) markers in farmed Atlantic salmon. *BMC Genomics* 13:244. doi: 10.1186/1471-2164-13-244
63. Houston, R. D., Haley, C. S., Hamilton, A., Guy, D. R., Mota-Velasco, J. C., Gheyas, A. A.,et al. (2010). The susceptibility of Atlantic salmon fry to freshwater infectious pancreatic necrosis is largely explained by a major QTL. *Heredity (Edinb.)* 105, 318–327. doi: 10.1038/hdy.2009.171
64. Houston, R. D., Haley, C. S., Hamilton, A., Guy, D. R., Tinch, A. E., Taggart, J. B.,et al. (2008a). Major quantitative trait loci affect resistance to infectious pancreatic necrosis in Atlantic salmon (*Salmo salar*). *Genetics* 178, 1109–1115. doi: 10.1534/genetics.107.082974
65. Houston, R. D., Gheyas, A., Hamilton, A., Guy, D. R., Tinch, A. E., Taggart, J. B.,et al. (2008b). Detection and confirmation of a major QTL affecting resistance to infectious pancreatic necrosis (IPN) in Atlantic salmon (*Salmo salar*). *Dev. Biol. (Basel)* 132, 199–204. doi: 10.1159/000317160
66. Houston, R. D., Taggart, J. B., Cézard, T., Bekaert, M., Lowe, N. R., Downing, A.,et al. (2014). Development and validation of a high density SNP genotyping array for Atlantic salmon (*Salmo salar*). *BMC Genomics* 15:90. doi: 10.1186/1471-2164-15-90
67. Johnson, N. A., Vallejo, R. L., Silverstein, J. T., Welch, T. J., Wiens, G. D., Hallerman, E. M.,et al. (2008). Suggestive association of major histocompatibility IB genetic markers with resistance to bacterial cold water disease in rainbow trout (*Oncorhynchus mykiss*). *Mar. Biotechnol.* 10, 429–437. doi: 10.1007/s10126-007-9080-7
68. Jones, P. G., Hammell, K. L., Gettinby, G., and Revie, C. W. (2013). Detection of emamectin benzoate tolerance emergence in different life stages of sea lice, *Lepeophtheirus salmonis,* on farmed Atlantic salmon, *Salmo salar* L. *J. Fish Dis.* 36, 209–220. doi: 10.1111/jfd.12022
69. Kjøglum, S., Grimholt, U., and Larsen, S. (2005). Non-MHC genetic and tank effects influence disease challenge tests in Atlantic salmon (*Salmo salar*). *Aquaculture* 250, 102–109. doi: 10.1016/j.aquaculture.2005.03.021
70. Kjøglum, S., Henryon, M., Aasmundstad, T., and Korsgaard, I. (2008). Selective breeding can increase resistance of Atlantic salmon to furunculosis, infectious salmon anaemia and infectious pancreatic necrosis. *Aquac. Res.* 39, 498–505. doi: 10.1111/j.1365-2109.2008.01904.x
71. Kjøglum, S., Larsen, S., Bakke, H. G., and Grimholt, U. (2006). How specific MHC class I and class II combinations affect disease resistance against infectious salmon anaemia in Atlantic salmon (*Salmo salar*). *Fish Shellfish Immunol.* 21, 431–441. doi: 10.1016/j.fsi.2006.02.001
72. Kolstad, K., Heuch, P. A., Gjerde, B., Gjedrem, T., and Salte, R. (2005). Genetic variation in resistance of Atlantic salmon (*Salmo salar*) to the salmon louse *Lepeophtheirus salmonis. Aquaculture* 247, 145–151. doi: 10.1016/j.aquaculture.2005.02.009
73. Langefors,Å., Lohm, J., Grahn, M., Andersen, Ø., and Von Schantz, T. (2001). Association between major histocompatibility complex class IIB alleles and resistance to *Aeromonas salmonicida* in Atlantic salmon. *Proc. Biol. Sci.* 268, 479–485. doi: 10.1098/rspb.2000.1378
74. Langevin, C., Blanco, M., Martin, S. A. M., Jouneau, L., Bernardet, J.-F., Houel, A.,et al. (2012). Transcriptional responses of resistant and susceptible fish clones to the bacterial pathogen *Flavobacterium psychrophilum. PLoS ONE* 7:e39126. doi: 10.1371/journal.pone.0039126

75. Leeds, T. D., Silverstein, J. T., Weber, G. M., Vallejo, R. L., Palti, Y., Rexroad, C. E.,et al. (2010). Response to selection for bacterial cold water disease resistance in rainbow trout. *J. Anim. Sci.* 88, 1936–1946. doi: 10.2527/jas.2009-2538
76. Lhorente, J. P., Gallardo, J. A., Villanueva, B., Araya, A. M., Torrealba, D. A., Toledo, X. E.,et al. (2012). Quantitative genetic basis for resistance to *Caligus rogercresseyi* sea lice in a breeding population of Atlantic salmon (*Salmo salar*). *Aquaculture* 324, 55–59. doi: 10.1016/j.aquaculture.2011.10.046
77. Lien, S., Gidskehaug, L., Moen, T., Hayes, B. J., Berg, P. R., Davidson, W. S.,et al. (2011). A dense SNP-based linkage map for Atlantic salmon (*Salmo salar*) reveals extended chromosome homeologies and striking differences in sex-specific recombination patterns. *BMC Genomics* 12:615. doi: 10.1186/1471-2164-12-615
78. Liu, Z., and Cordes, J. (2004). DNA marker technologies and their applications in aquaculture genetics. *Aquaculture* 238, 1–37. doi: 10.1016/j.aquaculture.2004.05.027
79. Lohm, J., Grahn, M., Langefors,Å., Andersen, Ø., Storset, A., and Von Schantz, T. (2002). Experimental evidence for major histocompatibility complex–allele–specific resistance to a bacterial infection. *Proc. Biol. Sci.* 269, 2029–2033. doi: 10.1098/rspb.2002.2114
80. Lund, T., Gjedrem, T., Bentsen, H., Eide, D., Larsen, H., and Røed, K. (1995). Genetic variation in immune parameters and associations to survival in Atlantic salmon. *J. Fish Biol.* 46, 748–758. doi: 10.1111/j.1095-8649.1995.tb01598.x
81. Martin, S. A., Blaney, S. C., Houlihan, D. F., and Secombes, C. J. (2006). Transcriptome response following administration of a live bacterial vaccine in Atlantic salmon (*Salmo salar*). *Mol. Immunol.* 43, 1900–1911. doi: 10.1016/j.molimm.2005.10.007
82. McClelland, E. K., and Naish, K. A. (2008). A genetic linkage map for coho salmon (*Oncorhynchus kisutch*). *Anim. Genet.* 39, 169–179. doi: 10.1111/j.1365-2052.2008.01699.x
83. Meuwissen, T. H. E., Hayes, B., and Goddard, M. (2001). Prediction of total genetic value using genomewide dense marker maps. *Genetics* 157, 1819–1829.
84. Meuwissen, T. H., Karlsen, A., Lien, S., Olsaker, I., and Goddard, M. E. (2002). Fine mapping of a quantitative trait locus for twinning rate using combined linkage and linkage disequilibrium mapping. *Genetics* 161, 373–379.
85. Miller, K. M., Winton, J. R., Schulze, A. D., Purcell, M. K., and Ming, T. J. (2004). Major histocompatibility complex loci are associated with susceptibility of Atlantic salmon to infectious hematopoietic necrosis virus. *Environ. Biol. Fishes* 69, 307–316. doi: 10.1023/B:EBFI.0000022874.48341.0f
86. Miller, K. M., and Withler, R. E. (1996). Sequence analysis of a polymorphic Mhc class II gene in Pacific salmon. *Immunogenetics* 43, 337–351. doi: 10.1007/BF02199802
87. Mitchell, S., and Rodger, H. (2011). A review of infectious gill disease in marine salmonid fish. *J. Fish Dis.* 34, 411–432. doi: 10.1111/j.1365-2761.2011.01251.x
88. Moen, T., Baranski, M., Sonesson, A. K., and Kjøglum, S. (2009). Confirmation and fine-mapping of a major QTL for resistance to infectious pancreatic necrosis in Atlantic salmon (*Salmo salar*): population-level associations between markers and trait. *BMC Genomics* 10:368. doi: 10.1186/1471-2164-10-368
89. Moen, T., Fjalestad, K. T., Munck, H., and Gomez-Raya, L. (2004a). A multistage testing strategy for detection of quantitative trait Loci affecting disease resistance in Atlantic salmon. *Genetics* 167, 851–858. doi: 10.1534/genetics.103.013227.
90. Moen, T., Hoyheim, B., Munck, H., and Gomez-Raya, L. (2004b). A linkage map of Atlantic salmon (*Salmo salar*) reveals an uncommonly large difference in recombination rate between the sexes. *Anim. Genet.* 35, 81–92. doi: 10.1111/j.1365-2052.2004.01097.x
91. Moen, T., Sonesson, A. K., Hayes, B., Lien, S., Munck, H., and Meuwissen, T. H. (2007). Mapping of a quantitative trait locus for resistance against infectious salmon anaemia in Atlantic salmon (*Salmo salar*): comparing survival analysis with analysis on affected/resistant data. *BMC Genet.* 8:53. doi: 10.1186/1471-2156-8-53
92. Mustafa, A., and MacKinnon, B. M. (1999). Genetic variation in susceptibility of Atlantic salmon to the sea louse *Caligus elongatus* Nordmann, 1832. *Can. J. Zool.* 77, 1332–1335. doi: 10.1139/cjz-77-8-1332

93. Nichols, K. M., Young, W. P., Danzmann, R. G., Robison, B. D., Rexroad, C., Noakes, M.,et al. (2003). A consolidated linkage map for rainbow trout (*Oncorhynchus mykiss*). *Anim. Genet.* 34, 102–115. doi: 10.1046/j.1365-2052.2003.00957.x
94. Norris, A., Foyle, L., and Ratcliff, J. (2008). Heritability of mortality in response to a natural pancreas disease (SPDV) challenge in Atlantic salmon, *Salmo salar* L., post-smolts on a West of Ireland sea site. *J. Fish Dis.* 31, 913–920. doi: 10.1111/j.1365-2761.2008.00982.x
95. Ødegård, J., Baranski, M., Gjerde, B., and Gjedrem, T. (2011). Methodology for genetic evaluation of disease resistance in aquaculture species: challenges and future prospects. *Aquac. Res.* 42, 103-114. doi: 10.1111/j.1365-2109.2010.02669.x.
96. Ødegård, J., Olesen, I., Gjerde, B., and Klemetsdal, G. (2006). Evaluation of statistical models for genetic analysis of challenge test data on furunculosis resistance in Atlantic salmon (*Salmo salar*): prediction of field survival. *Aquaculture* 259, 116–123. doi: 10.1016/j.aquaculture.2006.05.034
97. Ødegård, J., Olesen, I., Gjerde, B., and Klemetsdal, G. (2007a). Evaluation of statistical models for genetic analysis of challenge-test data on ISA resistance in Atlantic salmon (*Salmo salar*): prediction of progeny survival. *Aquaculture* 266, 70–76. doi: 10.1016/j.aquaculture.2007.02.012
98. Ødegård, J., Olesen, I., Gjerde, B., and Klemetsdal, G. (2007b). Positive genetic correlation between resistance to bacterial (furunculosis) and viral (infectious salmon anaemia) diseases in farmed Atlantic salmon (*Salmo salar*). *Aquaculture* 271, 173–177. doi: 10.1016/j.aquaculture.2007.06.006
99. Olesen, I., Hung, D., and Ødegård, J. (2007). Genetic analysis of survival in challenge tests of furunculosis and ISA in Atlantic salmon. Genetic parameter estimates and model comparisons. *Aquaculture* 272(Suppl. 1), S297–S298. doi: 10.1016/j.aquaculture.2007.07.155
100. Ozaki, A., Sakamoto, T., Khoo, S., Nakamura, K., Coimbra, M. R., Akutsu, T.,et al. (2001). Quantitative trait loci (QTLs) associated with resistance/susceptibility to infectious pancreatic necrosis virus (IPNV) in rainbow trout (*Oncorhynchus mykiss*). *Mol. Genet. Genomics* 265, 23–31. doi: 10.1007/s004380000392
101. Palti, Y., Gao, G., Miller, M. R., Vallejo, R. L., Wheeler, P. A., Quillet, E.,et al. (2014a). A resource of single-nucleotide polymorphisms for rainbow trout generated by restriction-site associated DNA sequencing of doubled haploids. *Mol. Ecol. Res.* 14, 588–596. doi: 10.1111/1755-0998.12204
102. Palti, Y., Gao, G., Liu, S., Kent, M. P., Lien, S., Miller, M. R.,et al. (2014b). The development and characterization of a 57K SNP array for rainbow trout. *Mol. Ecol. Res.* doi: 10.1111/1755-0998.12337 [Epub ahead of print].
103. Palti, Y., Genet, C., Gao, G., Hu, Y., You, F., Boussaha, M.,et al. (2012). A second generation integrated map of the rainbow trout (*Oncorhynchus mykiss*) genome: analysis of conserved synteny with model fish genomes. *Mar. Biotechnol.* 14, 343–357. doi: 10.1007/s10126-011-9418-z
104. Palti, Y., Genet, C., Luo, M.-C., Charlet, A., Gao, G., Hu, Y.,et al. (2011). A first generation integrated map of the rainbow trout genome. *BMC Genomics* 12:180. doi: 10.1186/1471-2164-12-180
105. Palti, Y., Nichols, K. M., Waller, K. I., Parsons, J. E., and Thorgaard, G. H. (2001). Association between DNA polymorphisms tightly linked to MHC class II genes and IHN virus resistance in backcrosses of rainbow and cutthroat trout. *Aquaculture* 194, 283–289. doi: 10.1016/S0044-8486(00)00526-3
106. Palti, Y., Parsons, J. E., and Thorgaard, G. H. (1999). Identification of candidate DNA markers associated with IHN virus resistance in backcrosses of rainbow (*Oncorhynchus mykiss*) and cutthroat trout (*O. clarki*). *Aquaculture* 173, 81–94. doi: 10.1016/S0044-8486(98)00471-2
107. Perry, G. M. L., Tarte, P., Croisetière, S., Belhumeur, P., and Bernatchez, L. (2004). Genetic variance and covariance for 0+ brook charr (*Salvelinus fontinalis*) weight and survival time of furunculosis (*Aeromonas salmonicida*) exposure. *Aquaculture* 235, 263–271. doi: 10.1016/j.aquaculture.2004.03.002

108. Phillips, R., Keatley, K., Morasch, M., Ventura, A., Lubieniecki, K., Koop, B.,et al. (2009). Assignment of Atlantic salmon (*Salmo salar*) linkage groups to specific chromosomes: conservation of large syntenic blocks corresponding to whole chromosome arms in rainbow trout (*Oncorhynchus mykiss*). *BMC Genet.* 10:46. doi: 10.1186/1471-2156-10-46.
109. Piepho, H.-P., and Koch, G. (2000). Codominant analysis of banding data from a dominant marker system by normal mixtures. *Genetics* 155, 1459–1468.
110. Pomp, D., Allan, M. F., and Wesolowski, S. R. (2004). Quantitative genomics: exploring the genetic architecture of complex trait predisposition. *J. Anim. Sci.* 82, E300–E312.
111. Porto-Neto, L., Kijas, J., and Reverter, A. (2014). The extent of linkage disequilibrium in beef cattle breeds using high-density SNP genotypes. *Genet. Select. Evol.* 46, 22. doi: 10.1186/1297-9686-46-22
112. Purcell, M. K., Nichols, K. M., Winton, J. R., Kurath, G., Thorgaard, G. H., Wheeler, P.,et al. (2006). Comprehensive gene expression profiling following DNA vaccination of rainbow trout against infectious hematopoietic necrosis virus. *Mol. Immunol.* 43, 2089–2106. doi: 10.1016/j.molimm.2005.12.005
113. Qureshi, S. T., Skamene, E., and Malo, D. (1999). Comparative genomics and host resistance against infectious diseases. *Emerging Infect. Dis.* 5, 36–47. doi: 10.3201/eid0501.990105
114. Råberg, L., Sim, D., and Read, A. F. (2007). Disentangling genetic variation for resistance and tolerance to infectious diseases in animals. *Science* 318, 812–814. doi: 10.1126/science.1148526
115. Rexroad, C. E. III, Palti, Y., Gahr, S. A., and Vallejo, R. L. (2008). A second generation genetic map for rainbow trout (*Oncorhynchus mykiss*). *BMC Genet.* 9:74. doi: 10.1186/1471-2156-9-74
116. Rise, M. L., Jones, S. R. M., Brown, G. D., Von Schalburg, K. R., Davidson, W. S., and Koop, B. F. (2004a). Microarray analyses identify molecular biomarkers of Atlantic salmon macrophage and hematopoietic kidney response to *Piscirickettsia salmonis* infection. *Physiol. Genomics* 15, 21–35. doi: 10.1152/physiolgenomics.00036.2004
117. Rise, M. L., Von Schalburg, K. R., Brown, G. D., Mawer, M. A., Devlin, R. H., Kuipers, N.,et al. (2004b). Development and application of a salmonid EST database and cDNA microarray: data mining and interspecific hybridization characteristics. *Genome Res.* 14, 478–490. doi: 10.1101/gr.1687304
118. Roberts, R. J., and Pearson, M. D. (2005). Infectious pancreatic necrosis in Atlantic salmon, *Salmo salar* L. *J. Fish Dis.* 28, 383–390. doi: 10.1111/j.1365-2761.2005.00642.x
119. Rodriguez, M. F., Lapatra, S., Williams, S., Famula, T., and May, B. (2004). Genetic markers associated with resistance to infectious hematopoietic necrosis in rainbow and steelhead trout (*Oncorhynchus mykiss*) backcrosses. *Aquaculture* 241, 93–115. doi: 10.1016/j.aquaculture.2004.08.003
120. Røed, K. H., Fjalestad, K. T., and Strømsheim, A. (1993). Genetic variation in lysozyme activity and spontaneous haemolytic activity in Atlantic salmon (*Salmo salar*). *Aquaculture* 114, 19–31. doi: 10.1016/0044-8486(93)90247-V
121. Rothschild, M. F., and Soller, M. (1997). Candidate gene analysis to detect traits of economic importance in domestic livestock. *Probe* 8, 13–22.
122. Rozas, M., and Enríquez, R. (2014). Piscirickettsiosis and *Piscirickettsia salmonis* in fish: a review. *J. Fish Dis.* 37, 163–188. doi: 10.1111/jfd.12211
123. Ruane, N. M., and Jones, S. R. M. (2013). "Amoebic Gill Disease (AGD) of farmed Atlantic salmon (*Salmo salar* L.)," in *Proceedings of the ICES Identification Leaflets for Diseases and Parasites of Fish and Shellfish.* Leaflet No. 60 (Copenhagen: International Council for the Exploration), 6.
124. Sakamoto, T., Danzmann, R. G., Gharbi, K., Howard, P., Ozaki, A., Khoo, S. K.,et al. (2000). A microsatellite linkage map of rainbow trout (*Oncorhynchus mykiss*) characterized by large sex-specific differences in recombination rates. *Genetics* 155, 1331–1345.
125. Salem, M., Vallejo, R. L., Leeds, T. D., Palti, Y., Liu, S., Sabbagh, A.,et al. (2012). RNA-Seq identifies SNP markers for growth traits in rainbow trout. *PLoS ONE* 7:e36264. doi: 10.1371/journal.pone.0036264

126. Salte, R., Bentsen, H. B., Moen, T., Tripathy, S., Bakke, T. A., Ødegård, J.,et al. (2009). Prospects for a genetic management strategy to control *Gyrodactylus salaris* infection in wild Atlantic salmon (*Salmo salar*) stocks. *Can. J. Fish. Aquat. Sci.* 67, 121–129. doi: 10.1139/F09-168
127. Salte, R., Gjøen, H. M., Norberg, K., and Gjedrem, T. (1993). Plasma protein levels as potential marker traits for resistance to furunculosis. *J. Fish Dis.* 16, 561–568. doi: 10.1111/j.1365-2761.1993.tb00892.x
128. Sanchez, C., Smith, T., Wiedmann, R., Vallejo, R., Salem, M., Yao, J.,et al. (2009). Single nucleotide polymorphism discovery in rainbow trout by deep sequencing of a reduced representation library. *BMC Genomics* 10:559. doi: 10.1186/1471-2164-10-559
129. Silverstein, J. T., Vallejo, R. L., Palti, Y., Leeds, T. D., Rexroad, C. E., Welch, T. J.,et al. (2009). Rainbow trout resistance to bacterial cold-water disease is moderately heritable and is not adversely correlated with growth. *J. Anim. Sci.* 87, 860–867. doi: 10.2527/jas.2008-1157
130. Shum, B., Mason, P., Magor, K., Flodin, L., Stet, R., and Parham, P. (2002). Structures of two major histocompatibility complex class I genes of the rainbow trout (*Oncorhynchus mykiss*). *Immunogenetics* 54, 193–199. doi: 10.1007/s00251-002-0450-z
131. Shum, B. P., Rajalingam, R., Magor, K. E., Azumi, K., Carr, W. H., Dixon, B.,et al. (1999). A divergent non-classical class I gene conserved in salmonids. *Immunogenetics* 49, 479–490. doi: 10.1007/s002510050524
132. Sonesson, A., and Meuwissen, T. (2009). Testing strategies for genomic selection in aquaculture breeding programs. *Genet. Select. Evol.* 41, 37. doi: 10.1186/1297-9686-41-37
133. Stear, M. J., Bishop, S. C., Mallard, B. A., and Raadsma, H. (2001). The sustainability, feasibility and desirability of breeding livestock for disease resistance. *Res. Vet. Sci.* 71, 1–7. doi: 10.1053/rvsc.2001.0496
134. Stet, R., De Vries, B., Mudde, K., Hermsen, T., Van Heerwaarden, J., Shum, B.,et al. (2002). Unique haplotypes of co-segregating major histocompatibility class II A and class II B alleles in Atlantic salmon (*Salmo salar*) give rise to diverse class II genotypes. *Immunogenetics* 54, 320–331. doi: 10.1007/s00251-002-0477-1
135. Sutherland, B., Koczka, K., Yasuike, M., Jantzen, S., Yazawa, R., Koop, B.,et al. (2014). Comparative transcriptomics of Atlantic *Salmo salar*, chum *Oncorhynchus keta* and pink salmon *O. gorbuscha* during infections with salmon lice Lepeophtheirus salmonis. *BMC Genomics* 15:200. doi: 10.1186/1471-2164-15-200
136. Taylor, J. F. (2014). Implementation and accuracy of genomic selection. *Aquaculture* 420–421(Suppl. 1), S8–S14. doi: 10.1016/j.aquaculture.2013.02.017
137. Taylor, R. S., Kube, P. D., Muller, W. J., and Elliott, N. G. (2009). Genetic variation of gross gill pathology and survival of Atlantic salmon (*Salmo salar* L.) during natural amoebic gill disease challenge. *Aquaculture* 294, 172–179. doi: 10.1016/j.aquaculture.2009.06.007
138. Thorgaard, G. H., Bailey, G. S., Williams, D., Buhler, D. R., Kaattari, S. L., Ristow, S. S.,et al. (2002). Status and opportunities for genomics research with rainbow trout. *Comp. Biochem. Physiol. B Biochem. Mol. Biol.* 133, 609–646. doi: 10.1016/S1096-4959(02)00167-7
139. Tsoi, S. M., Cale, J., Bird, I., Ewart, V., Brown, L., and Douglas, S. (2003). Use of human cDNA microarrays for identification of differentially expressed genes in Atlantic salmon liver during *Aeromonas salmonicida* infection. *Mar. Biotechnol.* 5, 545–554. doi: 10.1007/s10126-002-0112-z
140. Tsoi, S. C. M., Ewart, K. V., Penny, S., Melville, K., Liebscher, R. S., Brown, L. L.,et al. (2004). Identification of immune-relevant genes from Atlantic salmon using suppression subtractive hybridization. *Mar. Biotechnol.* 6, 199–214. doi: 10.1007/s10126-002-0101-2
141. Vallejo, R., Palti, Y., Liu, S., Evenhuis, J., Gao, G., Rexroad, C. III,et al. (2014a). Detection of QTL in rainbow trout affecting survival when challenged with *Flavobacterium psychrophilum*. *Mar. Biotechnol.* 16, 349–360. doi: 10.1007/s10126-013-9553-9
142. Vallejo, R. L., Palti, Y., Liu, S., Marancik, D. P., and Wiens, G. D. (2014b). Validation of linked QTL for bacterial cold water disease resistance and spleen size on rainbow trout chromosome Omy19. *Aquaculture* 432, 139–143. doi: 10.1016/j.aquaculture.2014.05.003

143. Vallejo, R. L., Wiens, G. D., Rexroad, C. E. III, Welch, T. J., Evenhuis, J. P., Leeds, T. D.,et al. (2010). Evidence of major genes affecting resistance to bacterial cold water disease in rainbow trout using Bayesian methods of segregation analysis. *J. Anim. Sci.* 88, 3814–3832. doi: 10.2527/jas.2010-2951
144. Verrier, E. R., Dorson, M., Mauger, S., Torhy, C., Ciobotaru, C., Hervet, C.,et al. (2013a). Resistance to a Rhabdovirus (VHSV) in rainbow trout: identification of a major QTL related to innate mechanisms. *PLoS ONE* 8:e55302. doi: 10.1371/journal.pone.0055302
145. Verrier, E. R., Ehanno, A., Biacchesi, S., Le Guillou, S., Dechamp, N., Boudinot, P.,et al. (2013b). Lack of correlation between the resistances to two rhabdovirus infections in rainbow trout. *Fish Shellfish Immunol.* 35, 9–17. doi: 10.1016/j.fsi.2013.03.369
146. Villanueva, B., Fernández, J., García-Cortés, L. A., Varona, L., Daetwyler, H. D., and Toro, M. A. (2011). Accuracy of genome-wide evaluation for disease resistance in aquaculture breeding programs. *J. Anim. Sci.* 89, 3433–3442. doi: 10.2527/jas.2010-3814
147. von Schalburg, K., Rise, M., Cooper, G., Brown, G., Gibbs, A. R., Nelson, C.,et al. (2005). Fish and chips: various methodologies demonstrate utility of a 16,006-gene salmonid microarray. *BMC Genomics* 6:126. doi: 10.1186/1471-2164-6-126
148. Walsh, B., and Henderson, D. (2004). Microarrays and beyond: what potential do current and future genomics tools have for breeders? *J. Animal Sci.* 82, E292–E299.
149. Weber, G. M., Vallejo, R. L., Lankford, S. E., Silverstein, J. T., and Welch, T. J. (2008). Cortisol response to a crowding stress: heritability and association with disease resistance to *Yersinia ruckeri* in rainbow trout. *N. Am. J. Aquac.* 70, 425–433. doi: 10.1577/A07-059.1
150. Weber, G. M., Wiens, G. D., Welch, T. J., Hostuttler, M. A., and Leeds, T. D. (2013). Comparison of disease resistance between diploid, induced-triploid, and intercross-triploid rainbow trout including trout selected for resistance to *Flavobacterium psychrophilum*. *Aquaculture* 410–411, 66–71. doi: 10.1016/j.aquaculture.2013.06.014
151. Wetten, M., Aasmundstad, T., Kjøglum, S., and Storset, A. (2007). Genetic analysis of resistance to infectious pancreatic necrosis in Atlantic salmon (*Salmo salar* L.). *Aquaculture* 272, 111–117. doi: 10.1016/j.aquaculture.2007.08.046
152. Wiens, G. D., Lapatra, S. E., Welch, T. J., Evenhuis, J. P., Rexroad, C. E. III, and Leeds, T. D. (2013a). On-farm performance of rainbow trout (*Oncorhynchus mykiss*) selectively bred for resistance to bacterial cold water disease: effect of rearing environment on survival phenotype. *Aquaculture* 388–391, 128–136. doi: 10.1016/j.aquaculture.2013.01.018
153. Wiens, G. D., Vallejo, R. L., Leeds, T. D., Palti, Y., Hadidi, S., Liu, S.,et al. (2013b). Assessment of genetic correlation between bacterial cold water disease resistance and spleen index in a domesticated population of rainbow trout: identification of QTL on chromosome Omy19. *PLoS ONE* 8: e75749. doi: 10.1371/journal.pone.0075749
154. Wientjes, Y. C. J., Veerkamp, R. F., and Calus, M. P. L. (2013). The effect of linkage disequilibrium and family relationships on the reliability of genomic prediction. *Genetics* 193, 621–631. doi: 10.1534/genetics.112.146290
155. Withler, R. E., and Evelyn, T. P. T. (1990). Genetic variation in resistance to bacterial kidney disease within and between two strains of coho salmon from British Columbia. *Trans. Am. Fish. Soc.* 119, 1003–1009. doi: 10.1577/1548-8659(1990)119<1003:GVIRTB>2.3.CO;2
156. Woram, R. A., Mcgowan, C., Stout, J. A., Gharbi, K., Ferguson, M. M., Hoyheim, B.,et al. (2004). A genetic linkage map for Arctic charr (*Salvelinus alpinus*): evidence for higher recombination rates and segregation distortion in hybrid versus pure strain mapping parents. *Genome* 47, 304–315. doi: 10.1139/g03-127
157. Yáñez, J. M., Bangera, R., Lhorente, J. P., Oyarzún, M., and Neira, R. (2013). Quantitative genetic variation of resistance against *Piscirickettsia salmonis* in Atlantic salmon (*Salmo salar*). *Aquaculture* 414–415, 155–159. doi: 10.1016/j.aquaculture.2013.08.009
158. Yáñez, J. M., Lhorente, J. P., Bassini, L. N., Oyarzún, M., Neira, R., and Newman, S. (2014a). Genetic co-variation between resistance against both *Caligus rogercresseyi* and *Piscirickettsia salmonis*, and body weight in Atlantic salmon (*Salmo salar*). *Aquaculture* 433, 295–298. doi: 10.1016/j.aquaculture.2014.06.026
159. Yáñez, J. M., Naswa, S., López, M. E., Bassini, L., Cabrejos, M. E., Gilbey, J.,et al. (2014b). "Development of a 200K SNP array for Atlantic Salmon: exploiting across continents

genetic variation," *in Proceedings of the 10th World Congress on Genetics Applied to Livestock Production*, Vancouver.

160. Yáñez, J. M., and Martínez, V. (2010). Factores genéticos que inciden en la resistencia a enfermedades infecciosas en salmónidos y su aplicación en programas de mejoramiento. *Arch. Med. Vet.* 42, 1–13.
161. Young, W. P., Wheeler, P. A., Coryell, V. H., and Keim, P. (1998). A detailed linkage map of rainbow trout produced using doubled haploids. *Genetics* 148, 839–850.
162. Zhang, Z., Niu, C., Storset, A., Bøgwald, J., and Dalmo, R. A. (2011). Comparison of *Aeromonas salmonicida* resistant and susceptible salmon families: a high immune response is beneficial for the survival against *Aeromonas salmonicida* challenge. *Fish Shellfish Immunol.* 31, 1–9. doi: 10.1016/j.fsi.2010.12.019

CHAPTER 12

Effects of Genotype by Environment Interaction on Genetic Gain and Genetic Parameter Estimates in Red Tilapia (*Oreochromis* spp.)

Nguyen H. Nguyen[1], Azhar Hamzah[2] and Ngo P. Thoa[3*]

[1]*School of Science and Engineering, University of the Sunshine Coast, Maroochydore, QLD, Australia*

[2]*National Prawn Fry Production and Research Center, Kota Kuala Muda, Malaysia*

[3]*Research Institute for Aquaculture No.1, Tu Son, Vietnam*

AUTHOR CONTRIBUTIONS

Conceived and designed the study: NN, and AH. Managed and oversaw the project: NN and AH. Analyzed the data and wrote the paper: NN, AH, and NT.

The extent to which genetic gain achieved from selection programs under strictly controlled environments in the nucleus that can be expressed in commercial production systems is not well-documented in aquaculture species. The main aim of this paper was to assess the effects of genotype by environment interaction on genetic response and genetic parameters for four body traits (harvest weight, standard length, body depth, body width) and survival in Red tilapia (*Oreochromis* spp.). The growth and survival data were recorded on 19,916 individual fish from a pedigreed population undergoing three generations of selection for increased harvest weight in earthen ponds from 2010 to 2012 at the Aquaculture Extension Center, Department of Fisheries, Jitra in Kedah, Malaysia. The pedigree comprised a total of 224 sires and 262 dams, tracing back to the base population in 2009. A multivariate animal model was used to measure genetic response and estimate variance and covariance components. When the homologous body traits in freshwater pond and cage were treated as genetically distinct traits, the genetic correlations between the two environments were high (0.85–0.90) for harvest weight and square root of harvest weight but the estimates were of lower magnitudes for length, width and depth (0.63–0.79). The heritabilities estimated for the five traits studied differed between pond (0.02 to 0.22) and cage (0.07 to 0.68). The common full-sib effects were large, ranging from 0.23 to 0.59 in pond and 0.11 to 0.31 in cage across all traits. The direct and correlated responses for four body traits were generally greater in pond than in cage environments (0.011–1.561 vs. −0.033–0.567 genetic standard deviation units, respectively). Selection for increased harvest body weight resulted in positive genetic changes in survival rate in both pond and cage culture. In conclusion, the reduced selection response and the magnitude of the genetic parameter estimates in the production environment (i.e., cage) relative to those achieved in the nucleus (pond) were a result of the genotype by environment interaction and this effect should be taken into consideration in the future breeding program for Red tilapia.

Keywords: red tilapia, harvest weight, genotype by environment interaction, genetic correlation, selection response

1. INTRODUCTION

Genotype by environment (G × E) interaction refers to the difference in the response of genotypes to different environments (Falconer and Mackay, 1996). There are two main forms of G × E interaction, due to scaling or re-ranking effects. The scaling effect is related to variance difference of traits between environments. A measure of the re-ranking G × E effect is the estimate of genetic correlations between performances in different environments (Falconer, 1952).

For Red tilapia (*Oreochromis* spp.), freshwater pond and cage are two prevailing culture environments in major producing countries, predominantly in Asia (Hamzah et al., 2008; Pongthana et al., 2010). In addition, a range of other production systems has been practiced including canals, mining pools, lakes, and reservoirs (Hamzah et al., 2008). In our breeding program for this species, selection was practiced for body weight in freshwater ponds and performance testing of siblings from each family was also conducted in freshwater cages over a grow period of about 3–4 months (Hamzah, unpublished). Recently, Thodesen et al. (2013) reported, in a different population of red tilapia, that the genetic correlation between body weight in freshwater earthen ponds and floating cages was high (0.92 ± 0.06), while that between freshwater earthen ponds and brackish water tanks was low (0.33 ± 0.14). The G × E interaction effect was reported for a range of fish species, including Nile tilapia (Bentsen et al., 2012; Trọng et al., 2013), rainbow trout (Kause et al., 2003), Atlantic cod (Kolstad et al., 2006), Asian and European sea bass (Dupont-Nivet et al., 2010; Domingos et al., 2013; Le Boucher et al., 2013). A synthesis of the literature review across farmed aquaculture species indicates that when the environments are similar (e.g., pond vs. cage), the G × E effect is not of commercial importance for body traits (e.g., Nguyen, 2016; Sae-Lim et al., 2016). However, when the environments in questions are remarkably dissimilar (e.g., freshwater vs. brackish water as reported in the study of Luan et al., 2008), the G × E interaction is significant especially for traits that are largely influenced by environmental factors, such as survival, sexual maturity, or fitness related traits.

To date, the extent to which genetic gain was realized in production systems (e.g., cage) when selection was conducted in a different environment (i.e., pond) is still not estimated in red tilapia. Quantification of such differences in the genetic gain and genetic parameter estimates between the testing and selection environments in the nucleus is needed because the aquaculture sector particularly for red tilapia is characterized by a diverse array of production systems from backyard farming to semi-medium and intensive large scales (Eknath et al., 2007; Trọng et al., 2013). In extreme cases if the G × E is significant, this should call the conduct of separate breeding programs for each production environment (Nguyen and Ponzoni, 2006; Sae-Lim et al., 2016). Understanding magnitude of the G × E effects on genetic gain and genetic parameter estimates (heritability and correlations) would assist not only the design but also the optimization of commercial genetic improvement programs for this species (Mulder et al., 2006; Sae-Lim et al., 2016).

The principal aim of our study was, hence, to evaluate selection response for four body traits in the two tested environments, namely pond and cage. In addition, we examined the magnitude of genotype by environment interaction for traits studied and reported the genetic parameters for body traits and survival in pond and cage culture.

2. MATERIALS AND METHODS

2.1 The Genetic Selection Line

The genetic selection line was established in 2008 from 103 full-sib families (one male × one female), each represented by 20–50 fish produced from a complete diallel cross involving three founder stocks (two imported stocks originated from Taiwan and Thailand, and another one from top ranking animals in a strain evaluation of 20 hatchery stocks in Malaysia; Hamzah et al., 2008). The founder stocks were reared under the same pond culture environments until they reached an average body weight of about 250–400 g before mating was initiated. In 2009, the progeny of the diallel cross experiment was produced in Malaysia, thus creating what we call the base population (G0). The selection line was formed with the 2009 progeny produced from 79 sires and 79 dams and it was selected for high breeding value for harvest body weight. A combined between and within family selection was practiced.

The average proportion of selected animals was 3.0% in females and 1.6% in males. Selection was practiced on breeding values for body weight but not by truncation (due to inability to reproduce of some selected breeders we had to resort to selecting lower ranking ones; also, the number of selected individuals contributed by each family was restricted to a maximum of two males and four females in order to avoid later inbreeding and mating of black spot fish was avoided. On average, the progeny was produced in each generation from 46 to 79 males and 53 to 79 dams. This design and experimental size were adhered throughout the course of the selection program (2010–2012) during which the data used for this study were collected and reported here.

2.2 Family Production

In each generation, production of full- and half-sib families was conducted in hapas installed in ponds, following the nested mating design (one male × two females) prepared from annual routine genetic evaluation and mate allocation analyses. However, in practice, the two females were not always successfully mated with a male, due to breeding failure or mortality loss. After 7 days of mating, fertilized eggs were collected from the mouth of the female and immediately transferred to hatching jars for artificial incubation. Fry were often hatched after about 5–7 days. As soon as after sac-yolk absorption, the hatched fry of each family were transferred from the incubators to the nursery hapas (1 m^3 with 2 mm mesh size), stocked at a density of 200 fry per m^3.

At least three nursery hapa replicates for each family were maintained in the same pond to reduce environmental differences between families. When the fingerlings reached an average weight of 5–10 g, about 100–150 individuals per family were randomly sampled and physically identified, using PIT (passive integrated transponder) tags. At tagging, the identification number, body weight (BW), standard length (L), body depth (D), and body width (W) were also recorded. After conditioning for about 3 days in fiberglass tanks without feeding, the tagged fingerlings were pooled together and representatives of each family were sent to communal grow-out testing in freshwater earthen pond and cage. The same producers were repeated in all generations during the course of the selection program for Red tilapia.

2.3 Testing Environments and Data Recording

The tagged fingerlings from each full-sib family were randomly allocated to two testing environments in cages or earthen ponds. Six cages (3.5 m width × 3.5 m length × 3 m depth) were positioned adjacent to each other in an irrigation canal (water depth of about 1.5 m) at Kodiang, Kedah, 22 km away from Jitra. An equal number of sibs per family was randomly assigned to each cage with an initial stocking density of 5 fish per square meter of surface water. The feeding rate was from 3 to 5% of their body weight on a commercial dry pellet feed with 32% protein content twice a day. Contemporaneously with freshwater cage culture, siblings from all families were also tested in earthen ponds of 0.01 ha located at the Aquaculture Extension Center, Department of Fisheries, Jitra, Kedah. The density in each pond was 2–3 fish per square meter of surface water. The same feeding, culturing and management practices were applied as used for the cages. In both environments, water quality parameters (temperature, pH, dissolved oxygen and total ammonia) were monitored once a week.

At the end of the grow-out period, fish were harvested after pond drainage to measure body traits and survival. Individual fish was weighted using a digital scale (nearest to 0.1 g). Standard length was also measured with a ruler. Body width and depth were measured at the mid-side of the fish, where they were largest, by an electronic digital caliper (0.01 mm). Survival was recorded as a binary response (coded as 1 for fish that were present at harvest and 0 for those were absent). Body weight data collected at harvest was then processed to determine estimated breeding values (EBVs) for all individuals in the pedigree. The best animals (highest EBV for body weight) were selected to become parents of the next generations. The same procedures regarding the family production, genetic evaluation and selection process were repeated in subsequent generations during the course of the study from 2009 to 2012. Table 1 shows the pedigree structure (number of sires, dams, and progeny) recorded over four generations from 2009 to 2012 (including the base population) in the two testing environments (i.e., freshwater pond and cage).

Table 1. Number of sire, dam and progeny in pond and cage environments.

Environment	Spawning year	Line	Sire	Dam	Progeny
Pond	2009	Base population	79	79	7,283
	2010	Selection	46	55	2,375
	2011	Selection	43	50	2,080
	2012	Selection	52	69	3,265
	Sub-total		220	253	15,003
Cage	2009	Base population	–	–	–
	2010	Selection	35	38	1,366
	2011	Selection	39	44	1,344
	2012	Selection	45	60	2,203
	Sub-total	Selection	119	142	4,913
Total			224	262	19,916

Total number of common sires and dams in both environments.

2.4 Statistical Analysis

Statistical analyses were performed on 19,916 data records (15,003 fish in ponds combined with 4,913 fish in cages) collected over three generations of selection (2010–2012), tracing back to the base population in 2009. Basic statistics given in Table 2 show that the fish were harvested at a similar weight and size in both pond and cage environments. All traits were evaluated for normality before undertaking further analyses and raw data were transformed when appropriate. Exploratory analyses using a general linear model (GLM) and PROC UNIVARIATE (SAS Institute Inc., 2007) were undertaken. All body traits followed approximate normal distribution, but square root transformation of body weight was also used to improve normality of the residuals. A series of statistical analyses was conducted separately for each environment as follows:

Table 2. Number of observations (n), mean, standard deviation (SD) and coefficient of variation (CV) for traits studied in in pond and cage environments.

Environment	Traits	Unit	*n*	Mean	SD	CV,%
Pond	Weight	g	4,206	245.7	96.7	39.4
	Length	cm	4,206	18.3	3.5	19.4
	Width	cm	4,206	7.7	1.4	18.7
	Depth	cm	4,206	3.4	0.7	20.6
	Survival	%	4,206	55.0	47.8	
Cage	Weight	g	2,122	251.4	138.5	55.1
	Length	cm	2,122	17.6	2.9	16.6
	Width	cm	2,122	7.9	1.7	21.7
	Depth	cm	2,121	3.2	0.7	21.3
	Survival	%	2,122	43.8	49.6	

2.5 Univariate Analysis Using Linear Mixed Model

After all the significant fixed effects were identified; the full model with the family effects was applied to analyse all traits studied. In matrix notation the mixed model can be written as:

$$y=Xb + Za + Wc + e \qquad (1)$$

Where y is the vector of observations for body traits and individual survival, b is the vector of the fixed effects including generation (or spawning seasons, 1–3), sex (female or male) and the two-way interactions between generation and sex. A linear covariate (age from birth to harvest) was fitted within sex by generations subclass as well-stocking weight ($P < 0.05$). Vector a is the random animal additive genetic effects ~ (0, A σ^2_a) where A is the additive genetic (numerator) relationship matrix among the animals, c is the vector of family effects (or maternal effects in addition to additive genetics) ~ (0, I σ^2_c) and e is the vector of residual effects ~ (0, I σ^2_e). The dam component ($\sigma 2_D$) is most likely a combination of maternal and common environmental effects (thus, σ^2_D=σ^2_M+CE, referred to as σ^2_C) caused by the separate rearing of full-sib families until individuals reached a suitable size for physical tagging. The log likelihood ratio test showed that the c^2 effects were significant for all the traits ($P < 0.05$). X, Z, and W are incidence matrices relating observations to fixed effects, additive genetic effect of the individual animal and common effects to full-sib included in the model, respectively.

The mixed model equation for the best linear unbiased estimator (BLUE) of estimable functions of b and the best linear unbiased prediction of a and c are:

$$\begin{bmatrix} \hat{b} \\ \hat{a} \\ \hat{c} \end{bmatrix} = \begin{bmatrix} X'X & X'Z & X'W \\ Z'X & Z'Z + A^{-1}\alpha_1 & Z'W \\ W'X & W'Z & W'W + I\alpha_2 \end{bmatrix}^{-1} \begin{bmatrix} X'y \\ Z'y \\ W'y \end{bmatrix} \quad (2)$$

where $\alpha 1=\sigma^2{}_e/\sigma^2{}_a$ and $\alpha 2=\sigma^2{}_e/\sigma^2{}_c$.

Under the linear mixed model [1], heritabilities for body traits and survival were calculated as $h^2 = \frac{\hat{\sigma}_a^2}{\hat{\sigma}_a^2+\hat{\sigma}_c^2+\hat{\sigma}_e^2}$ and the common full-sib effect as $c^2 = \frac{\hat{\sigma}_c^2}{\hat{\sigma}_a^2+\hat{\sigma}_c^2+\hat{\sigma}_e^2}$ where $\sigma^2{}_a\sigma^{a}{}_2$ is the additive genetic variance, the maternal variance ($\sigma^2{}_c\sigma^{c}{}_2$) and the residual variance ($\sigma^2{}_e\sigma^{e}{}_2$).

2.6 Generalized Linear Mixed Model

In addition to the linear mixed model as described above (Equation 1), survival was analyzed by generalized linear mixed model (GLMM), assuming that the data followed a binomial and underlying normal distributions with a logit link function (McCulloch et al., 2008). Under the logit model, the link function $(\hat{p} = e^x/(1+e^x))$ was used where p is the probability of fish survival recorded at harvest. The fixed effects were the same as defined in Equation (1), but the random effects were sire (s_k) and common full-sib group (c_l). With this logit threshold model, heritability was calculated using the variance of the logit link function, which implies a correction of the residual variance by factor $\pi^2/3$.

$$h^2 = \frac{4\sigma_s^2}{\sigma_s^2 + \sigma_c^2 + \sigma_e^2\frac{\pi^2}{3}} \quad (3)$$

where $\sigma^2{}_s$ is sire variance, $\sigma^2{}_c$ is common full-sib variance and $\sigma^2{}_e=1$.

For binomial observations, estimates of h^2 on the observed (0/1) scale can be transformed to the underlying liability scale (logit) using the formula of Robertson and Lerner (1949) as follows:

$$h_L^2 = \frac{h_O^2 p(1-p)}{z^2} \quad (4)$$

where $h^2{}_O$ is the heritability on the observed (0/1) scale, $h^2{}_L$ is the estimated heritability on the liability (logit or probit) scale, p is a proportion of survival in the data, and z is the height of the ordinate of normal distribution corresponding to a truncation point applied to p proportion of those traits. A similar transformation was also made for c^2.

3. MULTIVARIATE ANALYSIS

3.1 Genetic Correlation for the Same Trait between Pond and Cage Environments

Multivariate linear mixed model was applied to estimate the genetic correlation between homologous traits in pond and cage environments. In these analyses, body measurements in each environment were treated as different traits in order to examine whether there were differences associated with culture environment. The model was as described above, also note that the fixed effect of environment was not included. Genetic correlations between expression of body traits in pond and cage were estimated via numerator genetic relationships matrix **(A)** in the full pedigree. Since the traits were measured on animals in different environments, there is no environmental covariance between them. The phenotypic correlations also do not exist because any individual fish will only express the trait in one environment.

3.2 Genetic and Phenotypic Correlations among Traits in Each Environment

Multivariate linear mixed model analyses were also used to estimate the genetic and phenotypic correlations separately for each environment. The correlations were calculated from a series of bivariate analysis as the covariance divided by the product of the standard deviations of traits $r = \frac{\sigma_{12}}{\sqrt{\sigma_1^2}\sqrt{\sigma_2^2}}$ where σ_{12} was the estimated additive genetic or phenotypic covariance between the two traits, and $\sigma^2{}_1$ and $\sigma^2{}_2$ are the additive genetic or phenotypic variances of traits 1, 2, respectively. The multivariate analyses included final (harvest) body weight to minimize any possible bias in genetic parameter estimates (Kennedy, 1990).

All the statistical analyses were conducted in ASReml, including the full pedigree traced back to the base population.

3.3 Direct Selection Response and Correlated Responses

The direct response to selection for body weight in pond environment was measured as changes in estimated breeding value (EBVs) using the full animal model (Equation 1). This model was also used to estimate the correlated responses in standard length, body depth, body width, and survival in cage and pond environments. The same fixed effects as shown in Equation (1) were used in all analyses. The EBV estimates for body weight and other traits were expressed in an actual unit (g for weight, cm for length, width and depth, and % for survival) and in genetic standard deviation units (estimated breeding values in actual unit/the square root of genetic variance).

4. RESULTS

4.1 Genetic Parameters in Each Culture Environment

4.1.1 Heritabilities

Table 3 presents the variance components, heritability and common environmental effects for harvest weight, total length, body width, body depth, and survival in pond and cage environments. In either pond or cage environment, the estimated heritabilities for body traits were low to moderate, ranging from 0.05 to 0.39. Square root transformation of body weight slightly increased magnitude of the heritability in cage environment, whereas the estimate remained unchanged in pond. This is likely due to the improved distribution of the residuals in the latter (cage) relative to the former (pond) environments. Heritability for survival estimated from linear and nonlinear (generalized) mixed model using logit link function was low in pond (0.02–0.18), whereas the estimates in cage were moderate to high (0.19–0.68) (Table 3). When the observed heritability of linear survival (model 1) was transformed to the underlying liability scale, the h^2 estimates also differed between pond (0.013) and cage (0.119). The common full-sib effects (or maternal and common environmental effects, c^2) for body traits and survival were relatively large and the estimates were higher in pond than in cage (0.23 to 0.51 vs. 0.13–0.31, respectively.

Table 3. Variance components, heritability (h^2) and common full-sib effects (c^2) for body traits in pond and cage environments.

Environment	Traits	V_A	V_C	V_E	h^2	c^2
Pond	Weight$^{0.5}$	2.16	6.14	2.02	0.21 ± 0.10	0.59 ± 0.05
	Weight	1727.3	3920.6	2122.4	0.22 ± 0.10	0.51 ± 0.05
	Length	1.54	2.87	7.86	0.13 ± 0.08	0.23 ± 0.05
	Width	0.026	0.20	0.34	0.05 ± 0.08	0.36 ± 0.05
	Depth	0.37	0.83	0.69	0.19 ± 0.10	0.44 ± 0.06
	Survival^A	0.0051	0.075	0.175	0.02 ± 0.06	0.29 ± 0.04
	Survival ^{B}	0.075	0.61	1	0.18 ± 0.13	0.33 ± 0.04
Cage	Weight$^{0.5}$	1.33	1.43	4.36	0.19 ± 0.12	0.20 ± 0.06
	Weight	563.9	1141.7	5312.4	0.08 ± 0.08	0.16 ± 0.05
	Length	0.29	1.40	2.77	0.07 ± 0.10	0.31 ± 0.06
	Width	0.095	0.037	0.12	0.39 ± 0.13	0.15 ± 0.06
	Depth	0.52	0.19	0.72	0.36 ± 0.13	0.13 ± 0.06
	Survival^A	0.037	0.030	0.129	0.19 ± 0.08	0.15 ± 0.04
	Survival ^{B}	2.40	1.70	1	0.68 ± 0.18	0.11 ± 0.03

^{A}, Observed scale (0/1) and ^{B}, Liability scale using logit function.
V_A, V_C, and V_E, Genetic, common full-sib and residual variances, respectively.
Weight$^{0.5}$, Square root transformation.

4.2 Correlations among Traits Studied in Each Environment

Phenotypic and genetic correlations among traits studied in pond and cage environments are presented in Table 4. In general, the genetic correlations among body traits including live weight, total length, body width, and depth in cage were positive and higher than the phenotypic correlations in pond. The genetic correlations among body traits studied were high and close to unity in both pond and cage environments. The harvest weight showed non-significant genetic correlations with survival due to the high standard errors, although the estimates were positive in pond and negative in cage. Similarly, the phenotypic correlations between survival and body traits were in opposite directions between the two environments (Table 4). This was likely due to the greater mortality loss in cage than in pond and the dead fish in cage were those which had "good" size (big, thick, and round body shape). The genetic correlation between harvest weight and survival may have varied with the environments used or population-specific.

Table 4. Phenotypic (above) and genetic (below) correlations among traits in pond and cage environments.

Environment		Weight	Length	Width	Depth	Survival
Pond	Weight		0.65 ± 0.02	0.63 ± 0.03	0.84 ± 0.01	−0.01 ± .04
	Length	0.99 ± 0.03		0.43 ± 0.03	0.59 ± 0.02	0.18 ± 0.05
	Width	0.90 ± 0.11	0.94 ± 0.12		0.79 ± 0.02	0.32 ± 0.39
	Depth	0.96 ± 0.03	0.99 ± 0.03	0.98 ± 0.03		0.17 ± 0.04
	Survival	0.54 ± 0.51	0.21 ± 0.15	−0.23 ± .32	0.16 ± 0.29	
Cage	Weight		0.80 ± 0.01	0.70 ± 0.02	0.82 ± 0.01	−0.21 ± 0.04
	Length	0.92 ± 0.13		0.73 ± 0.02	0.81 ± 0.01	0.92 ± 0.04
	Width	0.99 ± 0.14	0.95 ± 0.14		0.74 ± 0.02	−0.75 ± 0.09
	Depth	0.97 ± 0.04	0.97 ± 0.05	0.99 ± 0.04		−0.55 ± 0.03
	Survival	−0.73 ± 0.88	0.75 ± 0.56	0.04 ± 1.98	0.15 ± 1.56	

4.3 Between-Environment Genetic Correlations for Body Traits

The genetic correlations between trait expressions in pond and cage environments were high (0.62 to 0.90; Table 5). The between-environment genetic correlation estimates for harvest weight were greater than those obtained for other body traits (r_g = 0.90 vs. 0.62–0.79). The estimated genetic correlations for length and survival between pond and cage environments were high but they were associated with large standard error (r_g ± SE = 0.63 ± 0.41 and 0.70 ± 0.61, respectively).

4.4 Selection Responses in Pond and Cage Environments

Genetic gain measured as EBVs and expressed in an actual unit (g) and in a genetic standard deviation unit (σ_A) is given in Table 6 for pond and cage environments. Note that selection was performed in pond and the indirect response was measured in cage. The direct and correlated responses estimated in this population were all in desired directions. Overall, the direct selection and correlated responses were greater in pond than those in cage, except for correlated changes in body width. The genetic trend increased steadily with generations of selection for all traits studied in both environments (pond and cage; Table 6). The magnitude of the selection response obtained for body weight in square root transformation scale was in good agreement with those estimated on original scale of measurement (0.390 vs. 0.310 σ_A). Selection for increased harvest body weight resulted in positive correlated genetic changes in survival rate with the gain of 0.567 and 1.561 genetic standard deviation units in cage and pond after three generations, respectively.

Table 5. Genetic correlations between trait expressions in pond and cage environments.

Traits	Genetic correlation	S.E
Weight$^{0.5}$	0.85	0.14
Weight	0.90	0.10
Length	0.63	0.41
Width	0.79	0.18
Depth	0.62	0.18
Survival	0.70	0.61

Table 6. Genetic responses measured as estimated breeding values in actual unit (g) and in genetic standard deviation unit (σ_A) in pond and cage environments.

Env.	Years	Genetic gain in harvest weight				Correlated responses in body traits and survival							
		Weight$^{0.5}$		Weight		Length		Depth		Width		Survival	
		Actual	σ_A	Actual	σ_A	Actual	σ_A	Actual	σ_A	Actual	σ_A	Actual	σ_A
Pond	2010	0.003	0.002	0.215	0.005	0.017	0.014	0.002	0.004	0.000	−0.003	0.022	0.308
	2011	0.568	0.387	17.784	0.428	0.270	0.218	0.157	0.257	0.017	0.104	0.071	0.996
	2012	1.188	0.809	34.686	0.835	0.479	0.386	0.337	0.554	0.035	0.216	0.111	1.561
Cage	2010	−0.036	−0.031	−0.871	−0.037	−0.011	−0.020	−0.024	−0.033	−0.007	−0.023	0.027	0.138
	2011	0.220	0.191	4.746	0.200	0.049	0.092	0.122	0.200	0.044	0.273	0.068	0.351
	2012	0.450	0.390	7.365	0.310	0.102	0.190	0.302	0.497	0.133	0.825	0.109	0.567

5. DISCUSSION

In Red tilapia, a prevailing culture system in commercial production is freshwater cage. In contrast, selection programs in the nucleus are often conducted in pond. The difference between the production and selection environments may have impacts on genetic gain and population parameter estimates. Our findings showed that the correlated responses in freshwater cages and the direct gain obtained in earthen ponds differed between pond (0.011–1.156 σ_A) and cage (−0.033 to 0.567 σ_A) environments. The gain in body weight was greater in pond than in freshwater cage across generations of selection from 2010 to 2012. The reduced selection response in cage relative to pond was likely due to the G × E interaction, which consists of two forms: scaling (heterogeneity of the variances) or re-ranking effects (low genetic correlation estimate for the same trait between the two environments). The scaling effect on genetic gain was generally not important (Ponzoni et al., 2008; Sae-Lim et al., 2016). Hence, the differences in the selection response in the present study may have been a result of the re-ranking effect. A measure of the re-ranking G × E effect is the genetic correlation for harvest weight between pond and cage environments; the estimate, albeit high, was not close to one. This is also because selection was practiced in pond not in cage or due to different environmental factors between the two culture systems. The smaller number of fish were tested in the latter (i.e., cage) than the former environment (pond) and the mortality rate as well fish loss due to diseases or environmental factors were greater in cage than in pond. It was likely due to the poor water quality in the irrigation canal or due to the limited space in cages. However, other factors may have been involved with the high mortality in cage, e.g., social competition. Overall, the selection response obtained for harvest weight in the present study was comparable with those reported in a different population of Red tilapia (Thodesen et al., 2013), Nile tilapia (Ninh et al., 2014; de Oliveira et al., 2016), common carp (Dong et al., 2015), marine finfish (Knibb et al., 2016) and/or giant freshwater prawn (Hung et al., 2013).

In a similar trend to body weight, correlated responses in other body traits (standard length, width, and depth) and survival were also greater in pond than in cage culture. The correlated genetic changes achieved for body traits were all in desired directions. The changes in body traits suggest that the Red tilapia line grew faster in the pond compared with the cage, whereas the fish reared in the cage was thicker than those in pond. Trọng et al. (2013) also observed that the Nile tilapia grown in cage develop a thicker body shape at the same length compared to fish grown in ponds. Further, the small positive correlated changes in survival were achieved in both environments, although its genetic associations with body weight were not significant, due to the high standard errors. This also reflects that correlation is not causation (cause and effect), especially for traits that are largely influenced by environmental factors as observed for survival in this study.

The discrepancy in the selection responses between the production and nucleus in this population of Red tilapia is consistent with the genetic correlation estimates between homologous traits recorded in freshwater pond and cage culture environments, which are different from one, especially for length, width and depth. Our estimates of the between-environment genetic correlations were, however, in accordance with those published for a range of different culture environments for harvest weight in Nile tilapia (Eknath et al., 2007; Bentsen et al., 2012; Khaw et al., 2012; Thodesen et al., 2013; Trọng et al., 2013; Thoa et al., 2016). In other farmed aquaculture species, the genetic correlation estimates reported for harvest weight between contrasting environments during grow-out were also very high and close to one, such as in Rainbow trout (Sylvén et al., 1991; Fishback et al., 2002; Kause et al., 2003), Atlantic cod (*Gadus morhua*) (Kolstad et al., 2006), Whitefish (*Coregonus lavaretus*; Quinton et al., 2007), Barramundi (Domingos et al., 2013), European sea bass (Dupont-Nivet et al., 2010; Le Boucher et al., 2013), Pacific white shrimp (*P. vannamei*) (Gitterle et al., 2005) and Oysters (*Crassostrea gigas*) (Swan et al., 2007). On the contrary, a low genetic correlation (r_g = 0.45 ± 0.09) between freshwater and salt water environments has been reported in Nile tilapia by Luan et al. (2008). A similar finding was also reported by Mas-Muñoz et al. (2013) for harvest weight (r_g = 0.56) and specific growth rate (r_g = 0.27) between intensive recirculation aquaculture system (RAS) and semi-natural environment (POND) in common sole (*Solea solea*), Arctic charr (*Salvelinus alpinus*) by Chiasson et al. (2013) and rainbow trout (Sae-Lim et al., 2013). These authors suggested that there was a strong genotype by environment interaction between these diverse environments used in their studies.

In addition to the selection responses and the G × E effects, the estimates of heritability for body traits and survival differed between pond and cage environments. Square root transformation used in this study reduced the magnitude of the difference in the heritability estimates for body weight between the two environments. The slight difference in heritability between grow-out testing environments was also reported in Nile tilapia or other fish species, such as Artic char (Chiasson et al., 2013) and rainbow trout (Sae-Lim et al., 2013). However, a meta-analysis of 26 studies in various aquaculture species indicated that the effect size of the difference in heritability for body weight between contrasting environments was not significant, $P > 0.05$ (Nguyen, unpublished results). Further, when analyses were conducted separately for each environment, the genetic correlation estimates among body traits were almost identical (close to one). The genetic correlations between harvest weight and survival, although not significant, had opposite signs between ponds and cages. Phenotypically, weight, width and depth had negative correlations with survival, likely because dead fish in cages are those which had "good" (thick and round) body shape. Overall our estimates of heritability and genetic correlations for body traits and survival are in line with the published results in Nile tilapia (Thoa et al., 2016) or rainbow trout (Sae-Lim et al., 2015).

On the other hand, the common full-sib effects (c^2) obtained for body traits in the present study were greater than those reported for other fishes (Nguyen et al., 2007) or crustacean species (Hung et al., 2014). This was mainly due to the separate rearing of each family in hapas for about 2 months before the fish were physically tagged to enable the communal grow-out of all families in ponds and cages. The c^2 effects were also larger in ponds than cages, which can be explained because of the management and environmental differences between the two culture systems.

Collectively, our results suggest that even though when the genetic correlations between trait expressions in environments are high, the amount of selection response would provide practical information to determine whether the G × E interaction is of biological significance or not. This is because in the context of aquaculture, the G × E interaction can occur, due to the large variation between the selection and production environments in terms of culture environments (Bentsen et al., 2012), scales of production (e.g., small vs. large commercial operations; Trọng et al., 2013), agro-climatic conditions (Sae-Lim et al., 2013, 2017), as well various farming and husbandry management practices (Eknath et al., 2007; Hamzah et al., 2014; Ninh et al., 2014).

In summary, a combination of the genetic parameter estimates and the realized selection responses achieved in the present study pointed out that there may be a need to conduct separate genetic evaluation of selection response for pond and cage culture in Red tilapia. However, multiple genetic improvement programs for different environments may not be fully justified when the predicted losses in production are smaller than the cost of running a new program (Nguyen, 2016). In farmed aquaculture species, a single breeding program is virtually always implemented for a range of environments especially in developing countries where resources and experience in managing breeding programs are limited. This is also due to a biosecurity issue, i.e., preventing disease transmission from commercial production populations into the nucleus. One strategy we practiced to minimize G × E effects in the present breeding program was that relatives of selection candidates were performance tested in both pond and cage, and a combined genetic evaluation of the data recorded in both environments alleviated the G × E effects, thus lessening the loss in genetic gain. Optimization of the selective breeding programs is also discussed by Sae-Lim et al. (2016).

6. CONCLUSION

The selection responses captured for body traits in the production environment (cage) were lower than those achieved in the selection environment (i.e., earthen ponds). The genetic correlation between pond and cage environments for body traits including harvest weight (the selection criterion), albeit high, was not close to one. There were also differences in the heritability and common full-sib effects between the two environments. However, running separate breeding programs for each environment is hardly justified, due to our limited resources. In the future genetic improvement for this population of Red tilapia, selection index approach should be applied by treating the expression in each environment as a separate trait, and weighing the traits in proportion to the relative importance of pond and cage culture. Accounting for the genotype by environment effects in the selective breeding program of Red tilapia will contribute to the sustained development of the aquaculture sector in major tilapia producing countries world-wide.

ETHICS STATEMENT

The study and the protocol were reviewed and approved by the animal ethics committee of the Department of Fisheries, Malaysia (the approval number ANE070312). They were standard commercial procedures.

ACKNOWLEDGMENTS

This study was funded by Department of Fisheries, Malaysia. We would like to thank Hoong Yip Yee and Mohd A. Aziz for their kindly assistance in the management of the animals. The Aquaculture Extension Center of the Malaysian Department of Fisheries in Jitra, Kedah provided facilities and additional manpower for the study.

REFERENCES

1. Bentsen, H. B., Gjerde, B., Nguyen, N. H., Rye, M., Ponzoni, R. W., Palada de Vera, M. S., et al. (2012). Genetic improvement of farmed tilapias: genetic parameters for body weight at harvest in Nile tilapia (*Oreochromis niloticus*) during five generations of testing in multiple environments. *Aquaculture* 338–341, 56–65. doi: 10.1016/j.aquaculture.2012.01.027

2. Chiasson, M. A., Quinton, C. D., Danzmann, R. G., and Ferguson, M. M. (2013). Comparative analysis of genetic parameters and quantitative trait loci for growth traits in Fraser strain Arctic charr (*Salvelinus alpinus*) reared in freshwater and brackish water environments. *J. Anim. Sci.* 91, 2047–2056. doi: 10.2527/jas.2012-5656
3. de Oliveira, C. A. L., Ribeiro, R. P., Yoshida, G. M., Kunita, N. M., Rizzato, G. S., de Oliveira, S. N., et al. (2016). Correlated changes in body shape after five generations of selection to improve growth rate in a breeding program for Nile tilapia *Oreochromis niloticus* in Brazil. *J. Appl. Genet.* 57, 487–493. doi: 10.1007/s13353-016-0338-5
4. Domingos, J. A., Smith-Keune, C., Robinson, N., Loughnan, S., Harrison, P., and Jerry, D. R. (2013). Heritability of harvest growth traits and genotype-environment interactions in barramundi, *Lates calcarifer* (Bloch). *Aquaculture* 402–403, 66–75. doi: 10.1016/j.aquaculture.2013.03.029
5. Dong, Z., Nguyen, N. H., and Zhu, W. (2015). Genetic evaluation of a selective breeding program for common carp *Cyprinus carpio* conducted from 2004 to 2014. *BMC Genet.* 16:94. doi: 10.1186/s12863-015-0256-2
6. Dupont-Nivet, M., Karahan-Nomm, B., Vergnet, A., Merdy, O., Haffray, P., Chavanne, H., et al. (2010). Genotype by environment interactions for growth in European seabass (*Dicentrarchus labrax*) are large when growth rate rather than weight is considered. *Aquaculture* 306, 365–368. doi: 10.1016/j.aquaculture.2010.05.025
7. Eknath, A. E., Bentsen, H. B., Ponzoni, R. W., Rye, M., Nguyen, N. H., Thodesen, J., et al. (2007). Genetic improvement of farmed tilapias: composition and genetic parameters of a synthetic base population of *Oreochromis niloticus* for selective breeding. *Aquaculture* 273, 1–14. doi: 10.1016/j.aquaculture.2007.09.015
8. Falconer, D. S. (1952). The problem of environment and selection. *Am. Nat.* 86, 293–298. doi: 10.1086/281736
9. Falconer, D. S., and Mackay, T. F. (1996). *Introduction to Quantitative Genetics.* Harlow: Longmans Green.
10. Fishback, A. G., Danzmann, R. G., Ferguson, M. M., and Gibson, J. P. (2002). Estimates of genetic parameters and genotype by environment interactions for growth traits of rainbow trout (*Oncorhynchus mykiss*) as inferred using molecular pedigrees. *Aquaculture* 206, 137–150. doi: 10.1016/S0044-8486(01)00707-4
11. Gitterle, T., Rye, M., Salte, R., Cock, J., Johansen, H., Lozano, C., et al. (2005). Genetic (co)variation in harvest body weight and survival in *Penaeus (Litopenaeus) vannamei* under standard commercial conditions. *Aquaculture* 243, 83–92. doi: 10.1016/j.aquaculture.2004.10.015
12. Hamzah, A., Nguyen, N. H., Ponzoni, R. W., Kamaruzzaman, N., and Suhba, B. (2008). "Performance and survival of three red tilapia strains (Oreochromis spp) in pond environment in Kedah state, Malaysia," *in 8th International Symposium on Tilapia in Aquaculture* (Cairo), 199–211.
13. Hamzah, A., Ponzoni, R. W., Nguyen, N. H., Khaw, H. L., Yee, H. Y., Nor, M., et al. (2014). Performance of the Genetically Improved Farmed Tilapia (GIFT) strain over ten generations of selection in Malaysia. *Pertanika J. Trop. Agric. Sci.* 37, 411–429.
14. Hung, D., Nguyen, N. H., Hurwood, D. A., and Mather, P. B. (2014). Quantitative genetic parameters for body traits at different ages in a cultured stock of giant freshwater prawn (*Macrobrachium rosenbergii*) selected for fast growth. *Mar. Freshw. Res.* 65, 198–205. doi: 10.1071/MF13111
15. Hung, D., Vu, N. T., Nguyen, N. H., Ponzoni, R. W., Hurwood, D. A., and Mather, P. B. (2013). Genetic response to combined family selection for improved mean harvest weight in giant freshwater prawn (*Macrobrachium rosenbergii*) in Vietnam. *Aquaculture* 412, 70–73. doi: 10.1016/j.aquaculture.2013.07.015
16. Kause, A., Ritola, O., Paananen, T., Mäntysaari, E., and Eskelinen, U. (2003). Selection against early maturity in large rainbow trout *Oncorhynchus mykiss*: the quantitative genetics of sexual dimorphism and genotype-by-environment interactions. *Aquaculture* 228, 53–68. doi: 10.1016/S0044-8486(03)00244-8

17. Kennedy, B. W. (1990). "Use of mixed model methodology in analysis of designed experiments," in *Advances in Statistical Methods for Genetic Improvement of Livestock*, eds D. Gianola and K. Hammond (Berlin: Springer-Verlag), 77–97.
18. Khaw, H. L., Ponzoni, R. W., Hamzah, A., Abu-Bakar, K. R., and Bijma, P. (2012). Genotype by production environment interaction in the GIFT strain of Nile tilapia (*Oreochromis niloticus*). *Aquaculture* 326–329, 53–60. doi: 10.1016/j.aquaculture.2011.11.016
19. Knibb, W., Miller, A., Quinn, J., D'Antignana, T., and Nguyen, N. H. (2016). Comparison of lines shows selection response in kingfish (*Seriola lalandi*). *Aquaculture* 452, 318–325. doi: 10.1016/j.aquaculture.2015.11.015
20. Kolstad, K., Thorland, I., Refstie, T., and Gjerde, B. (2006). Genetic variation and genotype by location interaction in body weight, spinal deformity and sexual maturity in Atlantic cod (*Gadus morhua*) reared at different locations off Norway. *Aquaculture* 259, 66–73. doi: 10.1016/j.aquaculture.2005.12.022
21. Le Boucher, R., Vandeputte, M., Dupont-Nivet, M., Quillet, E., Ruelle, F., Vergnet, A., et al. (2013). Genotype by diet interactions in European sea bass (*Dicentrarchus labrax* L.): nutritional challenge with totally plant-based diets. *J. Anim. Sci.* 91, 44–56. doi: 10.2527/jas.2012-5311
22. Luan, T. D., Olesen, I., Ødegård, J., Kolstad, K., and Dan, N. C. (2008). "Genotype by environment interaction for harvest body weight and survival of Nile tilapia (Oreochromis niloticus) in brackish and Fresh water ponds," in *The proceedings of the 8th International Symposium on Tilapia in Aquaculture* (Cairo).
23. Mas-Muñoz, J., Blonk, R., Schrama, J. W., van Arendonk, J., and Komen, H. (2013). Genotype by environment interaction for growth of sole (*Solea solea*) reared in an intensive aquaculture system and in a semi-natural environment. *Aquaculture* 410–411, 230–235. doi: 10.1016/j.aquaculture.2013.06.012
24. McCulloch, C. E., Searle, S. R., and Neuhaus, J. M. (2008). *Generalized, Linear, and Mixed Models.* Hoboken, NJ: Wiley Online Library.
25. Mulder, H., Veerkamp, R., Ducro, B., Van Arendonk, J., and Bijma, P. (2006). Optimization of dairy cattle breeding programs for different environments with genotype by environment interaction. *J. Dairy Sci.* 89, 1740–1752. doi: 10.3168/jds.S0022-0302(06)72242-1
26. Nguyen, N. H. (2016). Genetic improvement for important farmed aquaculture species with a reference to carp, tilapia and prawns in Asia: achievements, lessons and challenges. *Fish Fish.* 17, 483–506. doi: 10.1111/faf.12122
27. Nguyen, N. H., Khaw, H. L., Ponzoni, R. W., Hamzah, A., and Kamaruzzaman, N. (2007). Can sexual dimorphism and body shape be altered in Nile tilapia (*Oreochromis niloticus*) by genetic means? *Aquaculture* 272, S38–S46. doi: 10.1016/j.aquaculture.2007.08.013
28. Nguyen, N., and Ponzoni, R. (2006). Perspectives from agriculture: advances in livestock breeding-implications for aquaculture genetics. *NAGA* 29, 39–45.
29. Ninh, N. H., Thoa, N. P., Knibb, W., and Nguyen, N. H. (2014). Selection for enhanced growth performance of Nile tilapia (*Oreochromis niloticus*) in brackish water (15–20ppt) in Vietnam. *Aquaculture* 428, 1–6. doi: 10.1016/j.aquaculture.2014.02.024
30. Pongthana, N., Nguyen, N. H., and Ponzoni, R. W. (2010). Comparative performance of four red tilapia strains and their crosses in fresh-and saline water environments. *Aquaculture* 308, S109–S114. doi: 10.1016/j.aquaculture.2010.07.033
31. Ponzoni, R. W., Nguyen, N. H., Khaw, H. L., and Ninh, N. H. (2008). Accounting for genotype by environment interaction in economic appraisal of genetic improvement programs in common carp *Cyprinus carpio*. *Aquaculture* 285, 47–55. doi: 10.1016/j.aquaculture.2008.08.012
32. Quinton, C., Kause, A., Ruohonen, K., and Koskela, J. (2007). Genetic relationships of body composition and feed utilization traits in European whitefish (*Coregonus lavaretus* L.) and implications for selective breeding in fishmeal-and soybean meal-based diet environments. *J. Anim. Sci.* 85, 3198–3208. doi: 10.2527/jas.2006-792
33. PubMed Abstract | CrossRef Full Text | Google Scholar
34. Robertson, A., and Lerner, I. M. (1949). The heritability of all-or-none traits; viability of poultry. *Genetics* 34, 395–411.

35. SAS (2007). *SAS/STAT(r) 9.1. User's Guide.* Cary, NC: SAS.
36. Sae-Lim, P., Gjerde, B., Nielsen, H. M., Mulder, H., and Kause, A. (2016). A review of genotype-by-environment interaction and micro-environmental sensitivity in aquaculture species. *Rev. Aquaculture* 8, 369–393. doi: 10.1111/raq.12098
37. Sae-Lim, P., Kause, A., Janhunen, M., Vehviläinen, H., Koskinen, H., Gjerde, B., et al. (2015). Genetic (co) variance of rainbow trout (*Oncorhynchus mykiss*) body weight and its uniformity across production environments. *Genet. Sel. Evol.* 47, 46. doi: 10.1186/s12711-015-0122-8
38. Sae-Lim, P., Kause, A., Mulder, H., Martin, K., Barfoot, A., Parsons, J., et al. (2013). Genotype-by-environment interaction of growth traits in rainbow trout (*Oncorhynchus mykiss*): a continental scale study. *J. Anim. Sci.* 91, 5572–5581. doi: 10.2527/jas.2012-5949
39. Sae-Lim, P., Kause, A., Mulder, H., and Olesen, I. (2017). Breeding and genetics symposium: climate change and selective breeding in aquaculture. *J. Anim. Sci.* 95, 1801–1812. doi: 10.2527/jas.2016.1066
40. Swan, A. A., Thompson, P. A., and Ward, R. D. (2007). Genotype × environment interactions for weight in Pacific oysters (*Crassostrea gigas*) on five Australian farms. *Aquaculture* 265, 91–101. doi: 10.1016/j.aquaculture.2007.01.036
41. Sylvén, S., Rye, M., and Simianer, H. (1991). Interaction of genotype with production system for slaughter weight in rainbow trout (*Oncorhynchus mykiss*). *Livestock Prod. Sci.* 28, 253–263. doi: 10.1016/0301-6226(91)90146-H
42. Thoa, N. P., Ninh, N. H., Knibb, W., and Nguyen, N. H. (2016). Does selection in a challenging environment produce Nile tilapia genotypes that can thrive in a range of production systems? *Sci. Rep.* 6:21486. doi: 10.1038/srep21486
43. Thodesen, J., Rye, M., Wang, Y.-X., Li, S.-J., Bentsen, H. B., and Gjedrem, T. (2013). Genetic improvement of tilapias in China: Genetic parameters and selection responses in growth, pond survival and cold-water tolerance of blue tilapia (*Oreochromis aureus*) after four generations of multi-trait selection. *Aquaculture* 396–399, 32–42. doi: 10.1016/j.aquaculture.2013.02.010
44. Trọng, T. Q., Mulder, H. A., van Arendonk, J. A., and Komen, H. (2013). Heritability and
45. genotype by environment interaction estimates for harvest weight, growth rate, and shape of Nile tilapia (*Oreochromis niloticus*) grown in river cage and VAC in Vietnam. *Aquaculture* 384, 119– 127.doi: 10.1016/j.aquaculture.2012.12.022

CHAPTER 13

Artificial selection with traditional or genomic relationships: consequences in coancestry and genetic diversity

Silvia Teresa Rodríguez-Ramilo[1], Luis Alberto García-Cortés[1] and María Ángeles Rodríguez de Cara[2*]

[1]*Departamento de Mejora Genetica Animal, Instituto Nacional de Investigacion y Tecnologia Agraria y Alimentaria, Madrid, Spain*

[2]*Laboratoire d'Eco-anthropologie et Ethnobiologie, Museum National d'Histoire Naturelle, Paris, France*

Estimated breeding values (EBVs) are traditionally obtained from pedigree information. However, EBVs from high-density genotypes can have higher accuracy than EBVs from pedigree information. At the same time, it has been shown that EBVs from genomic data lead to lower increases in inbreeding compared with traditional selection based on genealogies. Here we evaluate the performance with BLUP selection based on genealogical coancestry with three different genome-based coancestry estimates: (1) an estimate based on shared segments of homozygosity, (2) an approach based on SNP-by-SNP count corrected by allelic frequencies, and (3) the identity by state methodology. We evaluate the effect of different population sizes, different number of genomic markers, and several heritability values for a quantitative trait. The performance of the different measures of coancestry in BLUP is evaluated in the true breeding values after truncation selection and also in terms of coancestry and diversity maintained. Accordingly, cross-performances were also carried out, that is, how prediction based on genealogical records impacts the three other measures of coancestry and inbreeding, and viceversa. Our results show that the genetic gains are very similar for all four coancestries, but the genomic-based methods are superior to using genealogical coancestries in terms of maintaining diversity measured as observed heterozygosity. Furthermore, the measure of coancestry based on shared segments of the genome seems to provide slightly better results on some scenarios, and the increase in inbreeding and loss in diversity is only slightly larger than the other genomic selection methods in those scenarios. Our results shed light on genomic selection vs. traditional genealogical-based BLUP and make the case to manage the population variability using genomic information to preserve the future success of selection programmes.

Keywords: genomicselection, coancestry, inbreeding, breedingvalue, geneticdiversity

1. INTRODUCTION

Best linear unbiased prediction (BLUP) is possibly the most common selection method in animal and plant breeding, where it is used to calculate estimated breeding values (EBVs). BLUP evaluations maximize the genetic gain given the data by increasing the accuracy of the predictions (Henderson, 1984). This method relies on both the additive relationship matrix between the individuals in the population, which are traditionally obtained from pedigree records, and on phenotypic records of the candidates to selection. Such is the power of BLUP that it is actually not only used in breeding programmes, but also in evolutionary ecology to estimate the strength of selection and evolutionary change (see Hadfield et al., 2010 for a review) and more recently in human genetics for the prediction of complex traits (Makowsky et al., 2011).

With the advent of high-throughput genotyping techniques and the development of chips containing thousands of single nucleotide polymorphisms (SNPs) at a reasonable cost, the implementation of genome-wide evaluations (Meuwissen et al., 2001; Goddard and Hayes, 2007) is routinely used in many breeding programs, and conventional BLUP selection based on pedigrees is now migrating to genomic selection.

Genome-based EBV (estimated breeding values based on high-density marker data across the genome) have generally yielded a higher accuracy than pedigree-based EBV (Meuwissen et al., 2001; Goddard, 2009; Hayes et al., 2009; Sonesson et al., 2012; Rodriguez-Ramilo et al., 2014). This is because genetic markers provide a more accurate relationship matrices than pedigree data (Goddard, 2009), which accounts for the expected genetic relationships. For example, while the genealogical relationship between two full-sibs is 0.5, using molecular markers like high-density SNP chips, a more accurate value can be obtained, thus showing that the true relationship deviates from 0.5 (Visscher et al., 2006) and varies among pairs of sibs, depending on the segregation of the parental chromosomes (Garcia-Cortes et al., 2013).

Genomic selection can therefore lead to high levels of accuracy at an early age and generation intervals can be shortened leading to faster genetic gains within a specific breeding program. Furthermore, genomic selection not only has increased the accuracy in the breeding values, but also the increase in inbreeding per generation is lower than that obtained with conventional pedigree-based BLUP selection (Daetwyler et al., 2007; Sonesson et al., 2012). However, both traditional and genomic selection increase the levels of both inbreeding and coancestry, thus decreasing the pool of genetic diversity. This has wide-ranging consequences, as it is clear that such variation is needed for selection but also to avoid leading the population into extinction (Frankham et al., 2002). A crucial issue thus is a thorough understanding of the measures of coancestry between individuals and how they are affected by the relationship matrix used in the selection process, i.e., pedigree or genomic-based coancestries.

Traditionally, genealogical measures from pedigree records were used to calculate coancestry. As molecular markers became commonly used, estimates of genealogical coancestry from these markers were developed (Weir et al., 2006). It is only with the high-density panels that replacing genealogical coancestry with marker-based coancestry has become accepted as leading to more accurate predictions (Meuwissen et al., 2001; Meuwissen, 2007; Solberg et al., 2008) and to maintain more diversity in conservation programmes (de Cara et al., 2011). However, while the increase in accuracy in the EBVs using different marker types and densities is well-understood (Solberg et al., 2008; Jannink, 2010), the effect of different measures of coancestries in genomic and traditional selection has not received as much attention (Sonesson et al., 2012; Bjelland et al., 2013; Luan et al., 2014). For instance, genomic selection to estimate marker effects and predict the breeding values from them exploits the linkage disequilibrium between the markers in the panel and the causal mutations or QTL (Habier et al., 2007; de los Campos et al., 2010). When selection is performed via BLUP based on genomic relationships, the genetic gain is superior based on these relationships as compared to BLUP based on pedigree based relationships (Villanueva et al., 2005; Meuwissen, 2007) when the number of candidates for selection is large (Bastiaansen et al., 2012; Sonesson et al., 2012). Furthermore, selection based on genomic relationships also leads to lower increases in inbreeding and maintains more diversity (Sonesson et al., 2012; Liu et al., 2014).

In this study we analyse the effect of BLUP selection with four measures of coancestry on the genetic gain and on the increase in coancestry and inbreeding. For this purpose, we carry out simulations with three different genome-based relationship matrices and the matrix of genealogical relationships when inferring breeding values using BLUP. The three genomic measures of coancestry were: (1) based on shared segments of homozygosity (Fisher, 1954; Stam, 1980; Gusev et al., 2009), (2) using identity by state, that is, marker-by-marker similarity (Eding and Meuwissen, 2001; Caballero and Toro, 2002) and (3) based on a marker-by-marker count corrected by allelic frequencies (VanRaden, 2008). We measured the performance of selection with BLUP based on these four coancestries by analysing the genetic gain as measured with the true breeding values (TBVs).

2. MATERIALS AND METHODS

2.1. Base Population

A base population was simulated with an effective size of 1000 individuals (half males, half females) during 10,000 generations until an equilibrium in the average genome-wide heterozygosity was reached. Every individual had a genome of 10 chromosomes of 1M with 10,100 biallelic positions each. Initially, every position in the genome carried alleles 0 or 1 at random, so that the average initial heterozygosity was 0.5. The mutation rate per position and generation was 2.5×10^{-3}. Every generation during the creation of the base population we firstly performed mutations in every individual, then chose a male and a female at random with replacement and produced an offspring with recombination. The number of recombinations per chromosome were sampled from a Poisson distribution and the recombination positions were drawn from a uniform distribution. The base populations were generated with a fortran 90 code available upon request.

2.2. Selection

We performed 100 replicates of each scenario here studied by selecting 1000 polymorphic positions from this base population to be later used as selective loci (also known as QTLs in the literature). We sampled these selective loci from positions with $0.05 < p_j < 0.95$, where p_j is the allelic frequency of allele 1 at locus j. Note thus that the 100 replicates are all created from one single base population by selecting different selected loci and different individuals in each replicate.

Founder individuals for each replicate were chosen at random from the base population without replacement, by drawing an equal number N of founder sires and dams from the base population to create generation 0. We then performed 6 generations of random mating to record the genealogy.

From generation 7 onwards we performed truncation selection for 15 generations (up to generation 21), by selecting the best 50% of the sires and 50% of the dams according to each individual's expected breeding value. These sires and dams were mated at random to produce N sires and N dams for the next generation.

The default parameters used in our simulations are N = 50, a marker density of 10,100 markers per chromosome and a trait with heritability of h^2 = 0.25. To have a thorough understanding of the dependence of the results on population size, heritability and marker density, we also studied the following scenarios: we evaluated population sizes N = 10 and N = 30, two other heritabilities of the quantitative trait (h^2 = 0.10 and 0.50) and two other lower marker densities (2525 and 5050 markers per chromosome). Table 1 shows a summary of the simulated scenarios.

Table 1. Parameters simulated for the different scenarios.

	N			h^2			SNPs		
	10	30	50	0.10	0.25	0.50	2525	5050	10,100
N	10	30	50	50	50	50	50	50	50
h^2	0.25	0.25	0.25	0.10	0.25	0.50	0.25	0.25	0.25
SNPs	10,100	10,100	10,100	10,100	10,100	10,100	2525	5050	10,100

2.3. Calculation of Phenotypic Values and True and Estimated Breeding Values

We calculated the TBV of individual i as

$$TBV_i = \sum_{j=1}^{n_S} a_j \left(x_{ij} - 1\right) \tag{1}$$

where x_{ij} is the number of copies of the allele 1 that individual i has at the j-th selective locus, a_j is the effect of the allele 1 at position j and n_S is the number of selective loci. The values of the effects a were drawn from a Gaussian distribution with mean zero and variance one. The phenotypic values (y_i) of individuals were simulated as

$$y_i = \mu + TBV_i + e_i, \tag{2}$$

where e_i is an error term for individual i, which was normally distributed with mean zero and variance $\sigma^2{}_e$. The phenotypic average μ was set arbitrarily to be equal to 100, although this value does not affect the EBV. The variance $\sigma^2{}_a$ was calculated as the empirical variance of the TBVs in the base population and $\sigma^2{}_e$ was adjusted so that the heritability was the desired h^2. We had the phenotypic values for all individuals in the population.

EBV were calculated by solving Henderson's mixed model equations (Henderson, 1984) as follows:

$$\begin{bmatrix} \mathbf{X'X} & \mathbf{X'Z} \\ \mathbf{Z'X} & \mathbf{Z'Z} + \frac{\sigma_e^2}{\sigma_a^2}\mathbf{A}^{-1} \end{bmatrix} \begin{bmatrix} \hat{\mu} \\ \hat{EBV} \end{bmatrix} = \begin{bmatrix} \mathbf{X'y} \\ \mathbf{Z'y} \end{bmatrix} \tag{3}$$

where **X** and **Z** are the incidence matrices for the fixed and random effects, respectively and **A** is the relationship matrix. We assumed the variance components to be known. Equation (3) provides the pedigree-based breeding values, while genomic based breeding values can be obtained by replacing **A** and $\sigma^2{}_a$ in Equation (3) by the following genomic relationships and variances.

2.3.1. Coancestry Estimates

The four following genetic relationship matrices, here defined as twice the coancestry coefficient, were used:

1) *Additive relationship matrix* (**A**): This was calculated using the coancestry coefficient between individuals i and k, $f_A(i, k)$ following (Malecot, 1948) as the probability that two alleles taken at random, one for each individual, are identical by descent (IBD).

2) *Marker-by-marker relationship matrix* (**G**): In this case, the coancestry coefficient between individuals i and k, $f_G(i, k)$, is the probability that two alleles at a given locus taken at random from each individual are equal (identical by state, IBS). In this study, $f_G(i, k)$ was calculated as $f_G(i,k) = \frac{1}{4M}\sum_{n=1}^{M}\sum_{l_i=1}^{2}\sum_{m_k=1}^{2} I_n(l_i, m_k)$ where M is the number of markers and $I_n(l_i, m_k)$ is the identity of gamete l from individual i with gamete m from individual k at marker n and takes a value of 1 if both alleles are identical and 0 otherwise.

3) *ROH-based relationship matrix* (**R**): Following the study by de Cara et al. (2013), the coancestry coefficient based on shared segments of the genome between individuals i and k was calculated as

$f_R(i,k) = \frac{1}{4L}\sum_j \sum_{a_i=1}^{2} \sum_{b_k=1}^{2} L_j(a_i, b_k)$, where $L_j(a_i, b_k)$ is the length of the j-th shared segment measured over the gametes a_i and b_k of individuals i and k and L is the length of the genome. For a region to be considered a shared segment, we used a minimum length of 100 shared contiguous markers. The idea behind this segment-based relationship is that a segment shared between parents is a potential run of homozygosity (ROH) in the offspring.

4) *Marker-by-marker corrected by allele frequencies relationship matrix* (**V**): Following VanRaden (2008) a measure of coancestry $f_V(i, k)$ between individuals i and k can be calculated as

$$f_V(i,k) = \frac{1}{M}\sum_{n=1}^{M} \frac{(g_{in} - p_n)(g_{kn} - p_n)}{p_n(1 - p_n)}. \quad (4)$$

where g_{in} refers to the gene frequency value genotypes 00, 01, and 11, coded as 1, 0.5, and 0, respectively, of individual i at locus n. Gene frequency is half the number of copies of the reference allele 1 and p_n is set at 0.5 (Forni et al., 2011).

Every generation we estimated the additive variance of the base population using restricted maximum likelihood (REML). We performed REML by using a Monte Carlo expextation-maximization (EM) algorithm (Guo and Thompson, 1991) to avoid the repeated matrix inversion required by exact algorithms (Meyer, 1991). Additive variances were estimated after six thousand iterations and discarding the first 1000. As for the base population, the fortran 90 code for the selection process is available upon request.

3. RESULTS

As summarized in Table 1, we studied a combination of three population sizes, three heritabilities of the trait and three marker densities. The default case unless otherwise stated is the case of 10,100 markers per chromosome, heritability $h^2 = 0.25$ and a population size with 50 males and 50 females per generation.

3.1. Distribution of Coancestries

Most likely, the differences in our results are going to be due to the distribution of coancestries, as the different selection strategies here performed are based on the matrix of relationships between individuals.

We show in Figure 1 the distributions for the four measures of relationships prior to selection and give the variance within each figure. There we can see how the shape of the distribution of the genealogical coancestry is multimodal, given the sparse nature of the genealogical coancestry matrix and its distribution has the largest variance of all coancestry matrices, as well as the lowest mean. The distribution of coancestries f_V and f_G are fairly similar, the first one having a lower mean and a slightly larger variance although both distributions have a very small variance. Lastly, the distribution of coancestries f_R has a mean considerably lower than the other genomic coancestries f_V and f_G and a substantially larger variance.

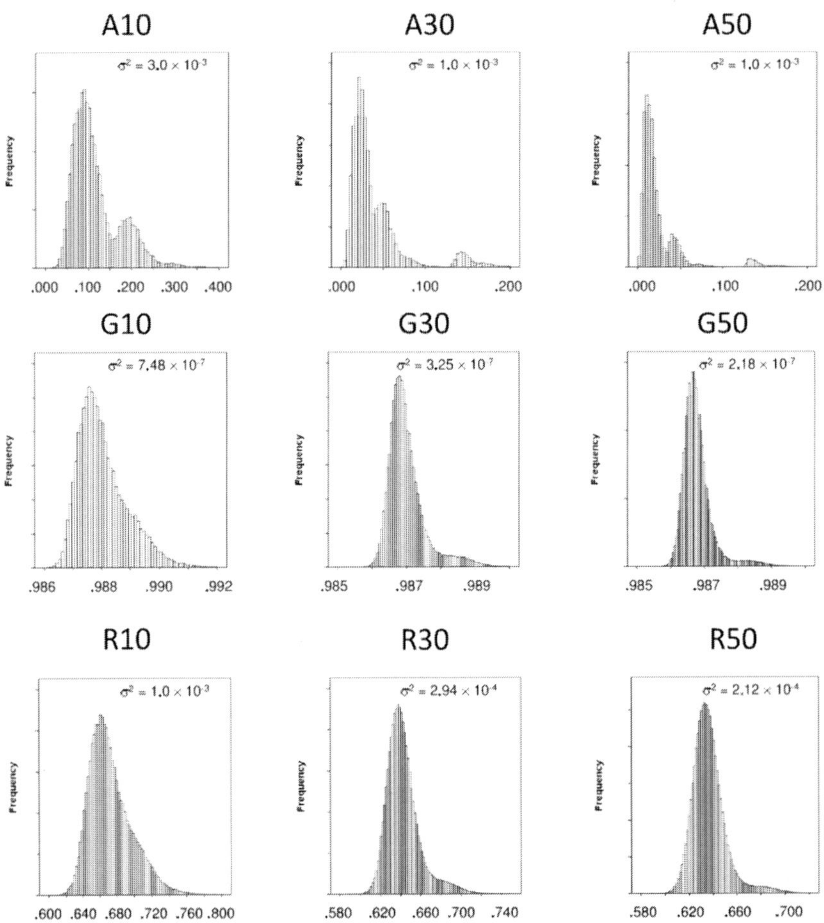

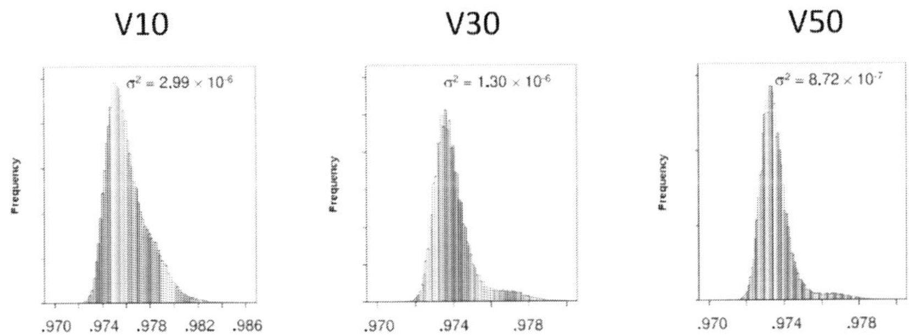

Figure 1. Histograms of the coancestries at generation 6 right before selection. Top row shows the histogram for genealogical coancestry f_A for 10, 30, and 50 individuals from left to right. Similarly, the second row shows the histogram for molecular marker-by-marker coancestry f_G. The third row shows the histograms for segment-based coancestry f_R, for $N = 10$, $N = 30$, and $N = 50$ from left to right. The bottom row shows the histogram for molecular marker-by-marker coancestry corrected by allelic frequencies f_V, for $N = 10$, $N = 30$, and $N = 50$ from left to right. The variance of each histogram is given within each plot.

3.2. Genetic Gain

Changes in TBVs obtained with the four relationship matrices for three population sizes $N = 10$, $N = 30$, and $N = 50$, three heritabilities $h^2 = 0.1$, $h^2 = 0.25$, and $h^2 = 0.50$, as well as three marker densities of 2525, 5050, and all 10,100 per chromosome are shown in Figure 2 vs. generations. We only show results after generation 7, when selection starts. For a better comparison between the different coancestries here used, we show the value at each generation minus the initial value right before selection (i.e., at generation 7). Overall, all four methods performed similarly in terms of genetic gain for the sizes here studied. As expected, the final TBV increased with the number of individuals and with the heritability of the trait. The density of markers had no effect when selecting with the genealogical coancestry f_A, as expected, and, within the range of densities here studied no differences were detected in the genetic gains achieved by the genomic based estimates f_V and f_G. The most surprising result is that for a low density of markers, the genetic gain is larger performing selection based on f_R. It must be noticed that the size for a region of homozygosity to be considered as such was kept constant and thus, a ROH of 100 contiguous markers covers a much longer stretch than for 10,100 marker per chromosome. This is also surprising as it has been pointed out that the longer the ROH, the more correlated ROH-based inbreeding is with genealogical inbreeding.

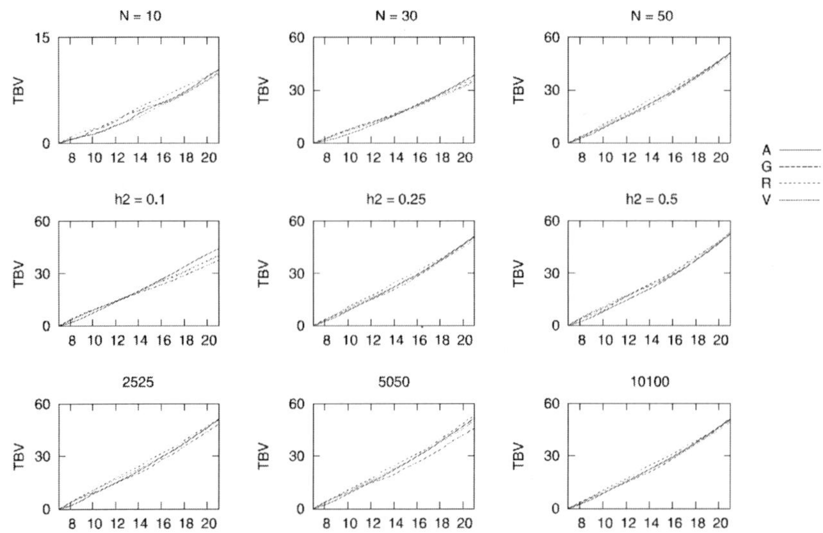

Figure 2. Mean true breeding values (tbv) for different marker densities (bottom row), heritability (middle row), and population size (top row) vs. Generations of selection. TBV values are shown minus the value right before truncation selection started. The default values are 50 sires and 50 dams, a heritability of 0.25 and 10,100 markers, unless the value at the top of the figure indicates otherwise.

3.3. Changes in Relatedness

We show in Figure 3 results for the changes in each of the four measures used of coancestry with each selection scenario. We have used a logarithmic scale as overall, the differences between genealogical based selection and genomic based selection were very large. That is, line "A" shows the results for genealogical coancestry resulting from selecting based on this coancestry f_A and so on for scenarios G, R, and V. The results for inbreeding are not shown as they display a very similar pattern. In order to better appreciate the differences between the four measures of coancestries, we show $log[(1 - f)/(1 - f_7)]$ in Figure 3. In this way, we compare the speed of increase in each average coancestry scaled with their values at generation 7 (f_7), right before selection started. The increase in genealogical coancestry (the decay in this log scale) is the largest, followed by ROH-based coancestry. Changes in f_V and f_G are hardly distinguishable and very similar to f_R for small heritability. The smaller the population, the larger the increase in any measure of coancestry. The differences in f_G and f_V are hardly different as heritability increases from $h^2 = 0.25$ to $h^2 = 0.5$.

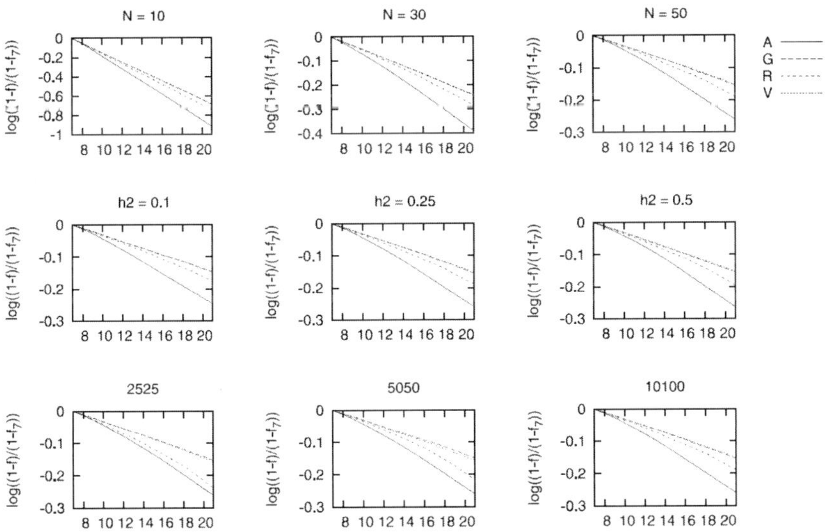

Figure 3. Change in each coancestry for different marker densities (bottom row), heritability (middle row), and population size (top row) vs. Generations of selection. The change in each coancestry is shown as log(1−f1−f7)log(1−f1−f7).

In Figure 4 we show a similar plot for the change in pedigree based coancestry obtained under each selection scenario. All cases studied showed that the three genomic based selection led to lower increases in pedigree-based coancestry and the differences between the selection based on genomic relationships are hardly noticeable. The results are very similar for f_V and f_G based BLUPs on genealogical coancestry and it seems that f_R-based BLUP leads to slightly larger genealogical coancestries.

3.4. Diversity Maintained

As a measure of the diversity maintained we used f_G, as this is directly related to observed heterozygosity. In Figure 5, we show the changes on this marker-by-marker relatedness over generations when selection was carried out using the four strategies analyzed. As previously done for all coancestries and for genealogical coancestry, we show its rate of decrease by plotting $log(1 - f_G)$ in Figure 5, minus this value right before starting selection $log(1 - f_G(7))$ to compare all selection processes. Therefore, in this scale, the largest decrease means the largest increase in f_G.

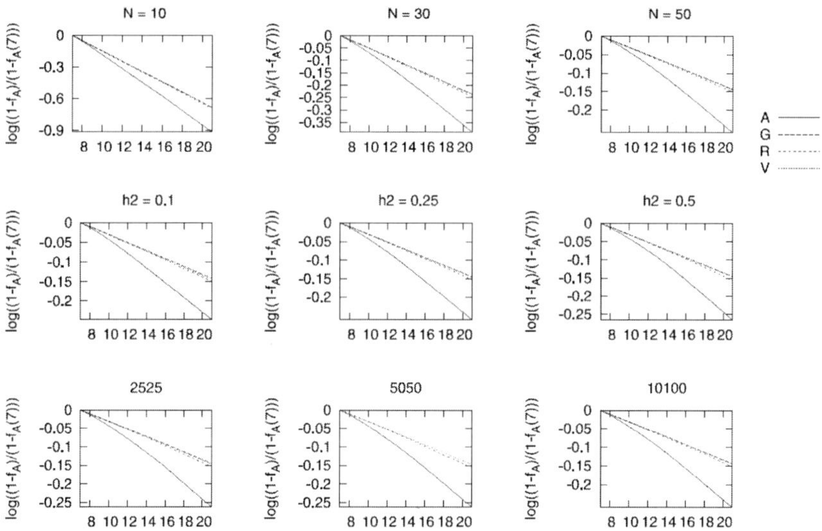

Figure 4. Changes in genealogical coancestry for different marker densities (bottom row), heritability (middle row), and population size (top row) vs. Generations of selection. The change in each coancestry is shown as log(1−fA1−fA(7))log(1−fA1−fA(7)), where f_A is the genealogical coancestry at each generation and $f_A(7)$ is the value before selection starts.

It is important to highlight that the highest loss in genetic diversity (the largest increase in f_G) was observed for the selection based on the additive relationship matrix without exception. The fastest decay is for the smallest population size of $N = 10$ and then for $N = 30$ and this decay is largest with decreasing population size than heritabilities or marker densities. Within each scenario, it seems that initially most diversity is maintained selecting with the genomic coancestries and the difference between f_G and f_V is small. The difference between f_G or f_V and f_R is small, though f_R-based BLUP can lead to slightly larger decreases in molecular coancestry than the other two genomic measures of relatedness, especially for small marker density. That is, f_R-based BLUP maintains slightly less genetic diversity than the other genomic based BLUPs.

4. DISCUSSION

We have shown here results for truncation selection performed with four different measures of coancestry: f_A, f_G, f_R and f_V. All results shown are selecting the top 50% of sires and dams and we have compared results with three different population sizes, three different heritabilities of the selected trait and three different number of markers per chromosome.

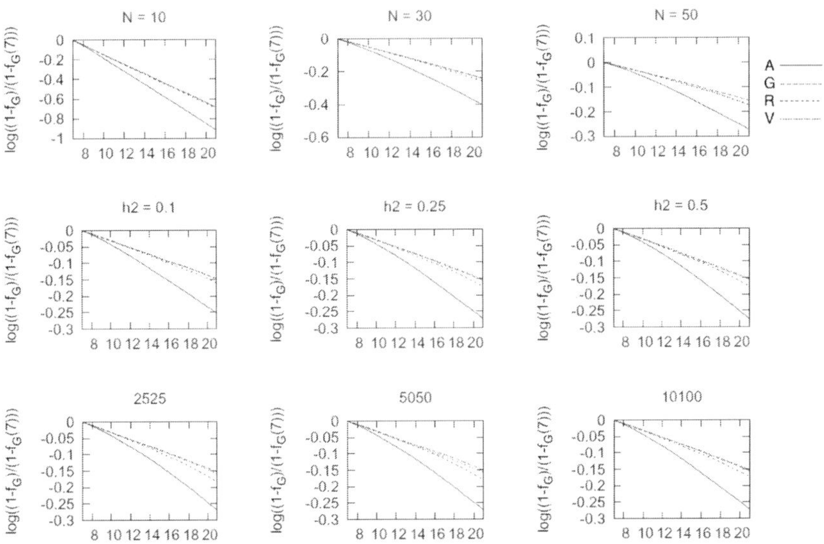

Figure 5. Changes in molecular coancestry as a measure of change in diversity for different marker densities (bottom row), heritability (middle row), and population size (top row) vs. Generations of selection. This change is shown as log(1−fG1−fG(7))log(1−fG1−fG(7)), where f_G is the value of molecular coancestry at each generation and $f_G(7)$ is the value before selection starts.

We have performed 6 initial generations of random mating to have a deeper pedigree and have a fairer comparison between molecular markers which record the whole population history and genealogies, which are usually only stored when the selection programme starts.

There seems to be currently a consensus that genomic BLUP selection, whereby we mean selection based on genomic measures of relatedness, is superior to traditional pedigree-based BLUP selection (Daetwyler et al., 2007, 2010; Sonesson et al., 2012) in terms of higher genetic gain and lower increase in inbreeding. However, few studies have paid attention to the loss of genetic variability caused by each selection strategies of selection (Jannink, 2010; Bastiaansen et al., 2012; Heidaritabar et al., 2014; Liu et al., 2014). We discuss our main conclusions and the differences with these previous studies below.

4.1. On Genetic Gain

One of the main properties of BLUP is that by definition, the largest gain is obtained when the additive genetic variance of the base population is known. This is a difficult task, as for a large number of loci under selection which may be linked, the standard formula of $\sigma_a^2 = \sum_{j=1}^{ns} 2p_j(1-p_j)a_j^2$ (Falconer and Mackay, 1996) does not apply. Furthermore, this variance is not appropriate when the performed BLUP relies on the genomic relationships f_G, f_R or f_V. Thus, we estimated the additive variance components using REML. While it is well-known that the estimates obtained with REML are more accurate for larger population sizes than the ones here studied, the differences between the four selection strategies here studied are small. We think that these differences are independent of whether the variance could have been better estimated. We believe that a more accurate estimate of the variance of the base population would lead to larger gains for all four BLUPs here performed and the differences in the trends would stay the same.

Overall, the genetic gain was very similar with the four relationship matrices, although BLUP based on f_R performed slightly better than the other BLUPs in terms of gain for lower marker densities. It also performed somewhat better for small population size and the intermediate heritability here studied, at least up to generation 18 (i.e., after 10 generations of selection). It is worth emphasizing that for the lower marker densities here studied, we kept the same threshold size of 100 consecutive markers for a ROH to be considered as such. That means that for 2525 markers per chromosome, such ROH would cover a section of about 4 cM, while for 10,100 a ROH of 100 consecutive markers covers 1 cM. Thus, for higher marker densities, it is likely that the gains could be increased by using a larger threshold for what is considered a ROH.

As expected, the final TBVs were larger for larger population size and for higher trait heritability. This is due to the larger genetic variance for larger population sizes in which selection can act upon, while the negative effects of inbreeding are reduced with higher population sizes. It is however somewhat surprising that the differences are small in genetic gain with marker densities for the genomic relationships matrices, particularly for f_V and f_G. This could indicate that a density of 2525 markers per chromosome would give the same correlation between the true genomic relationship if we had the whole sequence and that estimated with such marker density (Rolf et al., 2010).

It is likely that the lack of differences in genetic gain between the genomic and pedigree based relationships stems from the fact that we use the marker data to infer the relationships, but not to estimate the marker effects. It is in this later scenario where genomic selection seems considerably superior to traditional pedigree BLUP, although it depends on having enough training generations where both phenotypes and genotypes are recorded, as reviewed recently by Van Eenennaam et al. (2014).

In genomic selection, markers that densely cover the genome are expected to be in complete or partial linkage disequilibrium with the trait under selection. Genomic prediction based on IBS information uses the family structure of the population (Habier et al., 2007), since the markers capture the linkage disequilibrium that arises from the family structure. Recently, Luan et al. (2014) have proposed an approach to predict genomic estimated breeding values from runs of homozygosity. This study indicates that runs of homozygosity yield a multi-locus measure of linkage disequilibrium and thus can account for larger chromosomal distances to capture linkage disequilibrium than genomic prediction based on IBS information. It is worth noting that in their study, Luan et al. (2014) used a somewhat different definition of segment that we have used here. They obtained slightly better predictions for the ROH-based scenarios than for other genomic-based scenarios. Our results seem to be in line with those obtained by Luan et al. (2014), although a more thorough analysis of both methods is required for a better comparison. The measure of ROH used by Luan et al. (2014) does not seem to require a threshold size for a run of homozygosity, but it requires knowledge of the mutation rates and the effective population size.

No significant differences were detected between the genetic gain obtained with f_G and f_V. The reason is that with the $f_G$approach alleles that are IBD and IBS can not be distinguished and are both included in the coancestry (and inbreeding) measures. To express both pedigree- and genomic-based estimates in the same scale several methodologies have been proposed Toro et al. (2011). However, these methods are generally inaccurate and their performances are very similar to those for f_G Toro et al. (2002).

Sonesson et al. (2012) compared breeding schemes by simulating truncation or optimum contribution selection. They estimated breeding breeding values based on genome- or pedigree-based BLUP and recorded trait information on full-sibs of the candidates. This study concluded that to control inbreeding it is necessary to account for it on the same basis as what is used to estimate breeding values. Our results are in general agreement to those of Sonesson et al. (2012) regarding the genetic gain both with genomic- and pedigree-based selection procedures and with those of Bastiaansen et al. (2012), where higher accuracies were obtained for the genomic methods than for traditional pedigree-based BLUP.

4.2. On Coancestries and Inbreeding

As we have shown in Figure 4, the largest increases in coancestries, and similarly for inbreeding, is for the genealogical coancestry compared to other genomic measures of coancestry. At the same time, this increase in genealogical coancestry is larger with traditional pedigree-based BLUP than for any other BLUP here performed. This is in line with what Sonesson et al. (2012) obtained using BLUP combined with optimal contributions to control the increase in inbreeding, that the rate of increase in pedigree coancestry is higher for the pedigree-based selection scenario than for the genome-based selection approaches. This can be observed regardless the population size, the true heritability, or the density of markers. Bastiaansen et al. (2012) showed similar differences between traditional pedigree-based and genomic-based BLUP. They also showed how this difference built up with generations and was hardly noticeable after one round of selection. This study showed that the increase in inbreeding hardly depended on the genomic architecture of the selected trait, which is in line with what we observe in Figure 4, where the increase in coancestry seems independent of the marker density or the heritability of the trait. In agreement with Bastiaansen et al. (2012), we have also shown that genomic-based BLUPs can track Mendelian sampling within families, which is not possible with genealogical-based BLUP. Our results are apparently in contrast with the recent study of Liu et al. (2014), who obtained a lower increase in inbreeding for the larger heritability 0.25 in their study compared to that obtained for $h^2 = 0.05$. This is most likely due to the fact that they looked at the results after 8 generations of selecting the top 25% candidates each generation, while we have performed selection on the top 50% candidates and looked at the increase of coancestry after 14 generations of selection. This shows the importance of understanding the dynamics at different generation intervals.

Liu et al. (2014) debated whether using genealogical records would be a good measure of inbreeding, as it reflects expected relationships and not the actual ones. They proposed measuring inbreeding then based on runs of homozygosity, and obtained that genomic-based BLUPs lead to lower increases on genealogical inbreeding as compared to phenotype BLUP, but this was not the case for inbreeding measured with ROHs. Our results for f_R are very similar to those here presented for f_A, and thus in the scenarios here studied, all genomic measures lead to lower increases in inbreeding whether we measure it with genealogies or with ROHs.

Our results show that the increase in genealogical coancestry seems slightly larger for ROH-based BLUP as compared to the other genomic-based BLUPs, although the differences are small.

4.3. On Diversity Maintained

It is well-known that selection reduces variation around the selected loci due to hitchhiking (Maynard-Smith and Haigh, 1974; Heidaritabar et al., 2014; Liu et al., 2014). Thus, if we aim at maintaining diversity while selecting favorables variants, it is important to understand which selection strategy works better overall. We evaluated f_G as a measure of diversity maintained in the selection procedures simulated in the present study indicated that all genomic estimates maintained more variability than the pedigree-based ones. This result is in agreement with those also observed using simulated data but in the context of conservation programmes (de Cara et al., 2011), and with previous results in genomic selection (Liu et al., 2014).

An interesting study by Jannink (2010) showed that more variation could be maintained by placing more weight on favorable variants that are at low frequencies. This can potentially maintain more diversity both on the selected loci and on neutral loci. According to that study, this strategy leads to larger gains in the long-term, and thus this strategy could be optimal depending on how long is the long-term. Based on this study, it would be worthwhile studying whether placing weight on rare haplotypes could lead to a compromise between genetic gains and diversity maintained.

The study by Heidaritabar et al. (2014) has shown that changes in allelic frequencies are more localized around the selected loci with genomic based BLUP, while pedigree based BLUP leads to similar changes throughout the genome. Thus, it seems that genomic selection can lead to quick losses in genetic variation in specific regions of the genome, and thus great care is required if these regions provide potential adaptation of the breed.

In agreement with Liu et al. (2014), we have obtained that a larger heritability leads to larger decreases in diversity maintained when selecting with traditional BLUP. Similarly to what happened with genealogical inbreeding, the loss of diversity does not seem to depend on heritability when selecting with genomic-based BLUPs.

Interestingly, ROH-based BLUP seems to lead to slightly larger losses in diversity than the other genomic BLUPs, but massively smaller than pedigree BLUP. Consequently, a deep study of the factors involved in the definition of a ROH could help to improve the genetic gain obtained with this estimator while also keeping the a very high genetic variability.

In conclusion, in this study conventional pedigree based selection, which has been used for decades, results in similar genetic gains and does not maintain as much genetic variability as the genomic based selection methods. These results highlight the utility of genomic selection and also the need to manage the population variability using genomic information to preserve the future success of selection programs.

ACKNOWLEDGMENTS

STRR and LAGC were funded by Ministerio de Economía y Competitividad grant number CGL2012-39861-C02-02. MARdC was funded by LabEx grant ANR-10-LABX-0003-BCDiv from Agence Nationale de la Recherche Investissements programme ANR-11-IDEX-0004-02. The authors are grateful to Carmen Rodriguez Valdovinos and Luis Silio for access to their server. MARdC benefited from access to the computer cluster of the Genotoul bioinformatics platform Toulouse Midi-Pyrénées.

REFERENCES

1. Bastiaansen, J. W. M., Coster, A., Calus, M. P. L., van Arendonk, J. A. M., and Bovenhuis, H. (2012). Long-term response to genomic selection: effects of estimation method and reference population structure for different genetic architectures. *Genet. Sel. Evol.* 44:3. doi: 10.1186/1297-9686-44-3
2. Bjelland, D. W., Weigel, K. A., Vukasinovic, N., and Nkrumah, J. D. (2013). Evaluation of inbreeding depression in holstein cattle using whole-genome snp markers and alternative measures of genomic inbreeding. *J. Dairy Sci.* 96, 1–10. doi: 10.3168/jds.2012-6435
3. Caballero, A., and Toro, M. A. (2002). Analysis of genetic diversity for the management of conserved subdivided populations. *Cons. Genet.* 3, 289–299. doi: 10.1023/A:1019956205473
4. Daetwyler, H. D., Pong-Wong, R., Villanueva, B., and Woolliams, J. A. (2010). The impact of genetic architecture on genome-wide evaluation methods. *Genetics* 185, 1021–1031. doi: 10.1534/genetics.110.116855
5. Daetwyler, H. D., Villanueva, B., Bijma, P., and Woolliams, J. A. (2007). Inbreeding in genome wide selection. *J. Anim. Breed. Genet.* 124, 369–376. doi: 10.1111/j.1439-0388.2007.00693.x
6. de Cara, M. A. R., Fernandez, J., Toro, M. A., and Villanueva, B. (2011). Using genome-wide information to minimize the loss of diversity in conservation programmes. *J. Anim. Breed. Genet.* 128, 456–464. doi: 10.1111/j.1439-0388.2011.00971.x
7. de Cara, M. A. R., Fernandez, J., Toro, M. A., and Villanueva, B. (2013). Using genomic tools to maintain diversity and fitness in conservation programmes. *Mol. Ecol.* 22, 6091–6099. doi: 10.1111/mec.12560
8. de los Campos, G., Gianola, D., and Allison, D. B. (2010). Predicting genetic predisposition in humans: the promise of whole-genome markers. *Nat. Rev. Genet.* 11, 880–886. doi: 10.1038/nrg2898
9. Eding, H., and Meuwissen, T. H. E. (2001). Marker-based estimates of between and within population kinships for the conservation of genetic diversity. *J. Anim. Breed. Genet.* 118, 141–159. doi: 10.1046/j.1439-0388.2001.00290.x
10. Falconer, D. S., and Mackay, T. F. C. (1996). *Introduction to Quantitative Genetics*. Essex: Longman.
11. Fisher, R. A. (1954). A fuller theory of "junctions" in inbreeding. *Heredity* 8, 187–197.

12. Forni, S., Aguilar, I., and Misztal, I. (2011). Different genomic relationship matrices for single-step analysis using phenotypic, pedigree and genomic information. *Genet. Sel. Evol.* 43:1. doi: 10.1186/1297-9686-43-1
13. Frankham, R., Ballou, J. D., and Briscoe, D. A. (2002). *Introduction to Conservation Genetics.* Cambridge: Cambridge University Press.
14. Garcia-Cortes, L. A., Legarra, A., Chevalet, C., and Toro, M. A. (2013). Variance and covariance of actual relationships between relatives at one locus. *PLoS ONE* 8:e57003. doi: 10.1371/journal.pone.0057003
15. Goddard, M. E. (2009). Genomic selection: prediction of accuracy and maximisation of long term response. *Genetica* 136, 245–257. doi: 10.1007/s10709-008-9308-0
16. Goddard, M. E., and Hayes, B. (2007). Genomic selection. *J. Anim. Breed. Genet.* 124, 323–330. doi: 10.1111/j.1439-0388.2007.00702.x
17. Guo, S. W., and Thompson, E. A. (1991). Monte carlo estimation of variance component models for large complex pedigrees. *IMA J. Math. Appl. Med. Biol.* 8, 171–189.
18. Gusev, A., Lowe, J. K., Stoffel, M., Daly, M. J., Altshuler, D., Breslow, J. L., et al. (2009). Whole population, genome-wide mapping of hidden relatednesss. *Genome Res.* 19, 318–326. doi: 10.1101/gr.081398.108
19. Habier, D., Fernando, R. L., and Dekkers, J. C. M. (2007). The impact of genetic relationship information on genome-assisted breeding values. *Genetics* 177, 2389–2397. doi: 10.1534/genetics.107.081190
20. Hadfield, J. D., Wilson, A. J., Garant, D., Sheldon, B. C., and Kruuk, L. E. B. (2010). The misuse of blup in ecology and evolution. *Am. Nat.* 175, 116–125. doi: 10.1086/648604
21. Hayes, B. J., Visscher, P. M., and Goddard, M. E. (2009). Increased accuracy of artificial selection by using the realized relationship matrix. *Genet. Res.* 91, 47–60. doi: 10.1017/S0016672308009981
22. Heidaritabar, M., Vereijken, A., Muir, W. M., Cheng, T. M. H., Megens, H. J., Groenen, M. A. M., et al. (2014). Systematic differences in the response of genetic variation to pedigree and genome-based selection methods. *Heredity* 113, 503–513. doi: 10.1038/hdy.2014.55
23. Henderson, C. (1984). *Applications of Linear Models in Animal Breeding.* Guelph, ON: Guelph University Press.
24. Jannink, J.-L. (2010). Dynamics of long-term genomic selection. *Genet. Sel. Evol.* 42:35. doi: 10.1186/1297-9686-42-35
25. Liu, H., Sorensen, A. C., Meuwissen, T. H. E., and Berg, P. (2014). Allele frequency changes due to hitch-hiking in genomic selection programs. *Genet. Sel. Evol.* 46:8. doi: 10.1186/1297-9686-46-8
26. Luan, T., Yu, X., Dolezal, M., Bagnato, A., and Meuwissen, T. H. E. (2014). Genomic prediction based on runs of homozygosity. *Genet. Sel. Evol.* 46, 64. doi: 10.1186/s12711-014-0064-6
27. Makowsky, R., Pajewski, N. M., Klimentidis, Y. C., Vazquez, A. I., Duarte, C. W., Allison, D. B., et al. (2011). Beyond missing heritability: prediction of complex traits. *PLoS Genet.* 7:e1002051. doi: 10.1371/journal.pgen.1002051
28. Malecot, G. (1948). *Les Mathematiques de l'Heredite.* Paris: Masson et Cie.
29. Maynard-Smith, J., and Haigh, J. (1974). Hitch-hiking effect of a favorable gene. *Genet. Res.* 23, 23–35.
30. Meuwissen, T. H. E. (2007). Genomic selection: marker assisted selection on a genome wide scale. *J. Anim. Breed. Genet.* 124, 321–322. doi: 10.1111/j.1439-0388.2007.00708.x
31. Meuwissen, T. H. E., Hayes, B. J., and Goddard, M. E. (2001). Prediction of total genetic value using genome-wide dense marker maps. *Genetics* 157, 1819–1829.
32. Meyer, K. (1991). Estimating variances and covariances for multivariate animal models by restricted maximum likelihood. *Genet. Sel. Evol.* 23, 67–83.
33. Rodriguez-Ramilo, S. T., Garcia-Cortes, L. A., and Gonzalez-Recio, O. (2014). Combining genomic and genealogical information in a reproducing kernel hilbert spaces regression model for genome-enabled predictions in dairy cattle. *PLoS ONE* 9:e93424. doi: 10.1371/journal.pone.0093424
34. Rolf, M. M., Taylor, J. F., Schnabel, R. D., McKay, S. D., McClure, M. C., Northcutt, S. L., et al. (2010). Impact of reduced marker set estimation of genomic relationship matrices on

genomic selection for feed efficiency in angus cattle. *BMC Genet.* 11:24. doi: 10.1186/1471-2156-11-24

35. Solberg, T. R., Sonesson, A. K., Woolliams, J. A., and Meuwissen, T. H. E. (2008). Genomic selection using different marker types and densities. *J. Anim. Sci.* 86, 2447–2454. doi: 10.2527/jas.2007-0010
36. Sonesson, A. K., Woolliams, J. A., and Meuwissen, T. H. E. (2012). Genomic selection requires genomic control of inbreeding. *Genet. Sel. Evol.* 44:27. doi: 10.1186/1297-9686-44-27
37. Stam, P. (1980). The distribution of the fraction of the genome identical by descent in finite random mating populations. *Genet. Res.* 35, 131–155.
38. Toro, M. A., Barragan, C., Ovilo, C., Rodrigañez, J., Rodriguez, C., and Silio, L. (2002). Estimation of coancestry in iberian pigs using molecular markers. *Cons. Genet.* 3, 309–320. doi: 10.1023/A:1019921131171
39. Toro, M. A., Garcia-Cortes, L. A., and Legarra, A. (2011). A note on the rationale for estimating genealogical coancestry from molecular markers. *Genet. Sel. Evol.* 43:27. doi: 10.1186/1297-9686-43-27
40. Van Eenennaam, A. L., Weigel, K. A., Young, A. E., Cleveland, M. A., and Dekkers, J. C. M. (2014). Applied animal genomics: results from the field. *Ann. Rev. Anim. Biosci.* 2, 105–139. doi: 10.1146/annurev-animal-022513-114119
41. VanRaden, P. M. (2008). Efficient methods to compute genomic predictions. *J. Dairy Sci.* 91, 4414–4423. doi: 10.3168/jds.2007-0980
42. Villanueva, B., Pong-Wong, R., Fernandez, J., and Toro, M. A. (2005). Benefits from marker-assisted selection under an additive polygenic genetic model. *J. Anim. Sci.* 83, 1747–1752.
43. Visscher, P. M., Medland, S. E., Ferreira, M. A. R., Morley, K. I., Zhu, G., Cornes, B. K., et al. (2006). Assumption-free estimation of heritability from genome-wide identity-by-descent sharing between full siblings. *PLoS Genet.* 2:e41. doi: 10.1371/journal.pgen.0020041
44. Weir, B. S., Anderson, A. D., and Hepler, A. B. (2006). Genetic relatedness analysis: modern data and new challenges. *Nat. Rev. Genet.* 7, 771–780. doi: 10.1038/nrg1960

CHAPTER 14

Association between UCP Gene Polymorphisms and Growth, Breeding Value of Growth and Reproductive Traits in Mazandaran Indigenous Chicken

Seyed Mohammad Moazeni[1], Mohammadreza Mohammadabadi[2], Mostafa Sadeghi[3], Hosein Moradi Shahrbabak[3], Ali Esmailizadeh Koshkoieh[2], Farhad Bordbar[1]

[1]*MSc Student of Department of Animal Science, Shahid Bahonar University of Kerman, Kerman, Iran*

[2]*Associate Professor of Department of Animal Science, Shahid Bahonar University of Kerman, Kerman, Iran*

[3]*Assistant Professor of Department of Animal Science, Tehran University, Tehran, Iran*

ABSTRACT

The objective of this study was to estimate the allele and genotype frequencies of the UCP/Hha1 polymorphisms and to determine associations between these polymorphisms and growth traits, breeding value of growth and reproductive traits in Mazandaran Indigenous Chicken. To estimate breeding value of the reproductive traits, and phenotypic information of 18 successive generations of selection in breeding station of Mazandaran Indigenous Chicken (north of Iran), 190 birds were analyzed using a univariate animal model in ASREML procedure.. Genotyping of UCP gene polymorphism for all subjects was performed by the PCR?RFLP method. The evaluation of the association between this SNP with reproductive traits suggests a positive effect of TC genotype with age at first egg (ASM) compared with CC genotype. In addition, TC genotype was significantly associated with breeding value of age at first egg compared with CC genotype ($P < 0.05$). In conclusion, our results suggest that the TC genotype of the UCP gene is associated with age at sexual maturity (ASM) and breeding value of age at sexual maturity, and UCP polymorphisms may be used as DNA markers for selection in the breeding process of Mazandaran Indigenous Chicken.

Keywords: Reproductive Traits, Polymorphism, Chicken, UCP, PCR-RFLP

1. INTRODUCTION

Low energy expenditure, in addition to increased energy intake, has been a major cause of future weight gain, and variations in energy expenditure may be one of the underlying sources of variation in body weight [1]. Various determinants, including body composition, hormonal levels, activity of the sympathetic nervous system, and genetics are responsible for differences in metabolic rate among individuals [2]. A mitochondrial protein called uncoupling protein plays an important role in generating heat and burning calories by creating a pathway that allows dissipation of the proton electrochemical gradient across the inner mitochondrial membrane in brown adipose tissue, without coupling to any other energy-consuming process. This pathway has been implicated in the regulation of body temperature, body composition, and glucose metabolism [3]. The uncoupling protein (UCP) gene is one of the many genes that maybe related with growth traits, breeding value of growth and reproductive traits. UCP3 is the member of uncoupling proteins (UCPs) family of mitochondrial transporters, all of which are known to uncouple oxidative phosphorylation via proton leakage from the inner mitochondrial membrane. Thermoregulation and energy metabolism are potentially regulated by these mitochondrial proteins [4]. The uncoupling protein gene (UCP) is a candidate gene for growth traits, breeding value of growth and reproductive traits because it may contribute to the proton leak that occurs in mitochondria and partially uncouples the oxidative phosphorylation, thus releasing chemical energy as heat [5]. Uncoupling protein (UCP) gene family involved in energy metabolism of body, and had significant effects on energy balance related traits, such as body weight resting metabolic rate, and efficiency of food conversion etc. [6]. Liu *et al.* [7] found UCP-3 gene polymorphisms may contribute to body mass index (BMI) in the Caucasian population. Rudotsky *et al.* [8] found that human UCP-3 gene involved in energy balance and was associated with obesity and diabetes. UCP2 gene is expressed extensively in various tissues, like white adipose tissue, pancreas, skeletal muscle, and liver, in human [9]. Both environmental and genetic factors may have contribution toward susceptibility of growth traits, breeding value of growth and reproductive traits. Genetic diversity in indigenous breeds is a major concern considering the necessity of preserving what may be a precious and irreplaceable richness, regarding new productive demands. Conservation should be based on a deep knowledge of the genetic resources of the specific breed. Therefore, it is important to try to characterize genetically indigenous breeds [10]. A species without enough genetic diversity is thought to be unable to cope with changing environments or evolving competitors and parasites and also the ability of a population to respond adaptively to environmental changes depends on the level of genetic variability or diversity it contains. Therefore, studies of population genetic diversity and the structure of population within and between species may not only illustrate the evolutionary process and mechanism but also provide information useful for biological conservation of the Bovidae family, sheep, goats, birds and so on [11]. Molecular markers are increasingly used for the study of genetic diversity of populations in recent years [12]. Therefore, in the present study, we aimed to investigate association of polymorphism of UCP2 gene with growth traits, breeding value of growth and reproductive traits in subjects from Iran.

2. MATERIAL AND METHODS

2.1. Experimental Population and Sampling

Breeding station of Mazandaran Indigenous Chicken is located at 28 km far from Sari, the provincial capital of Mazandaran state, located in the north of Iran . In 1986 about 5000 cocks and hens were purchased from rural regions across the Mazandaran province and kept in a quarantine farm for a year. From those, about 2500 birds of two sexes were kept to produce hatching eggs and chicks produced from these eggs were transferred to the station in 1988. Since then birds have been individually tagged and trap nest has been used for pedigree recording. Parents of each generation (about 100 cocks and 800 hens) are selected among 6000 pedigreed and performance recorded birds produced each generation. In August 2009, a total of 205 blood samples from Mazandaran Indigenous Chicken, including 10 males and 195 females, were collected. Individuals were reared in Native chicken breeding station of Mazandaran and they belonged to generation 17 of the breeding station pedigreed animals. Individuals of this generation were developed by crossing 80 sires and 751 dams from generation16. Approximate 1 ml blood per chick from plumage veins was collected and kept in a tube containing anticoagulant EDTA (Ethylenediaminetetraacetic acid). All samples were delivered back to the laboratory in an ice box. The genomic DNA was extracted from white blood cells using standard salting out procedure described by Abadi *et al.* [13]. The DNA samples were dissolved in TE buffer which was made from 10 mMTris-Cl (pH 7.5) and 1 mM EDTA (pH 8.0) and were stored at 20°C for use.

2.2. Primer Synthesis and PCR-RFLP Reactions

The primers were designed on the basis of DNA sequence of the UCP (Accession: AF433170) using the oligonucleotide design tool Primer 5.0 software (F: 5'-TACCCCAAGCATGCAGAACTCA-3' and R: 5'- GGAACCGCA CCTTGACCAC-3'). Both PCR reactions were performed in a 20 µl mixture containing 10 pmol primers, 200 lMdNTP (deoxyribonucleotide triphosphate), 2 µl 10 × reaction buffer which contained 1.5 mM $MgCl_2$, 1 unit of Taq-DNA polymerase (Promega, Madison, WI), and 50 ng genomic DNA as template. PCR method was used to optimize the reaction accuracy: 94°C for 5 min, 35 cycles of 94°C for 30 s, annealing at 55°C for 60 s, 72°C for 60 s, and a final extension at 72°C for 7 min. PCR products were electrophoretically separated on 2% agarose gel (5 V/cm), stained with ethidium bromide. PCR products were digested by 10 units of Hha1 restriction enzymes (Fermentase, Lithuania), 6 ml of PCR product, 1.4 ml of Tango buffer and 2 ml nuclease-free water. The final volume of 10 ml was incubated in 37°C for 12h. The fragments were separated a 3.5% agarose gel stained with ethidium bromide.

2.3. Traits Measured

Whole information data file (18 successive generations) consisted of three fixed effects (generation, sex and hatch) and 11 recorded traits including body weight at hatch (BW1), body weight at age of 8 (BW8), 12 (BW12) weeks, body weight at sex maturation (WSM), age at first egg (ASM), egg number (EN), first egg weight EW1), average egg weight at age of 28 (EW28), 30 (EW30), 32 (EW32) weeks and average egg weight for the first 12 weeks of production (EW12). BW1, BW8 and BW12 have been measured in both male and female chicken. Also, three combined traits consisting of average of EW28, EW30 and EW32 (AV), intensity of egg production (EINT

= (egg number/days recording) × 100) and egg mass (EM = EN × EW12) were calculated to use in the analyses (**Table 1**). During 18 generations, genetic evaluation of the birds for body weight at 8 weeks, age of the hens at first egg, average egg weight and total number of eggs laid during first 12 weeks after flocks maturity (when 5% of the flock are in egg production) have been performed. Economic indexes are calculated for these traits and birds of two sexes are selected based on their aggregate genotypes for these traits. The goals of the breeding station on the one hand are to increase body weight, egg weight and egg number and on the other hand, to decrease age at first egg.

2.4. Statistical Analyses

Pedigree and data file were prepared using Visual FoxPro 9.0 software, the relational data base management system. SAS9.1 package was used to carry out descriptive statistics and fitting model. Fixed effects and their interactions were considered in animal model that provided having significant effect. Genetic analyses were performed using ASREML software [14]. Breeding values of growth and egg production traits was estimated using the BLUP based on an animal model with a relationship matrix. The used models in matrix notation were as follows:

$$Y = XB + ZA + e$$

where, Y is the vector of observations; b the vector of fixed effects of generation, sex and hatch; a the vector of random direct genetic effects; e the vector of random residual effects; X and Z are incidence matrices relating the observations to the respective fixed and direct genetic effects. Calculation of gene frequency was based on direct gene count method by: f (A) = (2nAA + n Aa) / 2n or f (G) = (2nGG + nGg) / 2n and standard error of frequency was calculated as $(p(1-p)/2n)^{1/2}$, where n is the sample size, p is the frequency of A or G allele. Marker-trait association analyses were conducted using following model in GLM procedure of SAS9.1 software. The significant differences of least squares means were tested with Tukey-Kramer's multiple range tests, and a P-value of ≤0.05 was considered statistically significant.

$$f(A) = (2nAA + n Aa) / 2n \text{ or } f(G) = (2nGG + nGg) / 2n$$

Y_{ijk} is the estimated breeding values of traits, μ the population mean, M_i the fix effect of genotypes, e_{ijk} the residual random error. There was no significant interaction between the genes additive effects.

3. RESULTS

3.1. Analysis of the Polymorphisms in the UCP Gene

Five variations in the UCP gene are shown in Table 1 with data obtained from 190 chickens. A single nonsynonymous SNP was a C > T substitution at nt position 1270 in exon-3 replacing Ala > Val. A PCR-RFLP test using HhaI restriction enzyme was designed to screen the UCP Ala118Val polymorphism (**Table 2**). **Figure 1** shows the electrophoretic gel pattern of the HhaI PCR-RFLP test where lanes 6 and 2 are Ala/Val heterozygotes with or without the 19-bp deletion, respectively. An additional restriction cut site was created within the PCR- amplified region by t > C silent substitution for cysteine at nt position 1316 in exon-3 that produced the second band from the top as seen in **Figure 1** for lane 5. The digested allele as allele "T", and the undigested allele was identified as allele "C". Three genotypes TC (Ala/Val), TT (Ala/Ala) and CC (Val/Val), were detected after PCR- RFLP analysis of the UCP gene. The allele frequencies were 0.21 for T and 0.79 for C, and the genotypic frequencies were 0.18, 0.12, and 0.70 for TC, TT and CC, respectively.

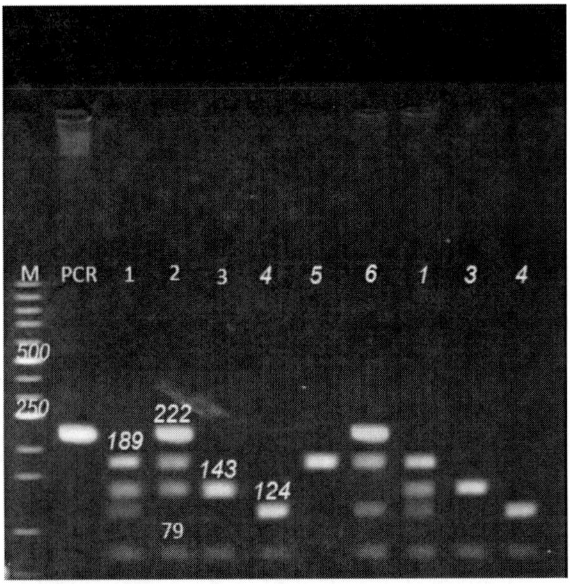

Figure 1. The electrophoretic gel patterns of the HhaI PCR- RFLP. M is Ladder, PCR is Fragment of PCR product, lanes 2 and 6 are heterozygote and other lanes are homozygote.

Table 1. Statistical description of data set for growth and egg production traits.

Traits	*No. of animal*	*Mean*	*Coefficient of variation*
BW1 (gr)	35.287	35.53	8.15
BW8 (gr)	43.067	563.7	17.09
BW12 (gr)	38.297	953.9	14.49
WSM (gr)	31.147	1694	11.90
ASM (day)	31.349	165.5	9.23
EN (number)	31.349	36.66	39.78
EW1 (gr)	27.294	41.21	15.74
EW28 (gr)	17.225	46.91	8.48
EW30 (gr)	19.031	48.12	8.50
EW32 (gr)	18.955	49.22	8.29
EW12 (gr)	18.847	46.62	9.28
AV (gr)	28.725	46.84	13.05
EM (gr)	28.725	1768	39.89
EINT (%)	31.349	57.07	33.29

BW1, BW8, BW12 = Body weight at birth, 8, 12 weeks of age. WSM = Body weight at sexual maturity. ASM = Age at first egg. EN = Egg number. EW1 = Weight of first egg. EW28, EW30 and EW32 = Average egg weight at 28, 30 and 32 weeks of age respectively. EW12 = Average egg weight for first 12 weeks of production. AV = Average for EW28, 30 and 32. EM = Egg mass (=EN × EW12). EINT = Egg production intensity (= Egg number/days recording) × 100).

Table 2. Summary of variations in UCP gene.

Location	*Position*	*Nucleotide changes*	*Amino acid changes*	*Subtype*
Exon 3	1270	C > T	Ala118Val	Missense
	1316	T > C	Cys133Cys	Silent

3.2. Relationship between Polymorphisms and Growth, Breeding Value of Growth and Reproductive Traits

The association between the AF433170: c. C1270T polymorphisms in the UCP gene and growth, breeding value of growth and egg production traits are shown in Table 3 and Table 4, respectively. UCP gene, the Missense mutations were significantly linked to ASM ($P < 0.05$) and also highly significantly linked to breeding value of ASM ($P < 0.05$). In this population, ASM was higher in the heterozygote (UCP Ala/Val) compared with the homozygote (UCP Val/Val).

4. DISCUSSION

The study of candidate genes is one of the primary methods to find specific genes related to economic traits in farm animals. UCP involved in regulation of energy metabolism has important physiological effects on economical important traits of farm animals. Gene polymorphisms can be used for improvement of the production traits by genetic selection, if the allelic association with the traits is determined. The chicken UCP gene was first discovered by Raimbault *et al.* [15]. Its function is still not clear, though there has been some research about it since its discovery. The nucleic acid sequence of avian UCP is highly homologous to both mammalian UCP2 and UCP3 [16]. The current hypothesis on the function of these two UCPs involves regulation of fatty acid oxidation, especially during metabolic stress [17] , possibly by transporting fatty acid ions out of the mitochondria [18]. Two significant associations ($P < 0.05$) were observed with growth, breeding value of growth and reproductive traits in this study (Table 3 and Table 4). Sherman et al. [19] observed significant association between a SNP (A/G) in bovine UCP-3 gene intron-3 and average daily gain and partial efficiency of growth in 9 Continental British hybrid beef steers. Han [20] reported the existence of three SNPs (G/A, C/T and G/T) in exon 3 and four genotypes in Chinese Qinchuan cattle. Association studies showed that the genotype AA had higher slaughter weight, carcass weight, carcass length, eye muscle area, water holding capacity, and marbling than other genotypes; the genotype AB had higher back fat than other genotypes. Li [21] found that the allele A of UCP-3/BglI polymorphism had higher frequencies than allele B in Nanyang, Luxi, and Yanbian cattle. Association analysis showed the superiority of the AA genotype over the AB genotype for the content of b-globin in Luxi cattle. Fasting, which is associated with a major reduction in total body energy expenditure, is associated with an increase in the expression of UCP2 and UCP3 [22]. However, it was also reported that UCP2 and UCP3 genes were unlikely to have a substantial effect on variation in obesity phenotypes, in particular US Caucasian human population [23]. In animal and human studies, UCP2 mRNA levels have been correlated with plasma-free fatty acid concentrations [24]. In adipose tissue of obesity-resistant mice, UCP2 expression is increased 2-fold when compared with obesity-prone animals [25]. Independent studies have identified quantitative trait loci for obesity or body fat in mice that coincide with the genomic location of the UCP2 gene [26] [27]. The SNP in the exon included synonymous and nonsynonymous polymorphisms. At any given position in a DNA sequence, a nucleotide can be substituted by any of the 4 nucleotide bases and may result in biallelic SNP. This occurs due to the low substitution rate of single nucleotides, estimated to be between 1×10^{-9} and 5×10^{-9} per nucleotide per year at neutral positions in mammals [28]. Based on these numbers, the probability of 2 independent base changes occurring at a single position is very low [28]. A large variation in UCP gene sequences in the studied population was found to reside in the first half of the gene. In the present study, a Missense polymorphism identified in the UCP gene at Ala118Val can be predicted to occur in the transmembrane region of the second domain close to the mitochondrial inner membrane space. The various biological functions of UCP are not known in detail, and based on the localization and nature of the codon UCP118 variant, it is difficult to predict if the encoded protein change may cause defects in the functionality of the UCP. Structural-functional studies are thus needed to elaborate on the effect of the UCP Ala118Val variant on UCP's protein structure as well as function. In a highly selected commercial population, changes in the allele and genotypic frequencies between generations may result from the net reproductive success of the individuals selected as breeding stock with the different genotypes and the selection norms adopted by the breeder. Variable associations of the identified polymorphisms may be a result of the differences in the population characteristics, sex, or both, indicating that the selection criteria may influence the production trait associations. This should be taken into consideration while selecting for the desired reproductive traits. Additional studies are required to expand the genetic, physiological, behavioral aspects involved in feed intake, digestion, and metabolism. The genomic diversity also has important implications in the evolutionary dynamics of species. Investigations of polymorphisms are useful for better understanding of the gene function, and those associated with commercially significant growth, breeding value of growth and reproductive traits have a potential for usage as molecular markers for selection programs. The identified polymorphisms and their associations with the traits of economic importance in present study provide greater insight into the role of gene involved in energy balance in poultry (UCP) and point toward the potential application of the findings for the enhancement of production traits by marker assisted selection.

Table 3. Association of the UCP genotypes at the growth and egg production traits (Mean ± S.E.).

Traits	Genotype TT	Genotype TC	Genotype CC
WSM	1795.28 ± 35.80	1842.07 ± 57.15	1721.10 ± 20.80
ASM	185.19 ± 3.45[ab]	189.29 ± 4.91[a]	176.08 ± 1.70[b]
EN	39.97 ± 1.36	42.58 ± 2.17	40.00 ± 0.79
EW28	48.33 ± 0.84	48.77 ± 1.46	47.86 ± 0.45
EW30	49.67 ± 0.74	50.30 ± 1.28	49.23 ± 0.43
AV	50.40 ± 0.65	51.48 ± 1.09	50.05 ± 0.37
EM	1981.90 ± 70.02	2157.50 ± 117.07	1985.76 ± 40.65
EINT	62.77 ± 2.37	65.76 ± 3.78	63.13 ± 1.37

[a,b]Values with different superscripts within the same row differ significantly ($P < 0.05$).

Table 4. Association of the UCP genotypes on breeding values of growth and egg production traits (Mean ± S.E.).

Traits	Genotype TT	Genotype TC	Genotype CC
WSM	13.74 ± 4.26	22.44 ± 6.86	−22.81 ± 6.80
ASM	−21.37 ± 0.98[ab]	−19.80 ± 1.61[a]	−24.57 ± 0.61[b]
EN	14.10 ± 0.28	14.26 ± 0.45	14.43 ± 0.17
EW28	0.85 ± 0.2	0.96 ± 0.33	0.94 ± 0.12
EW30	0.58 ± 0.24	0.38 ± 0.11	0.58 ± 0.15
AV	1.29 ± 0.17	1.50 ± 0.28	1.24 ± 0.10

[a,b]Values with different superscripts within the same row differ significantly ($P < 0.05$)

ACKNOWLEDGEMENTS

The authors thank all the team who worked on this study in the experiments and by providing technical assistance in the laboratory.

REFERENCES

1. Saltzman, E. and Roberts, S.B. (1995) The Role of Energy Expenditure in Energy Regulation: Findings from a Decade of Research. Nutrition Reviews, 53, 209-220. http://dx.doi.org/10.1111/j.1753-4887.1995.tb01554.x
2. Toubro, S., Sorensen, T.I.A., Ronn, B. and Astrup, A. (1996) Twenty Four-Hour Energy Expenditure: The Role of Body Composition, Thyroid Status, Sympathetic Activity, and Family Membership. Journal of Clinical Endocrinology and Metabolism, 81, 2670-2674.
3. Himms-Hagen, J. (1990) Brown Adipose Tissue Thermogenesis: Interdisciplinary Studies. Fatigue & Fracture of Engineering Materials and Structures, 4, 2890-2898.
4. Adams, S. (2002) Uncoupling Protein Homologs: Emerging Views of Physiological Function. Nutrition, 130, 711-714.
5. Krauss, S., Zhang, C.Y. and Lowell, B.B. (2002) A Significant Portion of Mitochondrial Proton Leak in Intact Thymocytes Depends on Expression of UCP2. Proceedings of the National Academy of Sciences of the United States of America, 99, 118-122. http://dx.doi.org/10.1073/pnas.012410699
6. Ricquier, D. and Bouillaud, F. (1997) The Mitochondrial Uncoupling Protein: Structural and Genetic Studies. Proceedings of the National Academy of Sciences of the United States of America-Biological Sciences, 56, 83-108. http://dx.doi.org/10.1016/s0079-6603(08)61003-x
7. Liu, Y.J., Liu, P.Y., Long, J., Lu, Y., Elze, L., Recker, R.R. and Deng, H.W. (2005) Linkage and Association Analyses of the UCP 3 Gene with Obesity Phenotypes in Caucasian Families. Physiological Genomics, 22, 197-203. http://dx.doi.org/10.1152/ physiolgenomics.00031.2005
8. Rudofsky, G.J., Schroedter, A., Schlotterer, A., Voron'ko, O.E., Schlimme, M., Tafel, J., Isermann, B.H., Humpert, P.M., Morcos, M., Bierhaus, A., Nawroth, P.P. and Hamann, A. (2006) Functional Polymorphisms of UCP 2 and UCP 3 Are Associated with a Reduced Prevalence of Diabetic Neuropathy in Patients with Type 1 Diabetes. Diabetes Care, 29, 89-94. http://dx.doi.org/10.2337/diacare.29.01.06.dc05-0757
9. Gimeno, R.E., Dembski, M., Gimeno, C.J., et al. (1997) Cloning and Characterization of an Uncoupling Protein Homolog: A Potential Molecular Mediator of Human Thermogenesis. Diabetes, 46, 900-906. http://dx.doi.org/10.2337/diab.46.5.900
10. Shojaei, M., Mohammad Abadi, M.R., Asadi Fozi, M., Dayani, O., Khezri, A. and Akhondi, M. (2011) Association of Growth Trait and Leptin Gene Polymorphism in Kermani Sheep. Journal of Cell and Molecular Research, 2, 67-73.
11. Askari, N., Mohammadabadi, M.R. and Baghizadeh, A. (2011) ISSR Markers for Assessing DNA Polymorphism and Genetic Characterization of Cattle, Goat and Sheep Populations. Iranian Journal of Biotechnology, 9, 222-229.
12. Zamani, P., Akhondi, M., Mohammadabadi, M.R., et al. (2013) Genetic Variation of Mehraban Sheep Using Two Intersimple Sequence Repeat (ISSR) Markers. African Journal of Biotechnology, 10, 1812-1817.
13. Abadi, M.M., Askari, N., Baghizadeh, A. and Esmailizadeh, A. (2009) A Directed Search around Caprine Candidate Loci Provided Evidence for Microsatellites Linkage to Growth and Cashmere Yield in Rayini Goats. Small Ruminant Research, 81, 146-151. http://dx.doi.org/10.1016/j.smallrumres.2008.12.012
14. Gilmour, A.R., Gogel, B.J., Cullis, B.R. and Thompson, R. (2006) ASReml User Guide. Release 2.0.
15. Raimbault, S., Dridi, S., Denjean, F., Lachuer, J., Couplan, E., Bouillaud, F., Bordas, A., Duchamp, C., Taouis, M. and Ricquier D. (2001) An Uncoulping Protein Homologue Putatively Involved in Facultative Muscle Thermogenesis in Birds. Biochemical Journal, 353, 441-444. http://dx.doi.org/10.1042/bj3530441
16. Evock-Clover, C.M., Poch, S.M., Richards, M.P., Ashwell, C.M. and McMurtry, J.P. (2002) Expression of an Uncoupling Protein Gene Homolog in Chickens. Comparative

Biochemistry and Physiology a-Molecular and Integrative Physiology, 133, 345-358. http://dx.doi.org/10.1016/S1095-6433(02)00113-7
17. Dulloo, A.G. and Samec, S. (2001) Uncoupling Proteins: Their Roles in Adaptive Thermogenesis and Substrate Metabolism Reconsidered. British Journal of Nutrition, 86, 123-139. http://dx.doi.org/10.1079/BJN2001412
18. Himms-Hagen, J. and Harper, M.E. (2001) Physiological Role of UCP3 May Be Export of Fatty Acids from Mitochondria When Fatty Acid Oxidation Predominates: A Hypothesis. Proceedings of the Society for Experimental Biology and Medicine, 226, 78-84.
19. Sherman, E.L., Nkrumah, J.D., Murdoch, B.M., Li, C., Wang, Z., Fu, A. and Moore, S.S. (2008) Polymorphisms and Haplotypes in the Bovine Neuropeptide Y, Growth Hormone Receptor, Ghrelin, Insulin-Like Growth Factor 2, and Uncoupling Proteins 2 and 3 Genes and Their Associations with Measures of Growth, Performance, Feed Efficiency, and Carcass Merit in Beef Cattle. Journal of Animal Science, 86, 1-16. http://dx.doi.org/ 10.2527/jas.2006-799
20. Han, R.H. (2008) Study on the Polymorphisms in UCP3, IGF2, CAPN1 Gene and Its Relationship with Carcass and Meat Quality Traits in Qinchuan Cattle and Its Hybrid Cattle (Dissertation). Northwest A & F University.
21. Li, Q.L. (2006) Study on the Relationship between Polymorphism of UCP3 Gene and Biochemical Index in Serum of Bovine (Dissertation). Northwest A & F University.
22. Boss, O., Hagen, T. and Lowell, B.B. (2000) Uncoupling Proteins 2 and 3, Potential Regulators of Mitochondrial Energy Metabolism. Diabetes, 49, 143-156. http://dx.doi. org/10.2337/diabetes.49.2.143
23. Guo, J.J., Liu, Y.J., Li, M.X., Yang, Y.J., Recker, R.R. and Deng, H.W. (2005) Linkage Exclusion Analysis of Two Candidate Regions on Chromosomes 7 and 11: Leptin and UCP2/UCP3 Are Not QTLs for Obesity in US Caucasians. Biochemical and Biophysical Research Communications, 332, 602-608. http://dx.doi.org/10.1016/j.bbrc.2005.04.160
24. Argyropoulos, G. and Harper, M.E. (2002) Uncoupling Proteins and Thermoregulation. Journal of Applied Physiology, 92, 2187-2198. http://dx.doi.org/10.1152/ japplphysiol.00994.2001
25. Fleury, C., Neverova, M., Collins, S., et al. (1997) Uncoupling Protein-2: A Novel Gene Linked to Obesity and Hyperinsulinemia. Nature Genetics, 15, 269-272. http://dx.doi.org/10.1038/ng0397-269
26. York, B., Truett, A.A., Monteiro, M.P., et al. (1999) Gene-Environment Interaction: A Significant Diet-Dependent Obesity Locus Demonstrated in a Congenic Segment on Mouse Chromosome 7. Mammalian Genome, 10, 457-462. http://dx.doi.org/10. 1007/ s003359901023
27. Taylor, B.A. and Phillips, S.J. (1996) Detection of Obesity QTLs on Mouse Chromosome-1 and Chromosome-7 by Selective DNA Pooling. Genomics, 34, 389-398. http://dx. doi.org/10.1006/geno.1996.0302
28. Vignal, A., Milan, D., San Cristobal, M. and Eggen, A. (2002) A Review on SNP and Other Types of Molecular Markers and Their Use in Animal Genetics. Genetics Selection Evolution, 34, 275-305. http://dx.doi.org/10.1186/1297-9686-34-3-275

CHAPTER 15

Revisiting demographic processes in cattle with genome-wide population genetic analysis

Pablo Orozco-terWengel [1*], MarioBarbato[1], EzequielNicolazzi [2], FilippoBiscarini [2], Marco Milanesi [3], WynDavies [4], DonWilliams [5], AlessandraStella [2], Paolo Ajmone-Marsan[3] and MichaelW.Bruford [1]

[1]*School of Biosciences, Cardiff University, Cardiff, UK*

[2]*Parco Tecnologico Padano, Lodi, Italy*

[3]*Faculty of Agriculture, Università Cattolica del Sacro Cuore, Piacenza, Italy*

[4]*Dinefwr, National Trust, Llandeilo, UK*

[5]*Independent Researcher, Swansea, UK*

The domestication of the aurochs took place approximately 10,000 years ago giving rise to the two main types of domestic cattle known today, taurine (*Bos taurus*) domesticated somewhere on or near the Fertile Crescent, and indicine (*Bos indicus*) domesticated in the Indus Valley. However, although cattle have historically played a prominent role in human society the exact origin of many extant breeds is not well known. Here we used a combination of medium and high-density Illumina Bovine SNP arrays (i.e., ~54,000 and ~770,000 SNPs, respectively), genotyped for over 1300 animals representing 56 cattle breeds, to describe the relationships among major European cattle breeds and detect patterns of admixture among them. Our results suggest modern cross-breeding and ancient hybridisation events have both played an important role, including with animals of indicine origin. We use these data to identify signatures of selection reflecting both domestication (hypothesized to produce a common signature across breeds) and local adaptation (predicted to exhibit a signature of selection unique to a single breed or group of related breeds with a common history) to uncover additional demographic complexity of modern European cattle.

Keywords: cattle, *Bos taurus*, *Bos indicus*, phylogeography, adaptation, SNP array

1. INTRODUCTION

Modern cattle comprise two species, *Bos taurus* and *Bos indicus*, both derived from the extinct aurochs (*B. primigenius*). The divergence between the two species has been dated ~250,000 years ago on the basis of mitochondrial DNA haplotypes (Bradley et al., 1996). The two species are interfertile and a number of hybrid breeds have been established (e.g., the American Beefmaster that was established from a Shorthorn × Hereford × Brahman cross). The details of cattle domestication have been a contentious, and arguments in favor of two or three domestication events have been put forward. The hypothesis arguing for two domestication sites suggests that taurine cattle were domesticated in the Fertile Crescent, likely from *B. primigenius,* while indicine cattle were domesticated in the Indus valley, likely from *B. primigenius namadicus* (Loftus et al., 1994; Troy et al., 2001; Lewis et al., 2011). The hypothesis arguing for three domestication sites adds a Northeast African center, that gave rise to the African taurine breeds deriving form *B. primigenius opisthonomous* (Wendorf and Schild, 1994; Payne and Hodges, 1997). Mitochondrial DNA evidence suggests that two main lineages were domesticated, one now typical of taurine breeds, and the other now typical of indicines (Loftus et al., 1994; Bradley et al., 1996). The domestication process is thought to have originated sometime between 8000 and 10,000 years ago (Ajmone-Marsan et al., 2010; Gautier et al., 2010; Lewis et al., 2011). Several potential routes have been hypothesized to explain the dispersal of taurine cattle from the Fertile crescent, mediated by Neolithic farmers, including alternative routes into Europe, occurring on at least two occasions (a Danubian route and a Mediterranean route Payne and Hodges, 1997; Pellecchia et al., 2007; Ajmone-Marsan et al., 2010; Taberlet et al., 2011). As aurochs only became extinct recently (Machugh et al., 1998; Beja-Pereira et al., 2006), it is possible that both taurine and indicine cattle crossed with their wild ancestor during their expansion out of the domestication centers, although the evidence supporting it is much debated (Troy et al., 2001; Gotherstrom et al., 2005; Beja-Pereira et al., 2006; Bollongino et al., 2008; Perez-Pardal et al., 2010; Edwards et al., 2011).

Several studies have addressed the population structure and phylogenetic relationships between taurine and indicine cattle breeds. Initially, studies were based on allozymes, mitochondrial DNA sequences, and microsatellites (Loftus et al., 1994; Medjugorac et al., 1994; Bradley et al., 1996; Machugh et al., 1998; Kantanen et al., 2000; Hanotte et al., 2002; Gautier et al., 2007) describing patterns of genetic variation in cattle, e.g., the two main lineages and differentiation between taurine and indicine breeds, albeit largely focusing on taurines. Nevertheless, such studies identified a clear differentiation between breeds from the two species, and between African and European taurines (Gautier et al., 2007). Interestingly, this handful of markers separated the African taurine breeds (e.g., Lagune, N'Dama and Somba) living in tsetse fly endemic regions from those that did not co-occur with these flies (e.g., Kuri and Borgou), suggesting that demographic signals among these breeds are strong and possibly related to properties of the environment they occupy. Recently, surveys of genome-wide genetic variation in many breeds have been carried out using SNP arrays (Bovine Hapmap et al., 2009; Mctavish et al., 2013; Decker et al., 2014). These more comprehensive analyses included tens of thousands of SNPs distributed across all chromosomes genotyped in several breeds from around the world. With these data, the differentiation between taurine and indicine breeds has been measured at approximately 10%, with approximately 5% for the divergence between African and non-African taurines (Gautier et al., 2010; Mctavish et al., 2013; Decker et al., 2014). Analyses using this type of large-scale data further identified a split between African longhorn taurines (N'Dama) and shorthorn taurines (e.g., Somba and Lagune—Gautier et al., 2010). Such analyses have also proved powerful enough to identify the mixed indicine/taurine genetic component of recently established hybrid breeds such as Santa Gertrudis and Beefmaster (Gautier et al., 2010; Mctavish et al., 2013).

Cattle are the most common large livestock species in the world; the global population size of which is approximately 1400 million animals (Felius et al., 2011; Taberlet et al., 2011). Among these, 1311 breeds have been recognized, 209 of which are now extinct (Rischkowsky and Pilling, 2007). Since domestication, cattle have spread around the world, with farmers selecting animals with desirable characteristics and establishing many breeds. This process, while not passive, was until recently slow and resulted in animals becoming adapted to the local conditions, e.g., feed types, local weather, and diseases (Wood, 1973; Cobb, 2006; Russell, 2007). The last two centuries saw a dramatic acceleration in the process of artificial/directional selection, largely as a consequence of the development of the breed concept, artificial insemination, and the development of statistical approaches used to estimate breeding values used to choose sires and dams for breeding purpose. This has limited the intercrossing between animals of divergent genetic background in order to increase or maintain breed purity, especially in dairy cattle [in beef cattle, conversely, cross-breeding is often used to exploit also non-additive genetic variation i.e., heterosis (Simm, 1998); still, pure lines for cross-breeding need to be selected and maintained]. Additionally, as a consequence of artificial insemination and marketing policies, few "champion" males with characteristics of interest have been used extensively to fertilize large proportions of females. The above developments have led to: (i) intense within-breed (or line) selection; (ii) importation/exportation of elite breeding stock to different environments/countries (migration); (iii) strong founder effect of "champion/top ranked" males both in breed formation or in already established breeds whose effective population size has thus been shrunken; (iv) larger genetic drift as a consequence of breed isolation. Generally, by trying to keep breeds pure, gene-flow among them has been reduced or even halted (unless done to improve breeds by upgrading Felius et al., 2011), and by using relatively few males to fertilize many females, inbreeding is increased (Taberlet et al., 2011). These approaches are used to manipulate the properties of a breed at the expense of the decrease in the genetic variability (Bulmer, 1971) which may be accompanied by a decrease in genetic health of the breed (e.g., reduction in fertility in Holstein cattle via the increase in frequency of deleterious variants through the process of inbreeding) (Pryce et al., 1997, 2004; Biffani et al., 2005). Overall, this change in farming practices resulted in a change from limited selection throughout much of the domestic history of cattle (~9800 years) to relatively stronger selection in the last ~200 years.

The relatively rapid distribution of cattle around the world over the last ~10,000 years suggests that populations have been potentially exposed to selective pressures deriving from the environmental variables that they had not been exposed previously, e.g., diseases occurring outside of the domestication centers (Hanotte et al., 2002; Felius et al., 2014; Xu et al., 2014). Consequently, while some cattle breeds represent recent introductions to new geographical areas such as the New World or Oceania, other breeds represent long established populations outside of the domestication center. The genetic makeup of such populations is a valuable source to understand the processes driving local adaptation, as the former represent genetic pools currently being shaped by local selective pressures (i.e., may serve as examples of selection in action) (Flori et al., 2012), while the latter represent established populations where the extant genetic variation reflects the outcome of local adaptation (Felius et al., 2014). In particular, long established local breeds are likely to represent the reservoirs of important genetic variation with adaptive potential that needs to be characterized (Taberlet et al., 2011; Felius et al., 2014) before it possibly disappears as a consequence of the widespread use of a handful of industrial breeds among others (Fao, 2007; Thornton, 2010; Herrero and Thornton, 2013; Perry et al., 2013).

Here, we collated a large-scale genome-wide dataset of SNP polymorphism in cattle breeds from around the world representing published taurine and indicine breeds, and new data for four European taurine breeds, all genotyped for a common set of ~35,000 SNPs in over 1300 animals. We used these data to characterize the distribution of genetic variation among cattle breeds around the world, and the genealogical relationships between these breeds. The dataset described here represents multiple breeds from various continents (e.g., multiple European and African taurines, as well as Asian and African indicines) providing a unique experimental set up that allows searching for signatures of selection that are consistent across multiple breeds from different continents. This approach enables to separate signatures of selection that may be breed specific from selection signals common for several breeds occurring in the same geographic region. We extend the published dataset with new SNP data to a case study where we compare white cattle breeds of conservation concern in the UK (White Park and Chillingham) with a candidate Roman ancestor, the Chianina, to help resolve the commonly held belief of a Roman origin of the Welsh White Park cattle (Felius et al., 2011).

2. MATERIALS AND METHODS

2.1 SNP Array Data

SNP array data was obtained for a total of 1346 animals representing 46 cattle breeds, and four species related to cattle (*B. gaurus*—Gaur, *B. javanicus*—Banteng, *B. gruniensis*—Yak, and *Bison bison*—Bison) used as outgroups (Table 1). The data comprises a combination of previously published datasets and new data collected for this study. The published data represents breeds published by Gautier et al. (2009); Gautier et al. (2010) and the Bovine HapMap project (Bovine Hapmap et al., 2009) genotyped with Illumina's BovineSNP50 v.1 and v.2 chip assay, respectively (Bovine Hapmap et al., 2009) (Table 1). Four new breeds were genotyped with the BovineHD Beach Chip for this project, namely, Welsh While Park (Dinefwr), Chianina, Romagnola and Chillingham. All SNP coordinates were converted to UMD3.1 bovine assembly (RefSeq: GCF_000003055.5). The datasets were merged using PLINK v1.7 (Purcell et al., 2007). The merged dataset was filtered to keep only individuals with 95% of their SNPs called and SNPs with 95% call rate across all samples. The SNPs and individuals that did not pass these filters were discarded. The total SNPs left in the dataset was 36,503 (data is available upon request from the authors). The data for each breed was phased with fastPHASE v1.4 (Scheet and Stephens, 2006).

Table 1. Breed abbreviations and species identification.

Abb	BN	Tax	*D*	*N*	Ho	*F*
ABO	Abondance	*Bos taurus*	2	22	0.308	0.089
ANG	Angus	*Bos taurus*	1	61	0.305	0.098
AUB	Aubrac	*Bos taurus*	2	22	0.301	0.112
BPN	Bretonne black pied	*Bos taurus*	2	18	0.317	0.064
BRU	French brown swiss	*Bos taurus*	2	18	0.288	0.150
BSW	Brown swiss	*Bos taurus*	1	24	0.280	0.171
CHA	Charolais	*Bos taurus*	1	20	0.314	0.071
CHL	Charolais	*Bos taurus*	2	26	0.320	0.054
GAS	Gascon	*Bos taurus*	2	22	0.305	0.100
GNS	Guernsey	*Bos taurus*	1	21	0.275	0.188
HFD	Hereford	*Bos taurus*	1	31	0.304	0.101
HO2	Holstein	*Bos taurus*	2	31	0.316	0.066
HOf	Holstein	*Bos taurus*	1	30	0.313	0.077
HOL	Holstein	*Bos taurus*	2	64	0.319	0.057
JE2	Jersey	*Bos taurus*	2	28	0.263	0.223
JEf	Jersey	*Bos taurus*	1	21	0.277	0.180
JER	Jersey	*Bos taurus*	2	28	0.263	0.223
LMS	Limousin	*Bos taurus*	1	44	0.309	0.086
MAN	Maine-Anjou (Rouge des Prés)	*Bos taurus*	2	16	0.303	0.105
MAR	Maraichine (Parthenaise)	*Bos taurus*	2	19	0.318	0.059
MON	Montbeliarde	*Bos taurus*	2	30	0.299	0.116
NOR	Normande	*Bos taurus*	2	30	0.307	0.094
NRC	Norwegian red cattle	*Bos taurus*	1	21	0.317	0.063
OUL	Oulmès Zaer	*Bos taurus*	2	27	0.288	0.149
PMT	Piedmontese	*Bos taurus*	1	24	0.321	0.053
PRP	French red pied lowland	*Bos taurus*	2	22	0.325	0.039
RGU	Red Angus	*Bos taurus*	1	15	0.305	0.100
RMG	Romagnola	*Bos taurus*	1	24	0.291	0.141
ROM	Romagnola	*Bos taurus*	3	13	0.293	0.134
CHI	Chianina	*Bos taurus*	3	14	0.285	0.158
CIL	Chillingham	*Bos taurus*	3	16	0.026	0.924
W_P	White park	*Bos taurus*	3	15	0.245	0.276
SAL	Salers	*Bos taurus*	2	22	0.285	0.157
TAR	Tarine	*Bos taurus*	2	18	0.300	0.113
VOS	Vosgienne	*Bos taurus*	2	20	0.312	0.078
BAO	Baoule	*Bos taurus*	2	29	0.216	0.362
LAG	Lagune	*Bos taurus*	2	30	0.183	0.460
NDA	N'Dama	*Bos taurus*	1	25	0.209	0.381
ND1	N'Dama	*Bos taurus*	2	14	0.235	0.307
ND2	N'Dama	*Bos taurus*	2	17	0.237	0.299
ND3	N'Dama	*Bos taurus*	2	25	0.210	0.381
SOM	Somba	*Bos taurus*	2	30	0.217	0.359
BRM	Brahman	*Bos indicus*	1	25	0.190	0.440
GIR	Gir	*Bos indicus*	1	24	0.160	0.528
NEL	Nelore	*Bos indicus*	1	21	0.161	0.524
ZBO	Zebu Bororo	*Bos indicus*	2	23	0.240	0.292
ZFU	Zebu Fulani	*Bos indicus*	2	30	0.241	0.289
ZMA	Zebu from Madagascar	*Bos indicus*	2	30	0.194	0.427
BMA	Beefmaster	*hybrid*	1	24	0.328	0.031
SGT	Santa Gertrudis	*hybrid*	1	24	0.313	0.075

BOR	Borgou	*hybrid*	2	30	0.263	0.222
KUR	Kuri	*hybrid*	2	30	0.261	0.229
SHK	Sheko	*hybrid*	1	20	0.250	0.260
SIM	Simmental	*Bos taurus*	1	3	NA	NA
GDV	Gelbvieh	*Bos taurus*	1	3	NA	NA
OBB	North American Bison	*Bos Bison*	2	4	NA	NA
OBJ	Banteng	*Bos javanicus*	1	2	NA	NA
OGR	Gaur	*Bos gaurus*	1	4	NA	NA
OYK	Yak	*Bos grunniens*	1	2	NA	NA

*The three letter code used to identify the breed throughout the manuscript is shown (**Abb**), along with the breed name (**BN**) and the taxonomic status indicating the species name or whether the breed is a B. taurus x B. indicus hybrid (**Tax**), the data set's origin (**D**), namely Bovine Hapmap Consortium (1), Gautier et al., 2010 (2), this study (3), the breed's sample size (**N**), the observed heterozygosity (**Ho**), and the inbreeding coefficient (**F**).*

2.2 Population Divergence

An analysis of population structure was carried out using the software Admixture v1.22 (Alexander et al., 2009), which uses a model-based estimation of indivdual ancestry for a range of prior values of K defined by the user. A cross-validation approach is used to determine the most likely number of populations (K) in the data. For each tested value of K, Admixture estimates the proportion of each individual's genotype deriving from each cluster. The values of K tested were in the range between 1 and 60 to accommodate potential population structure within breeds. For this analysis we excluded outgroup samples, breeds represented by fewer than 10 individuals, and SNPs with a minimum allele frequency lower than 0.01, linkage disequilibrium higher than 0.1 using a sliding window approach of 50 SNPs and a step size of 10 SNPs (Alexander et al., 2009). A principal components analysis was carried out using flashpca v1.2 (Abraham and Inouye, 2014) with default settings in order to investigate the ordinal relationships among breeds and individuals. Lastly, a NeighbourNet network was constructed using Reynold's distance among breeds using Splitstree v4.13.1 (Huson and Bryant, 2006), and a dendrogram using the same genetic distance and 100 bootstrap replicates to assess the statistical support of breed; clustering was performed using Phylip v3.69 (Felsenstein, 1993).

Within this framework, the Welsh White Park cattle breed was analyzed along the other cattle breeds to identify their relative position within the dendrogram and PCA explaining the similarities between cattle breeds. In particular, the common belief of a Roman origin of the Welsh White Park was addressed by comparing the position of this breed in relation to the Italian breeds, in particular the large Chianina breed, one of the oldest cattle breeds originating from the region of Valdichiana in central Italy, and Romagnola. Additionally, we estimated the relationships between Welsh White Park and other 17 breeds in the dataset using Treemix v1.12 (Pickrell and Pritchard, 2012) in order to determine the historical relationships between these breeds in terms of splits and migration (mixtures) between breeds. The breeds used for this analysis were Welsh White Park, the Italian

Chianina, Romagnola (ROM), and Piedmontese, the British breeds Chillingham, Hereford, Angus, Jersey (JER), as well as five other non-African taurines (Holstein (HOL), Brown Swiss, Charolais, Normande, Vosgienne), the four African taurines Baoule, Lagune, Somba and N'Dama (NDA), and the indicine breed Brahman as outgroup. Treemix was run iteratively for values of the migration parameter between 0 and 12. The f index representing the fraction of the variance in the sample covariance matrix ($\hat{W}$) accounted for by the model covariance matrix (W) was used to identify the number of modeled migration events that best fitted the data (Pickrell and Pritchard, 2012).

2.3 Demographic Estimation

Recent demographic history was measured by the trend in effective population size (Ne) change over the last 8000 years using the software SNeP v1.0 and default settings (Barbato et al., 2015). Loci with missing data were excluded from this analysis, as well as SNPs with a minimum allele frequency (MAF) smaller than 5% (Sved et al., 2008; Corbin et al., 2012). Linkage disequilibrium (LD) was calculated between each pair of SNPs separated by a minimum distance of 5 kbp and maximum distance of 1 Mbp using Hill and Robertson's r^2 (Hill and Robertson, 1968). LD values were grouped in 30 distances bins, such that each bin represented approximately the same number of pairwise LD estimates used to estimate Ne (Barbato et al., 2015). Within each bin the mean r^2 was calculated and used to estimate $N_t = 1/(4\,f(c))\,(1/r^2 - 1)$, where $f(c) = c\,[(1 - c/2)/(1 - 2)^2]$ (Sved, 1971); c is the linkage distance inferred from the physical distance between SNPs assuming $1\ Mb \sim 1\ cM$ and $N_t$represents the effective population size estimate at $t = 1/2c$ generations ago.

2.4 Signatures of Selection

As the dataset analyzed here comprised *B. taurus* breeds from Europe and Africa, as well as *B. indicus* breeds from Asia and Africa, we sought to identify signatures of selection in a replicated manner (Figure 1). Initially we compared four taurine breeds (two European—Gascon and Chianina; and two African—Baoule and N'Dama) with four indicine breeds (two African – Zebu Fulani and Zebu Bororo; and the Asian Gir and Brasilian Nelore; Nelore was recently derived from Indian Ongole cattle, therefore, we broadly refer to these breeds as Asian indicine) to identify signatures of selection between *B. taurus* and *B. indicus*. Within *B. taurus*, we randomly chose three European breeds to compare against three African breeds in order to identify selection signatures specific to breeds in these continents. In the same way, for *B. indicus* we compared three Asian breeds against three African breeds (Figure 1). In contrast to previous studies, this experimental design enabled us identifying potential signatures of selection reflecting adaption to the local environment, instead of breed specific signatures that potentially reflect the breed's particular history.

Figure 1. Pairwise comparisons for analysis of selection. Each box corresponds to one of the three types of pairwise comparisons carried out to detect signatures of selection. The blue box (comparison 1) indicates the comparison between four taurine and four indicine breeds used to detect signatures of selection between these breeds. The red box (comparison 2) shows the taurine breeds used in the comparison between African and not African breeds. The green box (comparison 3) shows the indicine breeds used in the comparison between African and not African breeds.

The test of selection used was the cross population extended haplotype homozygosity test (XP-EHH) calculated for each SNP in the dataset as implemented in Selscan v1.0 (Szpiech and Hernandez, 2014). Selscan was used with default settings except for the maximum gap in bp allowed between SNPs, which was set to 400,000 to address the BovineSNP50 SNP density. For the comparisons between taurine and indicine breeds the following pairwise combinations were used: N'Dama (African taurine)/Gir (Asiatic indice), Gascone (European taurine)/Zebu Bororo (African indicine), Baoule (African taurine)/Zebu Fulani (African indicine) and Chianina (European taurine)/Nelore (Asian indicine). For the within-taurine comparison three pairwise tests were done between a European breed and an African breed, and for the within-indicine comparison three pairwise comparisons were carried out between an Asiatic breed and an African breed (Figure 1). For each pairwise comparison, the XP-EHH results where standardized for each chromosome separately, and the 5 and 95% quantiles of the standardized XP-EHH distribution for all markers where used as threshold to identify outlier SNPs, i.e., SNPs with XP-EHH values equal or smaller (more negative) or equal or larger (more positive) than the chosen thresholds. As multiple XP-EHH were carried out for each comparison (e.g., four comparisons between taurine and indicine breeds) this analysis was the equivalent of a replicated test. We took advantage of this design to identify SNPs under selection as those that showed for several pairwise comparisons an extreme XP-EHH with the same sign (e.g., all pairwise comparisons should an extreme positive XP-EHH).

For the between species comparison we defined the consistent SNPs under selection as those showing the same XP-EHH trend in at least three of the four pairwise comparisons. We did not restrict ourselves to only those SNPs where the four tests showed the same result to avoid excluding SNPs that failed to pass the 5% MAF threshold within a particular breed but which showed a consistent signature of selection across the remaining three comparisons. For the within species comparison of breeds occurring on different continents we identified consistent SNPs under selection as those showing an extreme XP-EHH with the same sign in each of the three comparisons made. The SNPs showing consistent signatures of selection across multiple pairwise comparisons where linked to neighboring genes using a window approach of 50 K base pairs, and a Gene Ontology analysis was carried out on these SNPs using Gorilla (www.cbl-gorilla.cs.technion.ac.il; Eden et al., 2009).

3. RESULTS

Of the 36,503 SNPs analyzed in 1345 animals representing taurine and indicine breeds, the observed heterozygosity varied between a minimum of 0.026 (Chillingham) and a maximum of 0.33 (the hybrid breed Beefmaster; Table 1). The comparison of expected heterozygosity (unbiased to sample size) between the taurine and indicine breeds was not significant, as well as the comparison between breeds in and out of Africa (all Welch *t*-test *p*-values > 0. 2). However, the average observed heterozygosity in taurine breeds was slightly larger than that in indicine breeds [0.28 standard deviation (sd) 0.05 and 0.20 sd 0.04, respectively]. The average expected heterozygosity across breeds in both datasets was similar, i.e., 0.34 sd 0.0001. The inbreeding coefficient (F_{IS}) was significantly lower for taurine breeds (F_{IS}: 0.18 sd 0.16) than for indicine breeds (F_{IS}: 0.42 sd 0.11; Welch *t*-test *p*-value: 0.0011). Within *B. taurus*, the African breeds presented a significantly higher inbreeding coefficient than the non-African taurines (Welch *t*-test *p*-value: 1.54×10^{-7}). Similarly, the non-African taurines also presented significantly lower F_{IS} values than either of the groups of indicine breeds (Welch *t*-test *p*-value: 1.84×10^{-5} and 0.026, respectively). Among the non-African taurines, the Chillingham presented by far the highest inbreeding coefficient of the entire dataset (F_{IS}: 0.92), on average five-fold larger than other taurines, and two-fold larger than the indicine breeds. The Welsh White Park is thought to be related to the Chillingham breed, but presented an observed heterozygosity (0.245) only slightly smaller than the average value observed in the other non-African taurines. The F_{IS} value for Welsh White Park (0.27) was less than a third of that of Chillingham but still almost twice the average across the other non-African taurine breeds (0.14 sd 0.14). With the exception of Chillingham, the four breeds genotyped here presented typical observed heterozygosity and F_{IS} values for European taurine breeds (Table 1). Removing from the analysis the breeds with samples sizes smaller than 20 did not change the results (results not shown).

3.1 Admixture and Genetic Relationships

Admixture analysis between taurine and indicine breeds was initially run for values of the number of clusters (K) between 1 and 60. The cross validation (CV) statistic used to choose the most suitable number of clusters had its lowest value at $K = 60$. However, the shape of the CV curve suggested that higher K values may present lower CV. Consequently, we carried out runs of Admixture for larger values of K (e.g., 80). However, the CV values continued to reduce the larger the K numbers tested (SM Figure 1). While it may be possible that some family structure within breeds explains this trend, the substantially longer run time needed by the algorithm for such K values combined with the small decrease in CV suggests that clustering solutions larger than K ~60 potentially represent spurious results (a similar observation was made by Mctavish et al. (2013) who only achieved marginal increases on the likelihood scores using structure in their cattle dataset for values of K beyond 3).

The hierarchical clustering analysis with Admixture showed that the largest difference for $K = 2$ was between taurine and indicine breeds. Conditioning the data to $K = 3$ separated the taurine breeds with an African origin from the non-African, suggesting a large differentiation between taurine breeds in these two groups of samples (Figure 2). Contrastingly, for values up to $K = 10$, the African indicine breeds presented a remarkably similar distribution of genetic variation to that of the Asian indicine breeds, which did not form a separate cluster. Nonetheless, the African indicine breeds consistently presented a distinctive proportion of their genotype representing taurine ancestry throughout the analysis (a pattern almost entirely absent in the breeds of Asian origin).

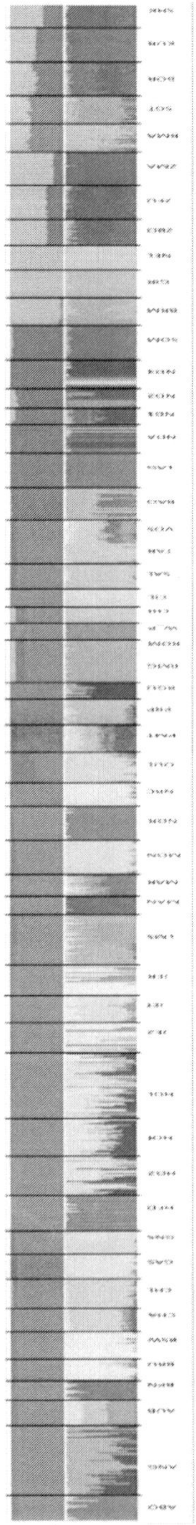

Figure 2. Analysis of Population Structure. Results of the analysis of population structure conditioning the dataset to 3 clusters (top row) and to 51 clusters (bottom row). Each animal is represented by a straight bar that is colored. The amount of a color reflects the individual's proportion of genetic variation originating in the cluster of that color. Each breed is labeled in the center of its box on the bottom of each plot. In the top row the non-African taurine breeds are labeled in blue, the African taurine in red and the indicine (both African and not African in green). The new datasets produced for this project are highlighted in green.

The *B. taurus*/*B. indicus* hybrids show a genetic make-up consisting of non-African taurine and *B. indicus* genetic components (Beefmaster and Santa Gertrudis), or African taurine and *B. indicus* (Borgou, Kuri, and Sheko).

The Admixture analysis was complemented with a principal component analysis (Figure 3). The first principal component (PC1) explained ~10% of the total variation in the data and separated taurine form indicine breeds. Besides of PC1, the only other component explaining a substantial part of the variance was PC2 (~5%), which separates the African taurine from the remaining taurine, as well as the African indicine from the Asian indicine breeds. PC3 to PC7 explained between less than 2 and 1.4% of the variance, while the remaining components explained less than 1% of the total variance, thus were not taken into consideration. PC3 separated Holstein from the remaining taurine breeds. PC4 identified a group of British breeds formed by Angus, Red Angus and Hereford, and PC5 identified a component of variation specific to Hereford separating it from the other taurines. PC6 identified the Chillingham cattle as well as the Jersey breed, and PC7 identified a specific component of variation to Chillingham separating it from the remaining taurine breeds. The combination of PC1 and PC2 identify the major groups of individuals described here in terms of species and geographic component (Figure 3A) with the hybrid breeds occurring between these. As expected from the Admixture analysis, the breeds Kuri, Borgou, and Sheko are placed with these principal components between the group of African taurine and indicine. Similarly, Beef Master and Santa Gertrudis are placed between the group of non-African taurine and indicine breeds, although from this analysis it is not clear whether their indicine genetic component is more similar to African indicine or Asian indicine.

The methods described above identified differences between breeds but did not allow us to determine how these relate to each other. We therefore estimated a NeighbourNet network with these breeds using Reynold's distance. The torso of the network shows multiple pathways of connection between breeds reflecting the relatively recent divergence between many populations (Figure 4).

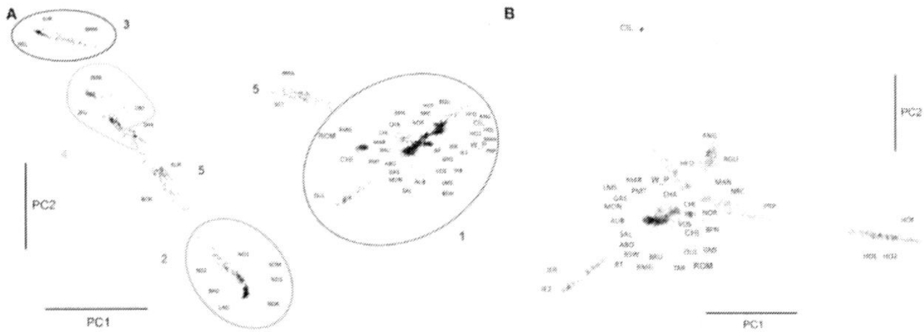

Figure 3. Principal Component Analyses. (A) This plots shows PC1 and PC2, which together explain ~15% of the variance in the dataset. The ellipses represent: Red=non-African taurine, Purple=African taurine, Dark Blue=Asian indicine, Light Blue=African indicine. Three letter codes refer to the breed abbreviations (Table 1). (B) Plot showing PC1 and PC2 of the analysis of the non-African taurine breeds, which together explain ~6.5% of the variance in that dataset. The new datasets produced for this project are highlighted in green in both (A,B).

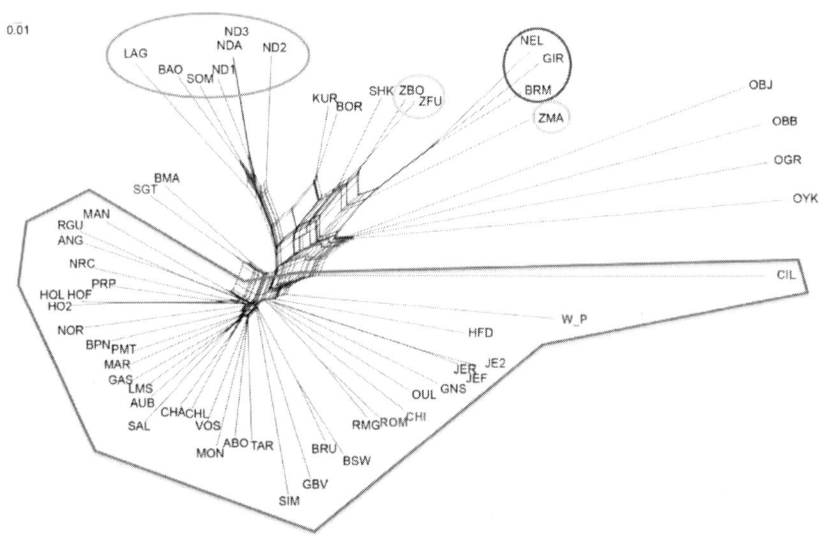

Figure 4. NeighbourNet depicting the relationships between cattle breeds. Three letter codes refer to the breed abbreviations (Table 1). The groups marked with ellipses in Figure 3A were marked with the same colors here as well. The new datasets produced for this project are highlighted in green.

However, the topology of the network resulted in the pattern expected from the previous analyses (Admixture and PCA), where a clear separation between taurine and indicine occurs, as well as the distinct separation between the African taurine breeds from the non-African taurines. Similarly to the PCA and in contrast to the Bayesian approach of Admixture, the network also resolved the African indicine breeds separately from their Asian counterparts. The dendrogram depicting the clustering of breeds using Reynold's distance recovered the grouping of breeds identified in the PCA with most branches (83%) supported with high statistical support (bootstrap values higher than 70%; SM Figure 2). The clustering of the outgroups was not well resolved, except for the grouping between bison and yak. The indicine breeds clustered together with the Asian indicine showing a greater similarity to each other, and then to the Madagascan zebu. The two African indicine clustered together and then clustered with the remaining indicines. The taurine breeds separated into two clusters, one representing the African taurines and the other the non-African taurines. The hybrids Beefmaster and Santa Gertrudis clustered together within the non-African taurine group reflecting their larger taurine genetic component. Contrastingly, the African hybrids (Sheko, Borgou, and Kuri) clustered with the indicine breeds suggesting a higher indicine genetic component in their genotype.

A comparison between the Welsh White Park cattle and the other British breeds (Chillingham, Hereford, and Jersey) and Italian breeds (Chianina, Romagnola, Piedmontese) was carried out to establish whether the UK white cattle have a genetic link with Roman white cattle, as is popularly believed. The Principal Components Analysis of this datasets identified the White Park breed clustering near the other British breeds (i.e., HFD, ANG, RGU), instead of closer to the Italian breeds (Figure 3B), as would be expected if White Park cattle had a strong Italian genetic component due to its speculated Roman origin. Similarly, in the dendrogram, the Italian breeds clustered together and separately from the Welsh White Park and Chillingham breeds, the latter two not clustering with any of the other taurine breeds suggesting their genetic distinctiveness. The Treemix analysis revealed a similar topology with the Welsh White Park and Chillingham clustering together and then clustering to Hereford and Angus, rather than clustering with the Italian breeds (Figure 5, SM Figures 3, 4). Adding up to seven or eight migration edges to the tree topology improved the amount of the variance explained by the phylogenetic model, albeit marginally (f statistic ~ 0.99965; Figure 5; Pickrell and Pritchard, 2012). The model without migration edges already presents an f statistic of 0.9981 (Figure 5 inset), a value above the threshold described previously, and above which the phylogram was not better explained by adding additional migration events (Pickrell and Pritchard, 2012; Decker et al., 2014). However, more migration edges did not seem to further increase this variance as the f statistic reached an asymptote between 7 and 8 migration edges (Figure 5 inset).

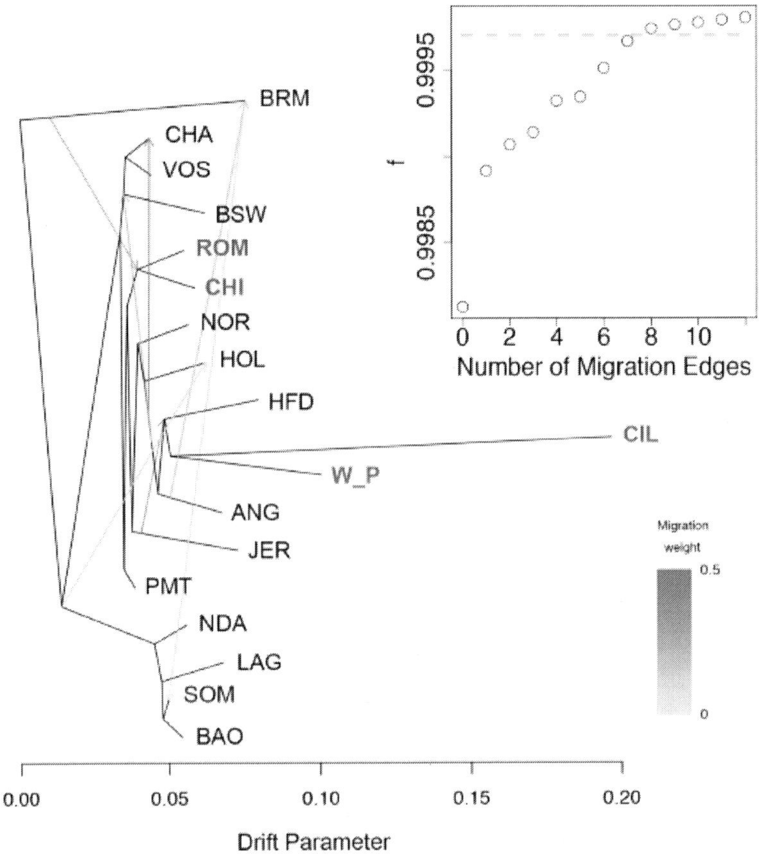

Figure 5. Phylogenetic network of the inferred relationships between 18 cattle breeds. The phylogenetic network inferred by Treemix of the relationships between Welsh White Park cattle breed and 17 other breeds is shown with Brahman (BRM) as outgroup. Migration edges between breeds are shown with arrows pointing in the direction toward the recipient breed of the migrants, and colored according to the ancestry percent received from the donor breed. The new datasets produced for this project are highlighted in green. The inset shows the f index representing the fraction of the variance in the sample covariance matrix ($\hat{W}$) accounted for by the model covariance matrix (W), as a function of the number of modeled migration events; the gray dashed line marks the number of migrant edges beyond which the f statistic asymptots.

The largest increase in variance occurred by adding the first migration edge to the graph from Brahman to the ancestor of Chianina and Romagnola, consistently with the Admixture results that suggest these Italian breeds have a minor (~10%) indicine genetic component. With the exception of two edges linking Brahman, Angus and Somba, the remaining migration edges were between taurine breeds. All migration edges had a weight less than ~0.2, with the exception of those from Jersey into Hereford, and from the ancestor of Hereford, White Park and Chillinghan, into Charolais, both of which have a weight above 0.4, suggesting the source population made a substantial genetic contribution to the recipient breeds.

3.2 Demographic History

The trend in historic effective population sizes for each breed was estimated using the distribution of linkage disequilibrium across the genome. All breeds showed a declining effective population size over the last ~2000 generations (i.e., approximately 8000 years assuming an average generation length of 4 years throughout these species history; Figure 6). Interestingly the African breeds showed an average higher effective population size for each species than the non-African breeds. Consistent with their hybrid nature, the five hybrid breeds presented a larger effective population size, probably reflecting the artificial increase in heterozygosity deriving from the admixture event (Figure 6).

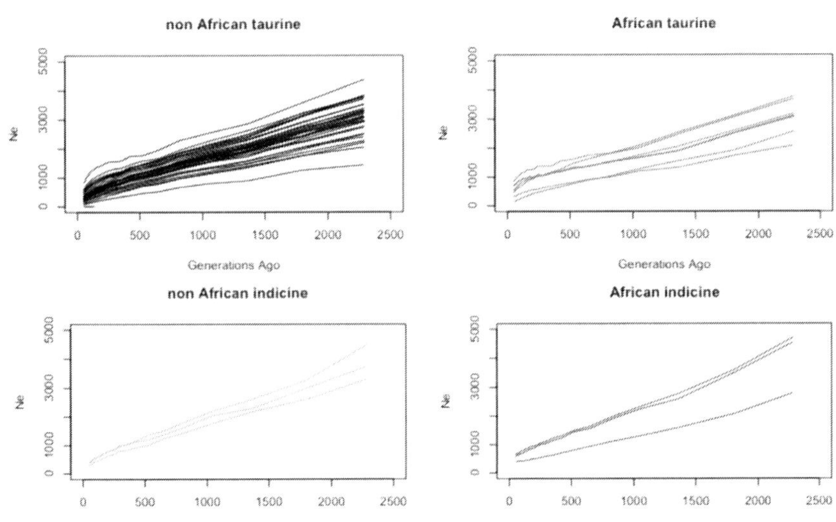

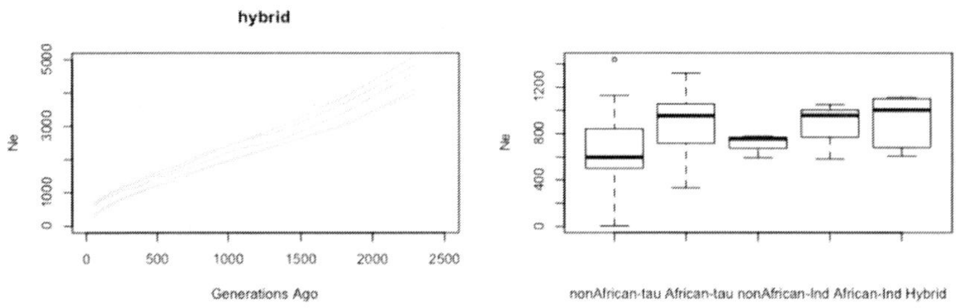

Figure 6. Demographic history reconstruction. Each plot shows the trend in effective population size for a set of cattle breeds between ~50 and ~2300 generations ago. The bottom right graph shows boxplots for the harmonic-mean of the effective population sizes trends estimated with SNeP for the various groups of cattle breeds analyzed.

3.3 Signals of Selection

A replicated approach using XP-EHH was used to identify consistent SNPs showing signatures of selection in the dataset (Figure 1). Four pairwise comparisons between taurine and indicine breeds were carried out resulting in 10,150 SNPs that passed the MAF threshold within population and which were tested in at least three out of the four pairwise comparisons. Among these SNPs, 3029 (~30%) showed consistent signatures of selection in the taurine breeds and 2385 (~24%) in the indicine breeds. The comparison between African and non-African indicine breeds resulted in 9388 SNPs tested, out of which 1768 (~19%) showed signatures of selection specific to the Asian breeds and 2049 (~22%) showed signatures of selection specific to the African breeds. For the comparison between taurine breeds, a similar number of SNPs was tested, 9938 SNPs. Of these 1872 (~19%) showed signatures of selection in European cattle breeds and 1739 (~17%) in African breeds. The SNPs showing consistent signatures of selection were linked to neighboring genes that occurred up to 50,000 bp downstream or upstream of the SNP position. No significantly enriched GO categories were found for any of the sets of breeds analyzed here. However, some patterns were observed, e.g., for the candidate SNPs under selection in Asian indicine several genes related to immunity were identified, e.g., LIPH (involved in platelet aggregation), Tespa1 (involve in thymocyte development), CCL14 (activates monocytes) (SM Table 1). However, interestingly, several RNA types were identified for several of the groups of breeds tested (Table 2). For the comparison between taurine and indicine breeds 6 snoRNA and 10 snRNA were identified among the taurine breeds, and two miRNA and 1 snoRNA in the indicine breeds. For the comparison between Asian and African indicine, a total of three miRNA, 1 snoRNA and 2 snRNA were identified for the Asian breeds and none for the African, while for the taurine comparison, the opposite was true, i.e., 1 snoRNA and 1 snRNA were identified for the African breeds and none for the European breeds. Overall this suggests that a substantial amount of selection signature occurred near non-coding RNAs that modify transcription of genes or of other RNAs.

Table 2. Classes of molecules under selection.

Molecule Class	Taurine vs. Indicine	
	Taurine	Indicine
Protein Coding	15	14
RNA (mi, sno, sn)	0,6,10	2,1,0
rRNA	0	0
Pseudogene	1	1
	Indicine	
	Asian	African
Protein coding	46	2
RNA (mi, sno, sn)	3,1,2	0
rRNA	3	0
Pseudogene	3	0
	Taurine	
	European	African
Protein coding	5	1
RNA (mi, sno, sn)	0	0,1,1
rRNA	0	0
pseudogene	0	0

DNA sequences coding for molecules in close proximity to SNPs under selection were grouped as Protein Coding, RNA types (mi, sno, and sn), rRNA and pseudogenes. The number of SNPs under selection close to each of these classes is shown for each of the three analyses presented.

4. DISCUSSION

This dataset represents one of the most comprehensive analyses of *B. taurus* and *B. indicus* cattle breed genetic data. Overall a higher observed genetic variation was observed in taurine breeds that originated or derived from European cattle breeds (~0.29), than in the other cattle breeds. Both African taurine and indicine breeds presented an intermediate level of observed heterozygosity (~0.22 in both sets) and the Asian indicine breeds presented the lowest observed heterozygosity (~0.17). While the observed difference could have a generic demographic explanation, e.g., a larger effective population size in the taurine breeds of European descent, such pattern is not consistent with what has been described with other types of molecular markers such as microsatellites (Rothammer et al., 2013). The most likely explanation for this pattern is ascertainment bias (Gautier et al., 2009; Matukumalli et al., 2009). In particular, for those breeds not included in the panel of breeds used to identify SNPs in the BovineSNP50 chip assay, minimum allele frequencies are expected to be substantially lower and the genetic variation captured with the chosen SNPs may not be reflective of the true underlying genetic variation in those breeds. In this case, SNP discovery was based on the genome sequence reads of five taurine breeds (Holstein, Angus, Limousin, Jersey, Norwegian Red) and one indicine (Brahman) breed compared to the reference genome of a Hereford cow (Bovine Hapmap et al., 2009). Ascertainment bias deriving from the design of this SNP array was previously described for a subset of the dataset analyzed here, where up to 30% of the SNPs had a MAF > 0.3 in taurine breeds, while only 19% of the SNPs had an equivalent MAF in indicine breeds (Nelore, Brahman, and Gir) (Bovine Hapmap et al., 2009). Similarly, the analysis of Holstein, Angus and Brahman samples carried out with a subset of the SNPs in the BovineSNP50 chip assay ascertained in each of the breeds separately, showed how Brahman animals presented a lower heterozygosity using the taurine SNPs (p-value < 0.05 and p-value = 0.07, respectively) than using the indicine SNPs (Neto and Barendse, 2010). Contrastingly, in the same analysis, for the Holstein and Angus derived SNPs, the taurine samples presented an observed heterozygosity typical to that of any other taurine breed. Furthermore, as no African taurine breeds and most indicine breeds were not included in the SNP ascertainment, the observed levels of genetic variation are likely to be an underestimate of the true genetic variation in these breeds (Matukumalli et al., 2009), and the associated inbreeding coefficient may consequently be an overestimate of the real ones. Thus, comparisons of genetic diversity discussed here are only likely to be meaningful within major genetic groups.

Among all cattle, the British Chillingham breed possessed unusually low observed heterozygosity (0.026) and an extremely high inbreeding coefficient (0.92). The Chillingham occurs in Northumberland, UK, and for the last ~300 years has lived in a feral state with almost no human intervention and no immigration (Visscher et al., 2001). Additionally, this breed passed through a bottleneck in 1947 where the herd reduced to five males and eight females. Previous analysis of genetic variation using 25 microsatellites identified only one marker with polymorphism (Visscher et al., 2001). Additionally, sequences of the full mitochondria of eight animals (sequencing depth ~2935x using next generation sequencing) revealed a single haplotype in the herd that differentiated them from other taurine breeds by three mutations, and one mutation (likely to be a recurrent mutation) that related them to indicine breeds and yak (Hudson et al., 2012). These studies suggest that Chillingham probably survives because the population has purged most deleterious mutations. The results shown here represent the largest study of genetic variation in this breed so far and did find similarly low levels of genetic variation in the breed (~900 polymorphic SNPs out of ~35,000). An extended analysis including more individuals and genetic markers in this breed may identify genetic variation of evolutionary importance not captured in this study, as well as shedding light on the distribution of homozygous and heterozygous tracts in the breed that could be used to monitor its genetic health.

4.1 Population Structure

The analyses of population structure easily identified the taurine and indicine genetic components in the dataset (Figure 2). However, when conditioning the number of clusters in the data to $K = 2$, the African taurine cattle presented a genetic composition similar to that of the American hybrid breeds Beef Master and Santa Gertrudis, the latter two are breeds of recent origin (less than 100 years old). Increasing the number of clusters to $K = 3$ clearly separates the African taurines in their own cluster, suggesting that despite the strong non-African taurine and indicine genetic component (as seen for $K = 2$), African taurines (both longhorn and shorthorn) have a substantial third genetic component specific to Africa which may derive from cross breeding with local aurochs (Mctavish et al., 2013). Consistent with previous observations, the African taurine N'Dama (ND1 and ND2) and Somba (SOM) breeds presented a minor indicine genetic component (Gautier et al., 2010). Among the hybrids, Beefmaster (BMA) and Santa Gertrudis (STG) presented a major genetic component of non-African taurine origin and the remaining of indicine origin, in accordance with their breed formation. Beefmaster and Santa Gertrudis are both hybrid breeds of North American origin resulting from the cross between European taurine cattle of British origin (Hereford and Shorthorn) and Brahman cattle (Mctavish et al., 2013). On the other hand, the remaining three hybrid breeds, the Borgou, Kuri, and Sheko present a genetic make-up of African taurine and indicine origin, with the first two having a somewhat larger taurine proportion than the Sheko breed. Among the taurine breeds, the Moroccan Oulmès Zaer (OUL) breed of unknown origin, presented approximately ~3/5 of its genotype of non-African taurine type and the remaining of African taurine type (Figure 2), consistent with previous results suggesting an influence of African longhorn taurine into the otherwise European taurine genetic background of this breed (Gautier et al., 2010; Decker et al., 2014). Interestingly, the non-African taurine cattle presented around 10% or less of their genetic make-up showing African taurine ancestry (Figure 2), possibly deriving form cross Mediterranean movement of cattle during the last millennia (Dürst, 1899; Do Vale, 1907; Felius et al., 2011). A further indicine genetic component was also detected in the French Charolais and Gascon breeds (<5%), and the Italian Romagnola and Chianina breeds (~10%). These results support observations by Decker et al. (2014) that suggest indicine admixture in European taurine breeds (the bulk of non-African taurines in this study) is rare instead of widespread (Mctavish et al., 2013). The genetic variation of African taurine is clearly distinct from that of other cattle breeds, and the PCA analysis suggests that the three main cattle groups (African and non-African taurine, and indicine) are almost as different from each other. While these results do not provide evidence for or against the hypothesis of a third domestication event in Africa (Felius et al., 2011), it suggests that African taurine cattle present a unique characteristics in their gene-pool.

The analysis of population structure using Admixture resulted in asymptoting CV values reflecting problems in identifying the true number of clusters in the data, possibly due to the presence of family structure within the breeds, or relatively low divergence between the estimated clusters. Thus, we also carried out principal components analysis, and as expected from the problems with Admixture, there is relatively low variation among cattle groups, i.e., the largest principal component (PC1) only explains ~10% of the variance in the dataset (separating taurine form indicine breeds), while PC2 explains ~5% of the variance (separating African from non-African taurines). Consistent with the Admixture result, the combination of PC1 and PC2 places the hybrid breeds Kuri, Borgou and Sheko between the African taurine and African indicine clusters, with the Sheko much closer to the indicine cluster reflecting its larger indicine background than Kuri and Borgou. The American Beef Master and Santa Gertrudis are placed between the non-African taurine and Asian indicine, but somewhat closer to the taurine cluster reflecting the larger taurine component of their genetic variation.

Lastly, the NeighbourNet analysis and the dendrogram identify the Madagascan zebu as more similar to the Asian indicine than to the other two African indicine. While this placing may be biased due to the ascertainment bias in the SNP array, it is tempting to suggest that the Madagascan population share a more recent history with Asian indicine breeds than the continental African breeds (e.g., a more recent establishment of the Madagascan population).

4.2 Roman Origin of Welsh White Park

Within the United Kingdom, traditional knowledge suggests that the Welsh White Park were brought to Britain by the Romans (Felius et al., 2011). While there is no evidence of the Romans having introduced cattle to Britain (Visscher et al., 2001), PCA places the White Park almost in the middle of the group of non-African taurine breeds (Figure 3A), consistently with the results by Decker et al. (2014) based on genotypes from five individuals. In particular, a comparison between Welsh White Park and the three Italian breeds in this dataset, Chianina, Romagnola and Piedmontese, revealed that the Italian breeds are more closely related to each other than to White Park, and the Welsh breed is more closely related to the other British breeds (Figures 3B, 5). Nevertheless, while this suggests that Welsh White Park are not similar to the Italian breeds it still does not answer the question regarding their origin. Both the NeighbourNet (Figure 4) and the phylogenetic network (Figure 5) show that the Welsh White Park and Chillingham cluster near each other (Figure 4). Such observation could be the outcome of genetic drift or inbreeding (or an interplay of these two parameters) in these breeds driving the divergence of their allelic frequencies (particularly in the case of Chillingham). Consistently with this hypothesis, the phylogenetic network analysis modeling genetic drift suggests that Welsh White Park and Chillingham have been through substantial drift in comparison to other taurines (Figure 5). Additionally, the Treemix analysis suggests that no migration events connect either the Welsh White Park or Chillingham with any of the Italian breeds, unless an additional edge that marginally increases the f-statistic by 0.01% (i.e., from 99.96 to 99.97% of the variance explained) is added; however, that additional migration edge is from Welsh White Park into Romagnola (contrary to common believe of an Italian ancestry of White Park) and has a weight near 0 (SM Figure 4) indicating that a very small fraction (if at all) of the Romagnola allelic variation may be of Welsh origin. While such idea could be appealing to some, it does not necessarily reflect a real historic event as the inferred edge provides an almost negligible contribution to the variance in the phylogram and corresponds to a minimal contribution of Welsh allelic variants into Romagnola. Additionally, the question of whether the inferred patterns of migration by Treemix are somehow dependent on the data analyzed could be raised. Further studies on the genetic history of these iconic breeds should include additional breeds such as the Podolian Hungarian Gray, or breeds close to the domestication center in order to better characterize the origin of this animals.

4.3 Demographic History

The demographic history of a population can be traced using the distribution of LD across the genome (e.g., Hill and Robertson, 1968), and if the cattle domestication event was marked by a strong bottleneck, it is expected that such a pattern should be visible with this dataset. However, it is important to note that the analyses presented here might provide a biased estimate of the true extent of demographic history, as LD calculated over SNP array markers is expected to decay at a lower rate than when the same estimates are obtained from whole-genome sequence data (Qanbari et al., 2014).

All the breeds analyzed present trends marked by a decline in their effective population size during the last ~2000 generations (Figure 6). Assuming an average generation length of 4 years, that suggests a constant decline across all breeds since the onset of the domestication process approximately 8000 years ago (Gautier et al., 2010). These results are consistent with observations describing a general decrease in breeds such as Angus, Holstein, Hereford, N'Dama and Brahman over the same period of time (Gautier et al., 2007; De Roos et al., 2008; Flury et al., 2010). Interestingly, overall, the harmonic-mean effective population size of the non-African taurine breeds was smaller than that of the non-African indicine, but both of those means were much smaller than that of the African taurines and indicines (Figure 6). These results suggest that either crossing between African taurine and indicine breeds has increased the effective population size in these breeds by moving genetic variation between breeds, and/or that more benign management practices in Africa have resulted in a milder reduction in effective population size. Although the second hypothesis cannot be ruled out, consistent with the first hypothesis, the hybrid breeds present overall a higher average effective population size.

4.4 Signatures of Selection

The unique nature of the data presented here provided us with a potentially powerful statistical framework to detect signatures of selection. In contrast to other studies, we examined multiple breeds for each species, accounting for their distribution in and out of Africa to search for signatures of selection in the equivalent of a replicated experimental design, where instead of identifying selection in a particular breed, we used several breeds of each species in each continent as biological replicas. The test used is an extension of Sabeti's extended haplotype homozygosity (Sabeti et al., 2002), which is based on the principle that a site under positive selection will rapidly increase in frequency, bringing neighboring variants to a high frequency in a process called hitchhicking (Smith and Haigh, 1974), thereby removing genetic variation around the site under selection (a selective sweep) and leaving a track of extended homozygosity. The magnitude of the effect of the selected site on the neighboring variation largely depends on how strong and recent the selective pressure is. If the process is slow, because selection is weak, recombination events separate the selected site from neighboring variants breaking down linkage and consequently resulting in a short area of extended homozygosity (Sabeti et al., 2002). On the contrary, if the process is fast because selection is strong, the area of extended homozygosity will be large, as there is not enough time for recombination to break down linkage. Consequently, identifying old signatures of positive selection using the EHH or any of its derived versions (e.g., XP-EHH) is rather difficult as the expected pattern of linkage disequilibrium is likely to have disappeared (Sabeti et al., 2007; Chen et al., 2010). However, despite the ancient divergence between *B. taurus* and *B. indicus*, i.e., approximately 250,000 years old (Bradley et al., 1996; Bovine Hapmap et al., 2009), a consistent signature of selection was found for several thousands of SNPs (5385 SNPs) probably reflecting relatively recent selective events rather than ancient ones.

Consistent with the expectation for the comparison between breeds in both species, a similar number of SNPs in each species showed signatures of selection (i.e., 3029 in taurine and 2385 in indicine). For the African vs. non-African comparison of taurine breeds 1872 of the SNPs under selection were on European breeds and 1739 were in African breeds. In contrast, for the comparison of African vs. non-African indicine breeds, 1768 of the SNPs under selection were on the Asian indicine breeds and 2049 were in the African breeds. The difference between the comparison of taurine breeds and indicine breeds in and out of Africa may reflect the shorter period of time that the later have spent in Africa since their introduction to that continent. The first domestic cattle to have entered Africa were taurines approximately 7000 years go (Payne and Hodges, 1997; Hassan, 2000; Felius et al., 2014). These domestic animals probably entered via a route crossing through Egypt and moving into northern Africa. In contrast, indicine breeds entered Africa much later via routes leading to East Africa via Egypt some time between ~3500 and 2500 years ago (Payne and Hodges, 1997; Hanotte et al., 2002). The timing of these migration events into Africa implies that taurine breeds have been in Africa between two and three times longer than indicine breeds.

At least three non-exclusive hypotheses can be derived from the taurine colonization history of Africa. Firstly, African taurine have become clearly differentiated from other non-African taurines as genetic drift exacerbated differences between the taurine propagules that spread from the domestication center into Europe and Africa. Secondly, African taurines had a longer time to hybridize with local aurochs, as possibly was the case for the European taurines, which hybridized with European aurochs (Decker et al., 2014). Consequently, African taurines are expected to carry unique genetic variation (deriving from local aurochs) that differentiates them from non-African taurines, as African aurochs were likely to genetically differ from European aurochs due to isolation by distance (Perez-Pardal et al., 2010; Mctavish et al., 2013). Thirdly, by having spent a longer time in Africa, African taurines are expected to have had a longer time to adapt to the local environment and although some of their signatures of local adaptation may have dissipated, they may have left a trace of genetic identity that helps differentiating African taurine form other cattle breeds. In contrast, *B. indicus* was introduced to Africa approximately ~2000 years ago, and during this short period of time African indicine breeds may have not had enough time to differentiate from other indicine breeds, as much as the African taurines did. However, as the presence of indicine breeds in Africa is recent it is reasonable to expect that these carry more signatures of selection than African taurines (~22 vs. ~17%, respectively).

The analysis of selection carried out here sought to identify SNPs under selection using the equivalent of a replicated experiment. For that purpose it was necessary that several pairwise comparisons identified the same SNP under selection showing the same sign in the XP-EHH analysis (e.g., a SNP showing selection for the European taurine variant in the three pairwise comparisons). The rationale for this is that genetic drift can cause extreme differences in allele frequencies between populations, however, the chance that the same allele changes allele frequency in the same direction in multiple populations is low (e.g., A gets fixed by chance in the three replicates). While our approach is expected to be robust as it combines the power of the track of linkage disequilibrium left after positive selection and its replicability across populations occupying similar environments (e.g., three European breeds tested against three African breeds), it is possible that weaker signatures of selection were not identified. Nevertheless, it is interesting to note that for the pairwise comparisons where several genes were identified to be in close proximity to the SNPs under selection some chromosomes seems to be more frequent under selection than others. For example, among the Asian indicine, eight genes occurred on the first chromosome, and 11 on the 19th chromosome (SM Table 1), while three out of the four genes identified among the European taurine breeds occurred on the 10th chromosome. Lastly, it is expected that short term adaptive processes may be mediated through modifications of regulatory pathways (Karim et al., 2011; Visscher and Goddard, 2011; David et al., 2013; Furusawa and Kaneko, 2013). Consistently with such hypothesis we find that several SNPs under selection occur in close proximity to RNA coding sequences involved in transcription regulation (e.g., miRNAs) and modification of other RNAs (e.g., snoRNAs).

The domestication of cattle in Eurasia had profound implications for society (Ajmone-Marsan et al., 2010; Groeneveld et al., 2010; Felius et al., 2014). Cattle are one of the major livestock species in the world and represent a large assemblage of widespread breeds as well as locally adapted breeds (Ajmone-Marsan et al., 2010; Felius et al., 2011, 2014). Studying the genetic resource of cattle breeds around the world is an on going work (e.g., Bovine Hapmap et al., 2009; Gautier et al., 2010; Mctavish et al., 2013; Decker et al., 2014) and approaches like the ones used here show the power that multiple comparisons have to derive particular trends and patterns that have characterized these breeds' histories. Additionally, identifying gaps representing regions where cattle has not yet been studied, as well as breeds that have not yet been genetically characterized is of utter importance to identify the genetic variation with adaptive value (e.g., local adaptation) and that should be the target of conservation efforts. Finally, whether Welsh White Park cattle were introduced to Britain by the Romans seems unlikely with the data described here, but cannot be entirely ruled out. The current results suggest that the breed presents a distinct genetic variation possible reflecting its ancient origin, but without some sort of calibration for the molecular phylogeny, an Approximate Bayesian computation approach that estimates times of divergence between breeds, or its comparison to local breeds close to the domestication center, this question is left partially unanswered (Csillery et al., 2010a,b; Sunnaker et al., 2013).

ACKNOWLEDGMENTS

The authors acknowledge the Chillingham Wild Cattle Association for the collection of samples for DNA analysis, as well as the NextGen consortium (http://nextgen.epfl.ch/) and Genomic-Resources (European Science Foundation; http://www.esf.org/index.php?id=7009) for funding, and two reviewers for their comments.

REFERENCES

1. Abraham, G., and Inouye, M. (2014). Fast principal component analysis of large-scale genome-wide data. *PLoS ONE* 9:e93766. doi: 10.1371/journal.pone.0093766
2. Ajmone-Marsan, P., Garcia, J. F., and Lenstra, J. (2010). On the origin of cattle: how aurochs became cattle and colonized the world. *Evol. Anthropol.* 19, 148–157. doi: 10.1002/evan.20267
3. Alexander, D. H., Novembre, J., and Lange, K. (2009). Fast model-based estimation of ancestry in unrelated individuals. *Genome Res.* 19, 1655–1664. doi: 10.1101/gr.094052.109

4. Barbato, M., Orozco-Terwengel, P., Tapio, M., and Bruford, M. (2015). SNeP: a tool to estimate trends in recent effective population size trajectories using genome-wide SNP data. *Front. Genet.* 6:109. doi: 10.3389/fgene.2015.00109
5. Beja-Pereira, A., Caramelli, D., Lalueza-Fox, C., Vernesi, C., Ferrand, N., Casoli, A., et al. (2006). The origin of European cattle: evidence from modern and ancient DNA. *Proc. Natl. Acad. Sci. U.S.A.* 103, 8113–8118. doi: 10.1073/pnas.0509210103
6. Biffani, S., Marusi, M., Biscarini, F., and Canavesi, F. (2005). Developing a genetic evaluation for fertility using angularity and milk yield as correlated traits. *Interbull. Bull.* 33, 63–66.
7. Bollongino, R., Elsner, J., Vigne, J. D., and Burger, J. (2008). Y-SNPs do not indicate hybridisation between European aurochs and domestic cattle. *PLoS ONE* 3:e3418. doi: 10.1371/journal.pone.0003418
8. Bovine Hapmap, C., Gibbs, R. A., Taylor, J. F., Van Tassell, C. P., Barendse, W., Eversole, K. A., et al. (2009). Genome-wide survey of SNP variation uncovers the genetic structure of cattle breeds. *Science* 324, 528–532. doi: 10.1126/science.1167936
9. Bradley, D. G., Machugh, D. E., Cunningham, P., and Loftus, R. T. (1996). Mitochondrial diversity and the origins of African and European cattle. *Proc. Natl. Acad. Sci. U.S.A.* 93, 5131–5135. doi: 10.1073/pnas.93.10.5131
10. Bulmer, M. G. (1971). The effect of selection on genetic variability. *Am. Nat.* 105, 201–211. doi: 10.1086/282718
11. Chen, H., Patterson, N., and Reich, D. (2010). Population differentiation as a test for selective sweeps. *Genome Res.* 20, 393–402. doi: 10.1101/gr.100545.109
12. Cobb, M. (2006). Heredity before genetics: a history. *Nat. Rev. Genet.* 7, 953–958. doi: 10.1038/nrg1948
13. Corbin, L. J., Liu, A. Y., Bishop, S. C., and Woolliams, J. A. (2012). Estimation of historical effective population size using linkage disequilibria with marker data. *J. Anim. Breed. Genet.* 129, 257–270. doi: 10.1111/j.1439-0388.2012.01003.x
14. Csillery, K., Blum, M. G. B., Gaggiotti, O. E., and Francois, O. (2010b). Invalid arguments against ABC: reply to AR Templeton. *Trends Ecol. Evol.* 25, 490–491. doi: 10.1016/j.tree.2010.06.011
15. Csillery, K., Blum, M. G., Gaggiotti, O. E., and Francois, O. (2010a). Approximate Bayesian Computation (ABC) in practice. *Trends Ecol. Evol.* 25, 410–418. doi: 10.1016/j.tree.2010.04.001
16. David, L., Ben-Harosh, Y., Stolovicki, E., Moore, L. S., Nguyen, M., Tamse, R., et al. (2013). Multiple genomic changes associated with reorganization of gene regulation and adaptation in yeast. *Mol. Biol. Evol.* 30, 1514–1526. doi: 10.1093/molbev/mst071
17. Decker, J. E., Mckay, S. D., Rolf, M. M., Kim, J., Molina Alcala, A., Sonstegard, T. S., et al. (2014). Worldwide patterns of ancestry, divergence, and admixture in domesticated cattle. *PLoS Genet.* 10:e1004254. doi: 10.1371/journal.pgen.1004254
18. De Roos, A. P., Hayes, B. J., Spelman, R. J., and Goddard, M. E. (2008). Linkage disequilibrium and persistence of phase in Holstein-Friesian, Jersey and Angus cattle. *Genetics* 179, 1503–1512. doi: 10.1534/genetics.107.084301
19. Do Vale, M. (1907). *Bovídeos Protuguezes* Lisbon: La Bécarre.
20. Dürst, J. (1899). *Die Rinder von Babylonien, Assyrien un Ägypten und ihr Zusammenhang mit den Rindern der Alten Welt.* Berlin: Gerog Reimber. doi: 10.5962/bhl.title.86398
21. Eden, E., Navon, R., Steinfeld, I., Lipson, D., and Yakhini, Z. (2009). GOrilla: a tool for discovery and visualization of enriched GO terms in ranked gene lists. *BMC Bioinformatics* 10:48. doi: 10.1186/1471-2105-10-48
22. Edwards, C. J., Ginja, C., Kantanen, J., Perez-Pardal, L., Tresset, A., Stock, F., et al. (2011). Dual origins of dairy cattle farming - evidence from a comprehensive survey of european Y-Chromosomal Variation. *PLoS ONE* 6:e15922. doi: 10.1371/journal.pone.0015922
23. Fao. (2007). "Global plan of action for animal genetic resources and the Interlaken Declaration," in *International Technical Conference on Animal Genetic Resources for Food and Agriculture* (Rome).
24. Felius, M., Beerling, M., Buchanan, D., Theunissen, B., Koolmees, P., and Lenstra, J. (2014). On the history of cattle genetic resources. *Diversity* 6, 705–750. doi: 10.3390/d6040705

25. Felius, M., Koolmees, P., Theunissen, B., Consortium, E. C. G. D., and Lenstra, J. (2011). On the breeds of cattle - historc and current classifictions. *Diversity* 3, 660–692. doi: 10.3390/d3040660
26. Felsenstein, J. (1993). *PHYLIP (Phylogeny Inference Package) version 3.5c.* 3.69 ed. (Seattle, WA: Department of Genetics, University of Washington).
27. Flori, L., Gonzatti, M. I., Thevenon, S., Chantal, I., Pinto, J., Berthier, D., et al. (2012). A quasi-exclusive European ancestry in the Senepol tropical cattle breed highlights the importance of the slick locus in tropical adaptation. *PLoS ONE* 7:e36133. doi: 10.1371/journal.pone.0036133
28. Flury, C., Tapio, M., Sonstegard, T. S., Drögemüller, T., Leeb, H., Simianer, H., et al. (2010). Effective population size of an indigenous Swiss cattle breed estimated fromlinkage disequilibirium. *J. Anim. Breed. Genet.* 127, 339–347. doi: 10.1111/j.1439-0388.2010.00862.x
29. Furusawa, C., and Kaneko, K. (2013). Epigenetic feedback regulation accelerates adaptation and evolution. *PLoS ONE* 8:e61251. doi: 10.1371/journal.pone.0061251
30. Gautier, M., Faraut, T., Moazami-Goudarzi, K., Navratil, V., Foglio, M., Grohs, C., et al. (2007). Genetic and haplotypic structure in 14 European and African cattle breeds. *Genetics* 177, 1059–1070. doi: 10.1534/genetics.107.075804
31. Gautier, M., Flori, L., Riebler, A., Jaffrezic, F., Laloe, D., Gut, I., et al. (2009). A whole genome Bayesian scan for adaptive genetic divergence in West African cattle. *BMC Genomics* 10:550. doi: 10.1186/1471-2164-10-550
32. Gautier, M., Laloe, D., and Moazami-Goudarzi, K. (2010). Insights into the genetic history of French cattle from dense SNP data on 47 worldwide breeds. *PLoS One* 5:e13038. doi: 10.1371/journal.pone.0013038
33. Gotherstrom, A., Anderung, C., Hellborg, L., Elburg, R., Smith, C., Bradley, D. G., et al. (2005). Cattle domestication in the Near East was followed by hybridization with aurochs bulls in Europe. *Proc. Biol. Sci.* 272, 2345–2350. doi: 10.1098/rspb.2005.3243
34. Groeneveld, L. F., Lenstra, J. A., Eding, H., Toro, M. A., Scherf, B., Pilling, D., et al. (2010). Genetic diversity in farm animals–a review. *Anim. Genet.* 41(Suppl. 1), 6–31. doi: 10.1111/j.1365-2052.2010.02038.x
35. Hanotte, O., Bradley, D. G., Ochieng, J. W., Verjee, Y., Hill, E. W., and Rege, J. E. (2002). African pastoralism: genetic imprints of origins and migrations. *Science* 296, 336–339. doi: 10.1126/science.1069878
36. Hassan, F. (2000). *The Origins and Development of African Livestock: Archaelogy, Genetics, Linguistics and Ethnography.* London: University College London Press.
37. Herrero, M., and Thornton, P. K. (2013). Livestock and global change: emerging issues for sustainable food systems. *Proc. Natl. Acad. Sci. U.S.A.* 110, 20878–20881. doi: 10.1073/pnas.1321844111
38. Hill, W. G., and Robertson, A. (1968). Linkage disequilibrium in finite populations. *Theor. Appl. Genet.* 38, 226–231. doi: 10.1007/BF01245622
39. Hudson, G., Wilson, I., Payne, B. I., Elson, J., Samuels, D. C., Santibanez-Korev, M., et al. (2012). Unique mitochondrial DNA in highly inbred feral cattle. *Mitochondrion* 12, 438–440. doi: 10.1016/j.mito.2012.05.003
40. Huson, D. H., and Bryant, D. (2006). Application of phylogenetic networks in evolutionary studies. *Mol. Biol. Evol.* 23, 254–267. doi: 10.1093/molbev/msj030
41. Kantanen, J., Olsaker, I., Holm, L. E., Lien, S., Vilkki, J., Brusgaard, K., et al. (2000). Genetic diversity and population structure of 20 North European cattle breeds. *J. Hered.* 91, 446–457. doi: 10.1093/jhered/91.6.446
42. Karim, L., Takeda, H., Lin, L., Druet, T., Arias, J. A., Baurain, D., et al. (2011). Variants modulating the expression of a chromosome domain encompassing PLAG1 influence bovine stature. *Nat. Genet.* 43, 405–413. doi: 10.1038/ng.814
43. Lewis, J., Abas, Z., Dadousis, C., Lykidis, D., Paschou, P., and Drineas, P. (2011). Tracing cattle breeds with principal components analysis ancestry informative SNPs. *PLoS ONE* 6:e18007. doi: 10.1371/journal.pone.0018007

44. Loftus, R. T., Machugh, D. E., Bradley, D. G., Sharp, P. M., and Cunningham, P. (1994). Evidence for two independent domestications of cattle. *Proc. Natl. Acad. Sci. U.S.A.* 91, 2757–2761. doi: 10.1073/pnas.91.7.2757
45. Machugh, D. E., Loftus, R. T., Cunningham, P., and Bradley, D. G. (1998). Genetic structure of seven European cattle breeds assessed using 20 microsatellite markers. *Anim. Genet.* 29, 333–340. doi: 10.1046/j.1365-2052.1998.295330.x
46. Matukumalli, L. K., Lawley, C. T., Schnabel, R. D., Taylor, J. F., Allan, M. F., Heaton, M. P., et al. (2009). Development and characterization of a high density SNP genotyping assay for cattle. *PLoS ONE* 4:e5350. doi: 10.1371/journal.pone.0005350
47. Mctavish, E. J., Decker, J. E., Schnabel, R. D., Taylor, J. F., and Hillis, D. M. (2013). New World cattle show ancestry from multiple independent domestication events. *Proc. Natl. Acad. Sci. U.S.A.* 110, E1398–1406. doi: 10.1073/pnas.1303367110
48. Medjugorac, I., Kustermann, W., Lazar, P., Russ, I., and Pirchner, F. (1994). Marker-derived phylogeny of european cattle supports demic expansion of agriculture. *Anim. Genet.* 25, 19–27. doi: 10.1111/j.1365-2052.1994.tb00399.x
49. Neto, L. R. P., and Barendse, W. (2010). Effect of SNP origin on analyses of genetic diversity in cattle. *Anim. Product. Sci.* 50, 792–800. doi: 10.1071/AN10073
50. Payne, W., and Hodges, J. (1997). *Tropical Cattle, Origins, Breeds and Breeding Policies.* Oxford: Blackwell Science Ltd.
51. Pellecchia, M., Negrini, R., Colli, L., Patrini, M., Milanesi, E., Achilli, A., et al. (2007). The mystery of Etruscan origins: novel clues from *Bos taurus* mitochondrial DNA. *Proc. Biol. Sci.* 274, 1175–1179. doi: 10.1098/rspb.2006.0258
52. Perez-Pardal, L., Royo, L. J., Beja-Pereira, A., Curik, I., Traore, A., Fernandez, I., et al. (2010). Y-specific microsatellites reveal an African subfamily in taurine (*Bos taurus*) cattle. *Anim. Genet.* 41, 232–241. doi: 10.1111/j.1365-2052.2009.01988.x
53. Perry, B. D., Grace, D., and Sones, K. (2013). Current drivers and future directions of global livestock disease dynamics. *Proc. Natl. Acad. Sci. U.S.A.* 110, 20871–20877. doi: 10.1073/pnas.1012953108
54. Pickrell, J. K., and Pritchard, J. K. (2012). Inference of population splits and mixtures from genome-wide allele frequency data. *PLoS Genet.* 8:e1002967. doi: 10.1371/journal.pgen.1002967
55. Pryce, J., Royal, M., Garnsworthy, P., and Maod, I. (2004). Fertility in high-producing dairy cow. *Livestock Product. Sci.* 86, 125–135. doi: 10.1016/S0301-6226(03)00145-3
56. Pryce, J., Veerkamp, R., Thompson, R., Hill, W. G., and Simm, G. (1997). Genetic aspects of common health disorders and measures of fertility in Holstein Friesian dairy cattle. *Anim. Sci.* 65, 353–360. doi: 10.1017/S1357729800008559
57. Purcell, S., Neale, B., Todd-Brown, K., Thomas, L., Ferreira, M. A., Bender, D., et al. (2007). PLINK: a tool set for whole-genome association and population-based linkage analyses. *Am. J. Hum. Genet.* 81, 559–575. doi: 10.1086/519795
58. Qanbari, S., Pausch, H., Jansen, S., Somel, M., Strom, T. M., Fries, R., et al. (2014). Classic selective sweeps revealed by massive sequencing in cattle. *PLoS Genet.* 10:e1004148. doi: 10.1371/journal.pgen.1004148
59. Rischkowsky, B., and Pilling, D. (2007). *The State of the World's Animal Genetic Resources for Food and Agriculture.* Rome: Food and Agriculture Organization of the United Nations.
60. Rothammer, S., Seichter, D., Forster, M., and Medugorac, I. (2013). A genome-wide scan for signatures of differential artificial selection in ten cattle breeds. *BMC Genomics* 14:908. doi: 10.1186/1471-2164-14-908
61. Russell, N. (2007). *Like Engend'ring Like. Heredity and Animal Breeding in Early Modern England.* Cambridge, MA: Cambridge University Press.
62. Sabeti, P. C., Reich, D. E., Higgins, J. M., Levine, H. Z., Richter, D. J., Schaffner, S. F., et al. (2002). Detecting recent positive selection in the human genome from haplotype structure. *Nature* 419, 832–837. doi: 10.1038/nature01140
63. Sabeti, P. C., Varilly, P., Fry, B., Lohmueller, J., Hostetter, E., Cotsapas, C., et al. (2007). Genome-wide detection and characterization of positive selection in human populations. *Nature* 449, 913–918. doi: 10.1038/nature06250

64. Scheet, P., and Stephens, M. (2006). A fast and flexible statistical model for large-scale population genotype data: applications to inferring missing genotypes and haplotypic phase. *Am. J. Hum. Genet.* 78, 629–644. doi: 10.1086/502802
65. Simm, G. (1998). *Genetic Improvement of Cattle and Sheep*. Wallingford: Farming Press.
66. Smith, J. M., and Haigh, J. (1974). The hitch-hiking effect of a favourable gene. *Genet. Res.* 23, 23–35. doi: 10.1017/S0016672300014634
67. Sunnaker, M., Busetto, A. G., Numminen, E., Corander, J., Foll, M., and Dessimoz, C. (2013). Approximate Bayesian computation. *PLoS Comput. Biol.* 9:e1002803. doi: 10.1371/journal.pcbi.1002803
68. Sved, J. A. (1971). Linkage disequilibrium and homozygosity of chromosome segments in finite populations. *Theor. Popul. Biol.* 2, 125–141. doi: 10.1016/0040-5809(71)90011-6
69. Sved, J. A., Mcrae, A. F., and Visscher, P. M. (2008). Divergence between human populations estimated from linkage disequilibrium. *Am. J. Hum. Genet.* 83, 737–743. doi: 10.1016/j.ajhg.2008.10.019
70. Szpiech, Z. A., and Hernandez, R. D. (2014). selscan: an efficient multithreaded program to perform EHH-based scans for positive selection. *Mol. Biol. Evol.* 31, 2824–2827. doi: 10.1093/molbev/msu211
71. Taberlet, P., Coissac, E., Pansu, J., and Pompanon, F. (2011). Conservation genetics of cattle, sheep, and goats. *C. R. Biol.* 334, 247–254. doi: 10.1016/j.crvi.2010.12.007
72. Thornton, P. K. (2010). Livestock production: recent trends, future prospects. *Philos. Trans. R. Soc. Lond. B Biol. Sci.* 365, 2853–2867. doi: 10.1098/rstb.2010.0134
73. Troy, C. S., Machugh, D. E., Bailey, J. F., Magee, D. A., Loftus, R. T., Cunningham, P., et al. (2001). Genetic evidence for Near-Eastern origins of European cattle. *Nature* 410, 1088–1091. doi: 10.1038/35074088
74. Visscher, P. M., and Goddard, M. E. (2011). Cattle gain stature. *Nat. Genet.* 43, 397–398. doi: 10.1038/ng.819
75. Visscher, P. M., Smith, D., Hall, S. J., and Williams, J. L. (2001). A viable herd of genetically uniform cattle. *Nature* 409, 303. doi: 10.1038/35053160
76. Wendorf, F., and Schild, R. (1994). Are the early holocene cattle in the eastern Sahara domestic or wild? *Evol. Anthropol.* 3, 118–128. doi: 10.1002/evan.1360030406
77. Wood, R. J. (1973). Robert Bakewell (1725-1795), pioneer animal breeder, and his influence on Charles Darwin. *Folia Mendeliana* 58, 231–242.
78. Xu, L., Bickhart, D. M., Cole, J. B., Schroeder, S. G., Song, J., Tassell, C. P., et al. (2014). Genomic signatures reveal new evidences for selection of important traits in domestic cattle. *Mol. Biol. Evol.* doi: 10.1093/molbev/msu333

CHAPTER 16

Computational Molecular Analysis of the Sequences of BMP15 Gene of Ruminants and Non-Ruminants

Bwaseh S. Bibinu[1*], Abdulmojeed Yakubu[2*], Steven B. Ugbo[1], Ndu I. Dim[1]

[1]*Department of Animal Breeding and Physiology, University of Agriculture Makurdi, Makurdi, Nigeria*

[2]*Department of Animal Science, Faculty of Agriculture, Nasarawa State University, Keffi, Shabu-Lafia Campus, Lafia, Nigeria*

ABSTRACT

Bone morphogenetic protein 15 (BMP15) is a member of the transforming growth factor β (TGFβ) super family that is expressed by oocytes and plays key roles in granulosa cell development and fertility in animal. This study investigated the molecular genetic variation of BMP15 gene of some selected mammalian species with a view to providing relevant genetic information for breeding and selection programmes in the studied species using computational methods. A total of thirty seven (37) BMP15 nucleotide sequences comprising goats (18), sheep (6), cattle (6), swine (4) and chicken (3) were retrieved from the GenBank. Sequence alignment, translation and comparison of the BMP15 gene of the various species were done with ClustalW. High degree of polymorphism of BMP15 gene was observed among the studied species. The significant value ($P < 0.01$) for relative proportions of non-synonymous substitutions per nonsynonymous site (dN) and the number of synonymous substitutions per synonymous site (dS) symbolized that non-synonymous sites evolved faster than the synonymous sites and positive selection effect over shadowed purifying selection. Functional analysis of missense mutations using PROVEAN showed that twelve amino acid substitutions (L10S, W13A, E20L, V28S, P31R, P31G, P40Q, L42W, Q46N, A52V, R58C and G64T) in goats, nine in sheep (H21R, S32R, I33A, A39W, Q46W, E51A, G54S, R61D and E72A), six in cattle (Q30M, T41W, E50R, I62R, H65E, and E72S), seven in swine (I7L, T9I, V33I, L35H, C40P, R46P and Q61R) and five in chickens (A20H, L27H, W43L, A47P and G50Y) appeared beneficial. The phylogenetic trees from nucleotide and amino acid sequences revealed the close relatedness of members of the bovidae family (goat, sheep and cattle). The present information could guide future efforts involving selection of markers of fecundity to improve genetically livestock species in Nigeria.

Keywords: BMP15, Sequences, SNPs, Phylogeny, Ruminants, Non-Ruminants

1. INTRODUCTION

One of the strides towards increased productivity of protein of animal origin is to have an in-depth knowledge of the genes associated with reproduction especially those related to ovulation and litter size. Identifying genes of major effect is also a prerequisite which offers the opportunity to improve production efficiency, product quality and product diversity in livestock industry, through utilizing them in breeding programs [1]. Most mutations in proteins are associated with diseases; only a small number of examples show the heterozygote advantage. A good example is the Bone morphogenetic protein 15 (BMP15) gene, which is identified to be related with female fecundity in several animals [2]. BMP15 is a member of the transforming growth factor β (TGFβ) superfamily that is expressed by oocytes and play key roles in granulosa cell development and fertility in animal models along with growth differentiation factor 9 (GDF9). It is one of the major causal mechanisms underlying either the highly prolific or infertile phenotypes of several sheep breeds [3] or other mammalian species including man [4] . Since BMP15 plays important role in prolificacy and fertility, it therefore means that attention should be paid to the sequences of the gene (using both dry and wet lab approaches) which are usually responsible for the observed phenotypic differences. A good knowledge of the sequences of this gene will help in identifying the variants responsible for various factors attributed to the gene. The objective of this study, therefore, was to investigate using *in-silico* method, the molecular genetic variation of BMP15 gene of some ruminants (cattle, sheep and goats) and non-ruminants species (swine and chicken) with a view to providing relevant genetic information for marker assisted selection in the studied species.

2. MATERIALS AND METHODS

2.1. Sequences of Species

A total of thirty seven (37) BMP15 nucleotide sequences comprising goats (18), sheep (6), cattle (6), swine (4) and chicken (3) were retrieved from the GenBank (NCBI) (www.ncbi.nlm.nih.gov). The Genbank accession numbers of the sequences are JF824149.1, JF824148.1, NM001285588.1, JQ350892.1, JQ350891.1, JQ350090.1, HM462258.1, HM462255.1, HM462254.1, FJ429281.1, GU732196.1, EU743938.1, JX860305, JX860304, JN655670.1, JN655669.1, EU095935.1 and EU888137.1 (Goat); KF644561.1, NM001114767.1, EU402923.1, FJ600405.1, JN655672.1 and JN655671.1 (Sheep); DQ463368.1, NM001031752.1, EU712722.1, AY572412.1, DQ489533.1 and AY304484.1 (cattle); NM,001005155.1, AY295074.1, AH012830.1, and AY295073.1 (swine); AY725199.1, NM001006589.2 and AY729025 (chicken).

2.2. Sequence Alignment and Translation

Sequence alignment, translation and comparison of the BMP15 gene of the various species were done with ClustalW as described by Larkin et al. [5] using IUB substitution matrix, gap open penalty of 15 and gap extension penalty of 6.66.

2.3. Substitution Analysis

The relative proportion of non-synonymous substitution per non synonymous site (dN) and the number of synonymous substitutions per synonymous site (dS) of the deduced amino sequences of BMP15 gene of the various species were computed by bootstrap method (1000 replicates) using the modified Nei-Gojobori (assumed transition/transversion bias = 2) method [6] . The ratio of nonsynonymous to synonymous divergence (dN/dS) was tested for departure from the neutral expectation of unity using the codon-based Z-distribution as by modified Nei-Gojobori, applying Jukes-Cantor correction.

2.4. Functional Analysis

In silico functional analysis of missense mutations was obtained using PROVEAN with threshold value of −2.5. PROVEAN collects a set of homologous and distantly related sequences from the NCBI NR protein database (released August 2011) using BLASTP (ver.2.2.25) with an E-value threshold of 0.1. The sequences are clustered based on a sequence identity of 80% to remove redundancy using the CD-HIT program (ver.4.5.5) [7] . If the PROVEAN score is smaller than or equal to a given threshold, the variation is predicted as deleterious [8] .

2.5. Phylogenetic Trees Analysis

Neighbor-Joining (NJ) trees were constructed using maximum composite likelihood method and pairwise deletion gap/missing data treatment as described by Saitou and Nei [9] and Tamura et al. [10]. The construction was done on the basis of genetic distances, depicting phylogenetic relationships among the BMP15 nucleotide and amino acid sequences of the investigated species. The reliability of the trees was calculated by bootstrap confidence values [11], with 1000 bootstrap iterations using MEGA 6.0 software [12] . However, the P-distance method of Nei and Kumar [13] was used for the amino acid sequences in place of maximum composite likelihood method. Similarly, UPGMA trees for the BMP15 gene were constructed with consensus nucleotide and amino acid sequences; using same models as those of the NJ trees. All the nucleotide sequences were trimmed to equal length of 222 bp corresponding to same region of the gene before generating the trees.

3. RESULTS

The variation in sequence length in base pair (bp) of BMP15 gene within and among species ranges between 222 bp and 6648 bp (Table 1). Across the species, five sequences (3 from goats and 2 from cattle) have 1185 bp, while two sequences each from goat and sheep have 1182 bp. Within species, 2, 3, 3 and 3 sequences have the same bp of 465, 1185, 1230 and 6648 respectively in goats, while 2 sequences each from sheep, cattle, swine and chicken have the same bp of 222, 1185, 988 and 1875. However, the rest of the sequences do not share the same bp within various species. In goats, the sequences JQ350892.1, JQ350891.1, JQ350890.1, GU732196.1, JN860304.1 and JN655670.1 are complete coding sequences (CDS) from DNA and had sequence length that are greater than six thousand base pair (>6000 bp) while the sequences HM462258.1, HM462254.1, HM462255,

JX860305.1, JX860304.1 and EU 095935.1 are partial coding sequences (partial CDS) whose length were less than one thousand base pair (<1000 bp). In sheep, cattle, swine and chicken the same reason holds.

Table 1. Accession number and sequence length variation of BMP15 gene of goats, sheep, cattle, swine and chicken.

Specie	Accession Number	Base pair (bp)	Sequence length variation
Goat	EU095935.1	393	393 - 6648
	HM462258.1	465	
	HM462254.1	465	
	HM462255.1	477	
	JX860305.1	762	
	JX860304.1	762	
	EU888137.1	1182	
	JQ350892.1	6495	
	JQ350891.1	6495	
	JQ350890.1	6495	
	GU732196.1	6494	
	JF824149.1	1230	
	JF824148.1	1230	
	NM001285588.1	1230	
	FJ429281.1	1436	
	EU743938.1	6648	
	JN655670.1	6648	
	JN655669.1	6648	
Sheep	KF644561.1	222	
	FJ600405.1	222	222 - 6641
	EU402923.1	260	
	NM001114767.1	1182	
	JN655671.1	6638	
	JN655672.1	6641	
Cattle	AY304484.1	720	
	DQ489533.1	953	720 - 1189
	EU712722.1	1184	
	AY572412.1	1185	
	NM001031752.1	1185	
	DQ463368	1189	
Swine	AY295074.1	988	
	AY295073.1	988	988 - 1316
	NM001005155.1	1194	
	AH012830.1	1316	
Chicken	AY729025.1	1053	
	AY725199.1	1875	1051 - 1875
	NM001006589.2	1875	

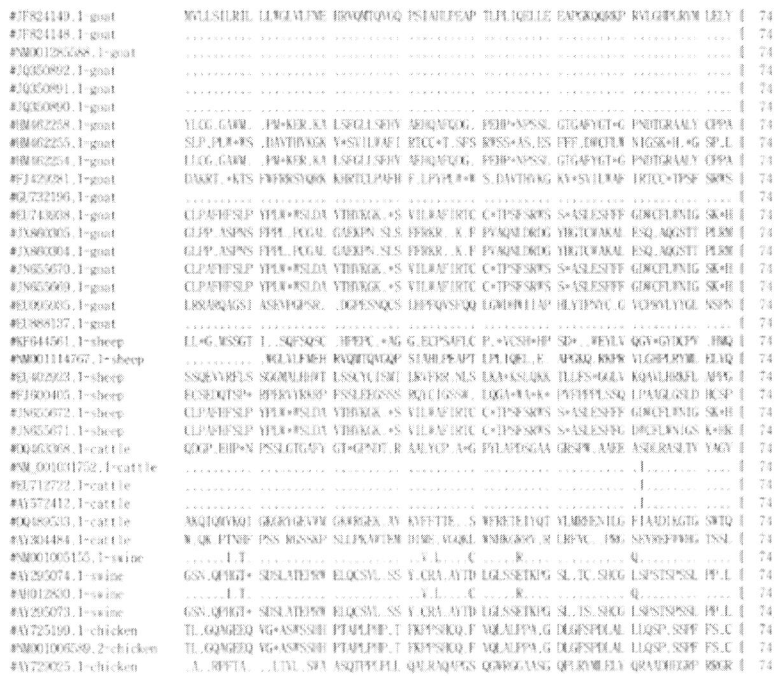

Figure 1. Amino acid prediction of goat, sheep, cattle, swine and chicken

The predicted amino acid sequences of goat, sheep, cattle, swine and chicken BMP15 showed varying degree of amino acid substitutions in the studied species (Figure 1).

Means of dS, dN and dN/dS with their corresponding Z-scores and P-values are presented in Table 2. Swine has the highest dN/dS value of 1.20 while sheep has the least value of 0.97. The Dn/dS pattern revealed positive selection for all species except sheep which has a dN/dS value that is less than one.

The results of functional analysis of coding nonsynonymous single nucleotide polymorphism (nsSNP) of BMP15 gene for goats, sheep, cattle, swine and chicken are presented in Tables 3-7, respectively.

Fourteen amino acid substitutions of the wild type alleles located in the putative peptide coding region of goats were obtained from the alignment of deduced amino acid sequences of goats. Out of these, twelve amino acid substitutions (L10S, W13A, E20L, V28S, P31R, P31G, P40Q, L42W, Q46N, A52V, R58C, G64T) were returned neutral indicating that the substitutions did not impair protein function, while the remaining two amino acid substitutions (M70F and Y74M) were returned as deleterious, an indication that the substitutions were harmful. In sheep, nine amino acid substitutions (H21R, S32R, I33A, A39W, Q46W, E51A, G54S, R61D and E72A) were returned neutral; an indication that they did not impair protein function while the remaining amino acid substitution (M70V) was predicted to be harmful. Out of the seven amino acid substitutions in cattle, six (Q30M, T41W, E50R, I62R, H65E, and E72S) appeared beneficial while the remaining one (Y74Q) appeared harmful.

For swine, seven amino acid substitutions (I7L, T9I, V33I, L35H, C40P, R46P and Q61R) were obtained and they all appeared beneficial (Table 6). Similarly, ten amino acid mutations were obtained from the deduced amino acid sequences of chicken, out of which, five (A20H, L27H, W43L, A47P and G50Y) appeared beneficial while the remaining five (Q31F, P38Q, E67G, P70F and R74C) appeared harmful.

The phylogeny based on nucleotide and amino acid sequences of BMP15 (Figure 2 and Figure 3) shows the clustering of sequences among the various species, although there was some intermingling between the species.

Table 2. Mean number of nucleotide substitutions per synonymous site (dS) and per nonsynonymous site (dN) with their ratio among goats, sheep, cattle, swine and chicken.

Specie	No. of codons	dS	dN	dN/dS	Z-score	P-value
Goat	340	2.28	2.42	1.06	10.77	0.00
Sheep	353	1.46	1.41	0.97	-1.232	1.00
Cattle	226	2.53	2.62	1.04	11.83	0.00
Swine	310	1.13	1.36	1.20	6.39	0.00
Chicken	347	0.70	0.77	1.10	7.85	0.00

dN/dS = ratio of nonsynonymous to synonymous divergence.

Table 3. Functional analysis of coding nsSNP of the BMP15 gene of goats using PROVEAN.

Variant	PROVEAN Score	Prediction
L10S	−0.947	Neutral
W13A	−0.722	Neutral
E20L	−0.959	Neutral
V28S	0.681	Neutral
P31R	−1.217	Neutral
P31G	−1.044	Neutral
P40Q	−2.283	Neutral
L42W	−2.283	Neutral
Q46N	−1.158	Neutral
A52V	−1.495	Neutral
R58C	−2.302	Neutral
G64T	−2.116	Neutral
M70F	−3.944	Deleterious
Y74M	−6.544	Deleterious

Default threshold is −2.5, that is: Variants with a PROVEAN score equal to or below −2.5 are considered "deleterious" while Variants with PROVEAN score above −2.5 are considered "neutral". G = glycine. A = Alanine. L = leucine. M = methionine. F = phenylalanine. W = tryptophan. Q = glutamine. E = glutamic acid. S = serine. P = proline. V = valine. Y = tyrosine. R = arginine. N = asparagine. T = threonine. C = cysteine.

Table 4. Functional analysis of coding nsSNP of the BMP15 gene of sheep using PROVEAN.

Variant	PROVEAN Score	Prediction
H21R	−1.206	Neutral
S32R	−1.171	Neutral
I33A	−0.057	Neutral
A39W	−1.871	Neutral
Q46W	−1.941	Neutral
E51A	−2.339	Neutral
G54S	−1.826	Neutral
R61D	−0.479	Neutral
M70V	−2.980	Deleterious
E72A	−2.313	Neutral

Default threshold is −2.5, that is: Variants with a PROVEAN score equal to or below −2.5 are considered "deleterious" while Variants with PROVEAN score above −2.5 are considered "neutral". G = glycine. A = Alanine. L = leucine. M = methionine. F = phenylalanine. W = tryptophan. K = lysine. Q = glutamine. E = glutamic acid. S = serine. P = proline. V = valine. I = isoleucine. C = cysteine. Y = tyrosine. H = histidine. R = arginine. N = asparagine. D = aspartic acid. T = threonine.

The genetic relationship of BMP15 of the studied species using UPGMA showed that members of the Bovidae family (sheep, goat and cattle) were closer at this locus compared to swine and chicken (Figure 4 and Figure 5).

BMP15 using ClustalW. Dots indicated identical amino acids and numbers on the right hand side represent site number and asterisks indicate indels.

Table 5. Functional analysis of coding nsSNP of the BMP15 gene of cattle using PROVEAN.

Variant	PROVEAN Score	Prediction
Q30M	−1.170	Neutral
T41W	−1.491	Neutral
E50R	−0.829	Neutral
I62R	−0.835	Neutral
H65E	−0.220	Neutral
E72S	−1.476	Neutral
Y74Q	−6.669	Deleterious

Default threshold is −2.5, that is: Variants with a PROVEAN score equal to or below −2.5 are considered "deleterious" while Variants with PROVEAN score above −2.5 are considered "neutral". M = methionine, Q = glutamine, T = threonine W = tryptophan, E = glutamic acid, R = arginine, I = isoleucine, H = histidine, S = serine, Y = tyrosine.

Table 6. Functional analysis of coding nsSNP of the BMP15 gene of swine using PROVEAN.

Variant	PROVEAN Score	Prediction
I7L	0.433	Neutral
T9I	0.329	Neutral
V33I	0.101	Neutral
L35H	−1.309	Neutral
C40P	3.229	Neutral
R46P	−1.053	Neutral
Q61R	0.187	Neutral

Default threshold is −2.5, that is: Variants with a PROVEAN score equal to or below −2.5 are considered "deleterious" while Variants with PROVEAN score above −2.5 are considered "neutral". I = isoleucine, L = leucine, T = threonine, V = valine, H = histidine, C = cysteine, P = proline, Q = glutamine, Y = tyrosine, R = arginine

Table 7. Functional analysis of coding nsSNP of the BMP15 gene of chicken using PROVEAN.

Variant	PROVEAN Score	Prediction
A20H	−0.392	Neutral
L27H	−1.971	Neutral
Q31F	−2.946	Deleterious
P38Q	−2.944	Deleterious
W43L	−2.161	Neutral
A47P	0.089	Neutral
G50Y	−2.034	Neutral
E67G	−3.811	Deleterious
P70F	−8.148	Deleterious
R74C	−4.981	Deleterious

Default threshold is −2.5, that is: Variants with a PROVEAN score equal to or below −2.5 are considered "deleterious" while Variants with PROVEAN score above −2.5 are considered "neutral". A = Alanine, H = histidine, L = leucine, Q = glutamine, F = phenylalanine, P = proline, W = tryptophan, G = glycine, Y = tyrosine, E = glutamic acid, R = arginine, C = cysteine.

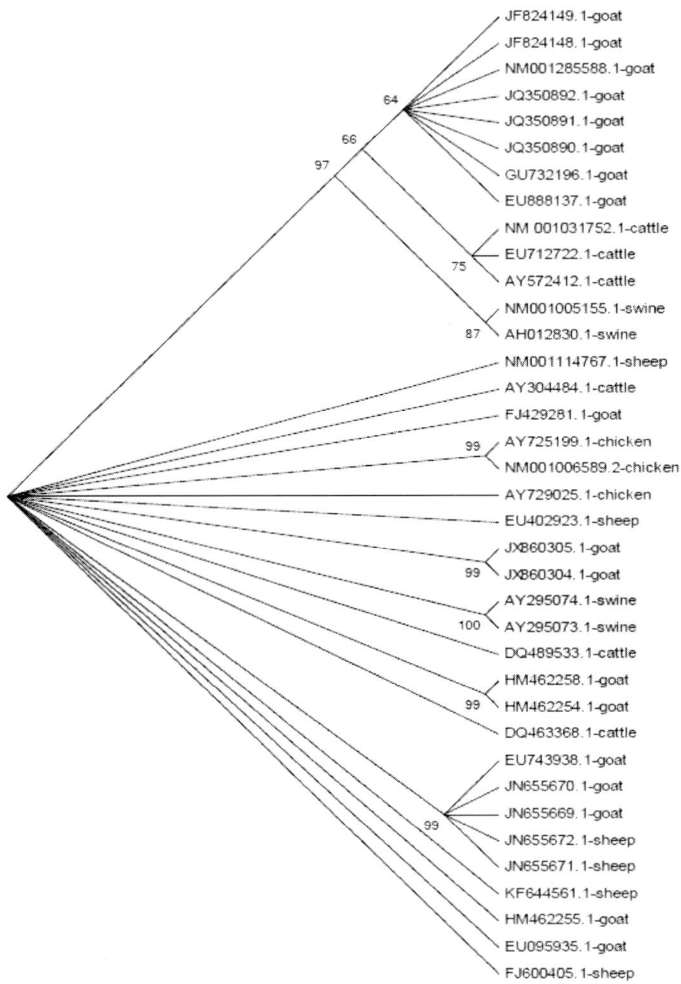

Figure 2. Phylogenetic relationships of goast, sheep, cattle, swine and chicken BMP15 nucleotide sequences.

4. DISCUSSION

BMP15, also known as GDF9B, is an X-linked gene in nature that is expressed in the oocyte and plays a key role in ovarian folliculogenesis. It is a polymorphic gene whose polymorphism has been proven to be associated with increased ovulation rate, sterility and litter size of farm animals [14]

[15] . The length variation observed within and across species in this study might be due to differences in the genomic region where the sequences were obtained from and differences due to complete coding sequences (CDS) or partial CDS. The variation in sequence length within and among species might result from evolution and differentiation [16] . There are cases where variability might result from DNA duplication, DNA rearrangement, short tandem repeat (STR), insertions or deletion of sequences [17] . This frameshift mutation often results in a completely different translation from the original protein and is also likely to cause a stop codon to be read which truncates further synthesis of protein [18] . This variability might initiate unique structures between individual members in conferring different biological activities [19] .

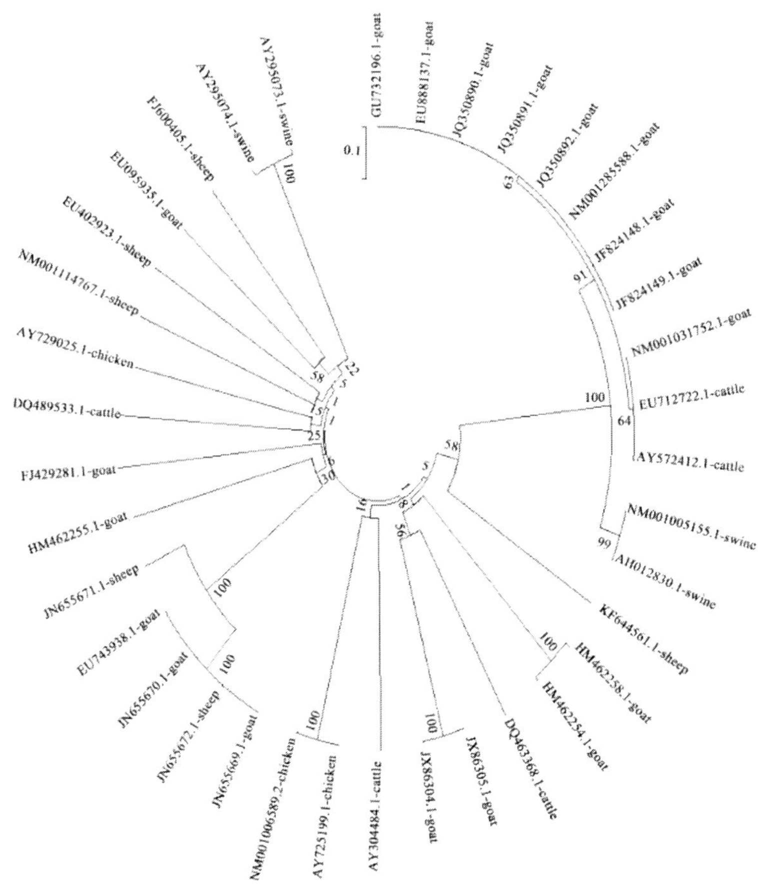

Figure 3. Phylogenetic relationships of goat, sheep, cattle, swine and chicken BMP15 amino acid sequences.

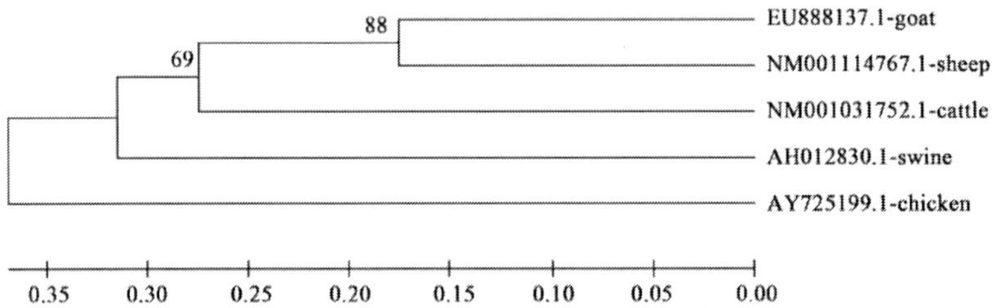

Figure 4. Phylogenetic tree of goat, sheep, cattle, swine and chicken BMP15 consensus nucleotide sequences using UPGMA.

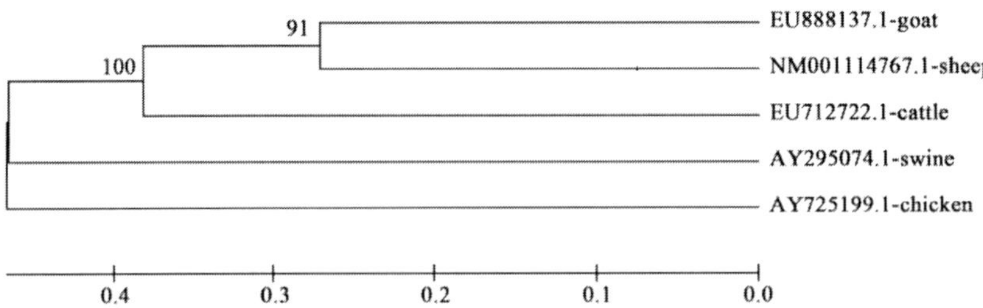

Figure 5. Phylogenetic tree of goat, sheep, cattle, swine and chicken BMP15 consensus amino acid sequences using UPGMA.

The ratio of the number of nonsynonymous substitution per nonsynonymous sites (dN; amino acid altering) to the number of synonymous mutations per synonymous sites (dS; silent mutation) also known as omega (ω = dN/dS), is a useful estimate of gene selective pressure [20] . An omega (ω) value greater than 1 ($\omega > 1$) is an indication of positive selection [21] . In the present study, the omega values obtained were greater than one (>1) except sheep which had a value of 0.97, indicating that there was excess of non-synonymous over synonymous nucleotide substitutions over the entire sequences across the species excluding sheep. This symbolizes that non-synonymous sites evolved faster than the synonymous sites and positive selection effect over-shadows purifying selection. It is an implication that balancing selection (or positive Darwinian section) favoured new variants and increased allelic polymorphism [22] which in turn might induce a change in the conformation of protein structure, thereby affecting the signaling pathway during follicle differentiation and ovulation [23] . The 0.97 omega value of sheep observed in this study, signifies purifying selection i.e. selection against non-synonymous mutations because of their deleterious effect.

One of the key issues in biology, is understanding how natural selection drives gene functional diversification across different species and lineages [24] . The varying substitutions of amino acids within and across species might be as a result of separate divergence from their common ancestor. This agrees with earlier submissions [20] [25] that as orthologs diverge from their most recent common ancestor, their different evolutionary trajectories lead to divergence in the selective constraints on homologous sites. This information may aid selection of superior animals within and between breeds for genetic improvement of desired traits.

The main goal in animal breeding is to select individuals that have high breeding values for traits of interest as parents to produce the next generation and to do so as quickly as possible. To date, most programs rely on statistical analysis of large data bases with phenotypes on breeding populations by linear mixed model methodology to estimate breeding values on selection candidates. However, there is a long history of research on the use of genetic markers to identify quantitative trait loci and their use in marker-assisted selection but with limited implementation in practical breeding programs [26] .

The neutral or beneficial amino acid substitutions are those substitutions that help in maintaining the structural integrity of cells and tissues. Also, they affect positively the functional roles of proteins involved in signal transduction of visual, hormonal, and other stimulants. However, the harmful amino acid substitutions could cause amino acid change further altering protein function which may lead to susceptibility to disease. They may modify enzyme activity, destabilize protein structures or disrupt protein interactions. The beneficial nsSNPs obtained in the present study, therefore, offer hope for future genetic improvement of goats, sheep, cattle, swine and chicken at the BMP15 locus. This is due to the fact that nsSNPs have been reported to be linked to economically important traits and disease development [17] . Some nsSNP are beneficial because the altered proteins result in increased sensitivity of granulosa cells to follicle stimulating hormone (FSH), which leads to accelerated follicular development and precocious ovulation of small follicles [27] while the reverse is the case for harmful or deleterious amino acid mutations [28] . According to Tariq et al. [29] , the prediction of SNPs status is promising in modern genetics analysis and breeding programmes as they have been used to identify those animals with higher breeding value. Gene polymorphism between and among species in the present study appeared to be diverse because BMP15 is a fecundity gene [14] with a very strong divergence and a rapid evolution [30] [31] . Therefore, BMP-15 gene as a genetic marker closely linked to the litter size trait can be used in marker- assisted selection (MAS) for high litter size production and productivity in ruminants and non-ruminants.

The dendrograms were constructed to compare the common ancestral nucleotide and amino acid sequences of the species since each tree may give a useful information for proper understanding of the evolutionary relationships. The phylogenetic tree revealed that clustering was species wise with some level of intermingling between species. This is an evidence of trans-species evolution which might be attributed to the coding nature of the sequences. The UPGMA consensus trees based on nucleotide and amino acid sequences revealed that goat, sheep and cattle were closer followed by swine while chicken was farther apart. This is in accordance with classical classification as goat, sheep and cattle are members of the bovidae family, and swine shares the order artiodactyla with the ruminants. The findings of this study also agrees with the submissions of Misra et al. [32] and Ugbo et al. [33] who observed similar clustering of members of the bovidae family and order artiodactyla [2] .

5. CONCLUSION

The study revealed that the gene BMP15 is a polymorphic gene that has numerous mutations which can be beneficial or harmful. However, the beneficial ones were more than the deleterious ones in the present study. The omega values obtained symbolized that non-synonymous sites evolved faster than the synonymous sites and positive selection effect over-shadowed purifying selection. Also, the phylogeny based on nucleotide and amino acid sequences revealed the close relatedness of members of the bovidae family (goat, sheep and cattle), an indication of their closeness in the evolutionary timescale. The information emanating from this study would be relevant in performing further genotype-phenotype research especially the selection of markers of fecundity to improve genetically livestock species in Nigeria, Sub-Saharan Africa.

REFERENCES

1. Moradband, F., Rahimi, G. and Gholizadeh, M. (2011) Association of Polymorphisms in Fecundity Genes of GDF9, BMP15 and BMP15-1B with Litter Size in Iranian Baluchi Sheep. Asian-Australasian Journal of Animal Science, 24, 1179-1183. http://dx.doi.org/10.5713/ajas.2011.10453
2. Ran, X., Lin, J.-B., Du, Z.-Y., Qing, C. and Wang, J.-F. (2009) Diversity of BMP15 and GDF9 Genes in White Goat of Guizhou Province and Evolution of the Encoded Proteins. Zoological Research, 30, 593-602. http://dx.doi.org/10.3724/SP.J.1141.2009.06593
3. Demars, J., Fabre, S., Sarry, J., Rossetti, R., Gilbert, H., Persani, L., Tosser-Klopp, G., Mulsant, P., Nowak, Z., Drobik, W., Martyniuk, E. and Bodin, L. (2013) Genome Wide Association Studies Identify Two Novel BMP15 Mutations Responsible for an Atypical

Hyperprolificacy Phenotype in Sheep. PLOS Genetics, 9, e1003482. http://dx.doi.org/10.1371/journal.pgen.1003482

4. Palmer, J.S., Zhao, Z.Z., Hoekstra, C., Hayward, N.K., Webb, P.M., Whiteman, D.C., Martins, N.G., Doomsma, D.I., Duffy, D.L. and Montgomery, G.W. (2006) Novel Variants in Growth Differentiation Factor 9 in Mothers of Dizygotic Twins. Journal of Clinical Endocrinology and Metabolism,91,4713-4716. http://dx.doi.org/10.1210/jc.2006-0970
5. Larkin, M.A., Blackshields G., Brown, N.P., Chenna, R., McGettigan, P.A., McWilliam, H., Valentin, F., Wallace, I.M., Wilm, A., Lopez, R., Thompson, J.D., Gibson, T.J. and Higgins, D.G. (2007) Clustal W and Clustal X Version 2.0. Bioinformatics, 23, 2947-2948. http://dx.doi.org/10.1093/bioinformatics/btm404
6. Zhang, J., Rosenberg, H.F. and Nei, M. (1998).Positive Darwinian Selection after Gene Duplication in Primate Ribonuclease Genes. Proceedings of the National Academy of Sciences of the United States of America, 95, 3708-3713. http://dx.doi.org/10.1073/pnas.95.7.3708
7. Li, W. and Godzik, A. (2006) Cd-Hit: A Fast Program for Clustering and Comparing Large Protein or Nucleotide Sequences. Bioinformatics, 22, 1658-1659. http://dx.doi.org/10.1093/bioinformatics/btl158
8. Choi, Y., Sims, G.E., Murphy, S., Miller, J.R. and Chan, A.P. (2012) Predicting the Functional Effect of Amino Acid Substitutions and Indels. PLoS ONE, 7, e46688. http://dx.doi.org/10.1371/journal.pone.0046688
9. Saitou, N. and Nei, M. (1987) The Neighbor-Joining Method: A New Method for Reconstructing Phylogenetic Trees. Molecular Biology and Evolution, 4, 406-425.
10. Tamura, K., Nei, M. and Kumar, S. (2004) Prospects for Inferring Very Large Phylogenies by Using the Neighbor-Joining Method. Proceedings of the National Academy of Sciences of the United States of America, 101, 11030-11035. http://dx.doi.org/ 10.1073/pnas.0404206101
11. Felsenstein, J. (1985) Confidence Limits on Phylogenies: An Approach Using the Bootstrap. Evolution, 39, 783-791. http://dx.doi.org/10.2307/2408678
12. Tamura , K., Stecher, G., Peterson, D., Filipski, A. and Kumar, S. (2013) MEGA6: Molecular Evolutionary Genetics Analysis Version 6.0. Molecular Biology and Evolution, 30, 2725-2729. http://dx.doi.org/10.1093/molbev/mst197
13. Nei, M. and Kumar S. (2000) Molecular Evolution and Phylogenetics. Oxford University Press, New York.
14. Wang, Y., Yuanxiao, L., Nana, Z., Zhanbin, W. and Junyan, B. (2011) Polymorphism of Exon 2 of BMP15 Gene and Its Relationship with Litter Size of Two Chinese Goats. Asian-Australian Journal of Animal Science, 24, 905-911.
15. Hanrahan, J.P., Gregan, S.M., Mulsant, P., Mullen, M. and Davis, G.H. (2004) Mutations in the Genes for Oocyte-Derived Growth Factors GDF9 and BMP15 Are Associated with Both Increased Ovulation Rate and Sterility in Cambridge and Belclare Sheep (Ovisaries). Biology of Reproduction, 70, 900-909. http://dx.doi.org/10.1095/biolreprod.103.023093
16. Yakubu, A., Alade, E.D. and Dim, N.I. (2014) Molecular Analysis of Solute Carrier Family 11 a1 (Slc11a1) Gene in Ruminants and Non-Ruminants Using Computational Method. Genetika, 46, 925-934. http://dx.doi.org/10.2298/GENSR1403925Y
17. Vincent, S.T., Momoh, O.M. and Yakubu, A. (2014) Bioinformatics Analysis of Beta Casein Gene in Some Selected Mammalian Species. Research Opinions in Animal and Veterinary Sciences, 4, 564-570.
18. Williams, L.E. and Eernegreen, J.J. (2013) Sequence Context of Indel Mutations and Their Effect on Protein Evolution in a Bacterial Endosymbiont. Genome Biology and Evolution, 5, 599-605. http://dx.doi.org/10.1093/gbe/evt033
19. Bhattacharya, T., Kumar, P. and Sharma, A. (2008) Molecular Cloning and Characterization of CSN2 in Indian Riverine Buffalo (Bubalus bubalis). Buffalo Bulletin, 27, 222-230.
20. Yakubu, A., Salako, A.E., De Donato, M., Takeet, M.I., Peters, S.O., Adefenwa, M.A., Okpeku, M., Wheto, M., Agaviezor, B.O., Sanni, T.M., Ajayi, O.O., Onasanya, G.O., Ekundayo, O.J., Ilori, B.M., Amusan, S.A. and Imumorin, I.G. (2013) Genetic Diversity in Exon 2 at the Major

Histocompatibility Complex DQB1 Locus in Nigerian Indigenous Goats. Biochemical Genetics, 51,954-966. http://dx.doi.org/10.1007/s10528-013-9620-y
21. Zhang, J., Nielsen, R. and Yang, Z. (2005) Evaluation of an Improved Branch-Site Likelihood Method for Detecting Positive Selection at the Molecular Level. Molecular Biology and Evolution,22,2472-2479. http://dx.doi.org/10.1093/molbev/msi237
22. Bergstrom, T. and Gyllensten, U. (1995) Evolution of the Mhc Class II Polymorphism: The Rise and Fall of Class II Gene Function in Primates. Immunological Reviews, 143, 13-31. http://dx.doi.org/10.1111/j.1600-065X.1995.tb00668.x
23. Cui, H.X., Zhao, S.M., Cheng, M.L., Guo, L., Ye, R.Q., Liu, W.Q. and Gao, S.Z. (2009) Cloning and Expression Levels of Genes Relating to the Ovulation Rate of Yunling Black Goat. Biology of Reproduction 80,219-226. http://dx.doi.org/10.1095/biolreprod.108.069021
24. Toll-Riera, M., Laurie, S. and Alba, M.M. (2011) Lineage-Specific Variation in Intensity of Natural Selection in Mammals. Molecular Biology and Evolution, 28, 383-398. http://dx.doi.org/10.1093/molbev/msq206
25. Marini, N.J., Thomas, P.D. and Rine, J. (2010) The Use of Orthologous Sequences to Predict the Impact of Amino Acid Substitutions on Protein Function. PLoS Genetics, 6, e1000968. http://dx.doi.org/10.1371/journal.pgen.1000968
26. Dekkers, J.C.M. (2012) Application of Genomics Tools to Animal Breeding. Current Genomics, 13,207-212. http://dx.doi.org/10.2174/138920212800543057
27. Moore, R.K. and Shimasaki, S. (2005) Molecular Biology and Physiological Role of the Oocyte Factor, BMP-15. Molecular and Cellular Endocrinology, 234, 67-73. http://dx.doi.org/10.1016/j.mce.2004.10.012
28. Rossetti, R., Di Pasquale, E., Marozzi, A., Bione, S., Toniolo, D., Grammatico, P., Nelson, L.M., Beck-Peccoz, P. and Persani, L. (2009) BMP15 Mutations Associated with Primary Ovarian Insufficiency Cause a Defective Production of Bioactive Protein. Human Mutation, 30, 804-810. http://dx.doi.org/10.1002/humu.20961
29. Tariq, A.M., Al-Shammari, A.S., Al-Muammar, M.N. and Alhamdan, A.A. (2013) Evaluation and Identification of Damaged SNPs in COL1A1 Gene Involved in Osteoporosis. Archives of Medical Science, 9, 899-905.
30. Auclair, S., Rossetti, R., Meslin, C., Monestier, O. and Di Pasquale, E. (2013) Positive Selection in Bone Morphogenetic Protein 15 Targets a Natural Mutation Associated with Primary Ovarian Insufficiency in Human. PLoS ONE, 8, e78199. http://dx.doi.org/10.1371/journal.pone.0078199
31. Monestier, O., Servin, B., Auclair, S., Bourquard, T., Poupon, A., Pascal, G. and Fabre, S. (2014) Evolutionary Origin of Bone Morphogenetic Protein 15 and Growth and Differentiation Factor 9 and Differential Selective Pressure between Mono- and Polyovulating Species. Biology of Reproduction, 91,Article83. http://dx.doi.org/10.1095/biolreprod.114.119735
32. Misra, S.S., Ganai, T.A.S., Mir, S.A. and Kirmani, M.A. (2011) Molecular Characterization of Partial Exon-2 of the Bone Morphogenetic Protein 15 (BMP15) Gene in Indian Buffalo (Bubalus bubalis): Its Contrast with Other Species. Buffalo Bulletin, 30, 24-54.
33. Ugbo, S.B., Yakubu, A., Omeje, J.N., Bibinu, B.S., Musa, I.S., Egahi, J.O. and Dim, N.I. (2015) Assessment of Genetic Relationship and Application of Computational Algorithm to Assess Functionality of Non-Synonymous Substitutions in DQA2 Gene of Cattle, Sheep and Goats. Open Journal of Genetics, 5, 145-158. http://dx.doi.org/10.4236/ ojgen.2015.54011

CHAPTER 17

Some Peculiarities of Horse Breeding

Marcilio Dias Silveira da Mota[1] and
Luciana Correia de Almeida Regitano[2]

[1] Department of Animal Breeding and Nutrition, School of Veterinary Medicine and Animal Science, University of Sao Paulo State, Botucatu/SP, Brazil

[2] Animal Molecular Genetics, Embrapa Southeast Cattle, Sao Carlos/SP, Brazil

1. INTRODUCTION

There are approximately 60 million horses in the world, most of them living in America, Asia and some European countries. Currently China has the largest herd (around 8 million), followed by the United States (7 million), Mexico (a little more than 6 million) and Brazil (a little less than 6 million). Together, these four countries have close to 45% of the world's equine population [1].

The horse population's growth rate has been either constant or decreasing in most of the countries, with only some regions in Central America, Asia and Europe keeping positive growth rates (Figure 1). From 2003 to 2007, Puerto Rico presented the highest growth rate (45.4%), while the biggest Decrease was in Benin, Africa (-12.9%) (Figure 2).

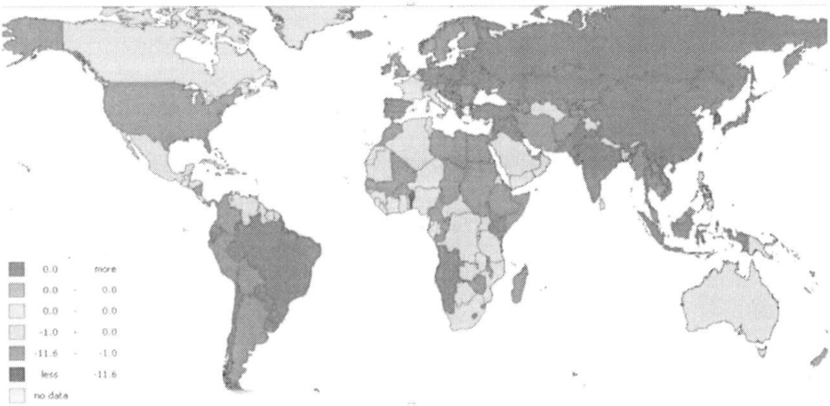

Figure 1. Growth rate of horse population in the world from 1997 to 2007 (Source [1])

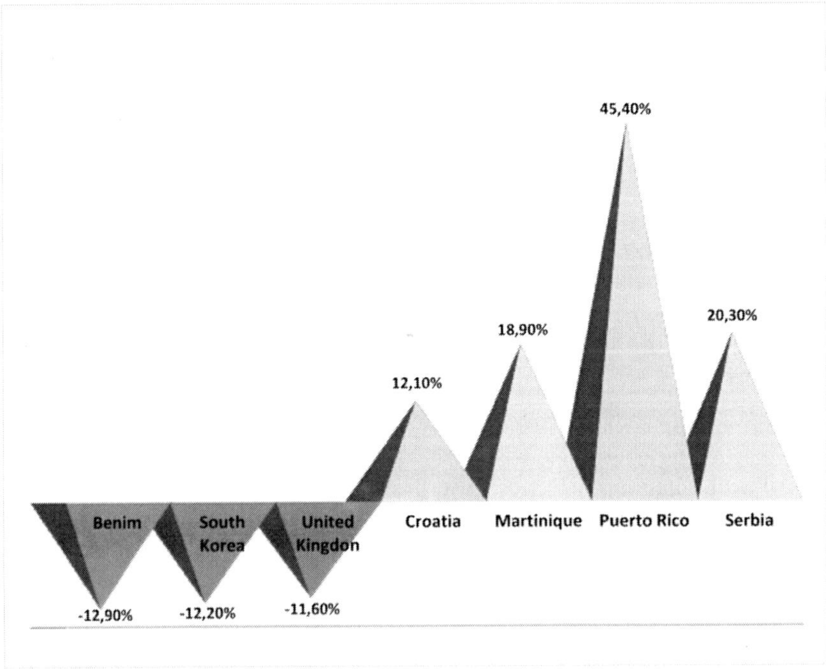

Figure 2. Countries with higher and lower equine population growth rates from 2003 to 2007

The world's export trade of live horses is concentrated in Europe and America, together representing 84% (58% and 26% respectively) of 2.1 billion dollars moved around yearly [2]. According to this source, the United States were the main exporter (148,472 animals, representing 48.8% of the world trade), raising around 474 million, in average $ 3,200 per animal.

Similar to the exportation, the world trade of live horses in relation to imports is highly supported by Europe (49.8% of the world trade), followed by Asia (26%) and Mexico (82,854), Canada (63,240) and Italy (46,333) are the main importers in number of animals, although the highest global expenditure in this aspect occurred in the United Kingdom (U.S. $ 498 million), UAE ($ 236 million) and Ireland (U.S. $ 233 million), representing together 46% of annual turnover - [2].

2. HORSE BREEDING

The expression "improvement of equine species," according to [3] dates back to the French articles about horses and donkeys in the *"Histoire Naturelle"* by Buffon in 1753. Despite the strong creationist dogmas of that time, these articles anticipated evolutionist ideas, while they described the concept of race degeneration due to the influence of location or climate. Adopted by the French Veterinary schools, it was believed that in order to return to an ideal type of creation, it would be necessary to mate mares and stallions with opposite types, to compensate traits that differed from the ideal. This meant basically to look for stallions on distant lands or regions.

Only decades later, after Darwin's evolution theory introduced the concept of selection is that the improvement of native populations emerged to counterbalance the situation, and effectively replace the desire to keep local types. Later, and gradually, the management of purebred animals predominated over crossbreeding, ending in the Thoroughbred Studbook's creation (first worldwide). It was a very important step for the use of animal production techniques in the 19th century [4]. According to the author, these techniques required precise identification procedures records of races held at different times and places in order to relate them to the same horse. The Stud Book, from the animal lineage certification, not only made it possible to relate the own horse performance information, but also their relatives. This fact prepared the way for the pedigree selection and progeny tests.

In addition, until the beginning of last century, horses were the focuses of experimental tests of inheritage theories. In this sense, historically the main concepts introduced by the horses, especially by the Thoroughbred breed, were [4]: performance selection in a purebred, introducing of a precise method for identifying an animal and its relatives, thinking about the male and female roles in the herd and the widespread use of planned matings.

Currently, although research in several countries are published every year in the literature involving some kinds of horse breeding study, few of them have ongoing consistent selection program. It means, in a way that most of research results in this area don't generate practical application and therefore it adds little to the species' development.

In most cases, this fact occurs because the breeders don't show interest in using the research results and not because of the research quality. The studies, depending on the availability and quality of information, usually do not consider the breeders' interests. Thus, in most countries there is wide gap between research institutes / universities and breeder associations. This gap is particularly due to the fact that more than in other species, horse breeders consider other breeders as potential competitors and give little importance to joint actions needed in breeding programs.

On the other hand, closer relation in Japan, Canada and some European countries, have allowed major advances in horse breeding. In Germany, for example, a country that stands out for jumping and dressage competitions and export of horses, the opinion, awareness and cooperation among breeders gave associations set guidelines that would meet not only their needs, and those of the research institutes / universities, but also the interest of the country. So the Breeders' Associations, supported by these institutions, publishes annually a Sire Handbook containing productive traits considered of general interest. This liaison has been very productive, resulting in19 gold, 8 silver and 12 bronze medals only in the last two Olympics[1] - , besides over 50 medals in some recent world championships [5].

3. SOME ADVANTAGES AND DIFFICULTIES IN HORSE BREEDING

Comparing to other farm animal species, there are some advantages and difficulties specifically related to researches on horse breeding, when quantitative genetics principles are used.

The advantages are fundamentally the amount of performance information and pedigree extension.

- The depth of the genealogy in most breeds is high

Traditionally, horse breeders associations consider the "pedigree" as a key factor to select their animals, so that in most of them, herd control is efficient. This fact goes back to the world's first Stud Book opening in 1791 (General Stud Book) for Thoroughbred animals. It was the basis for the other *Stud*Books, not only for horses, but also for all other domestic species.

Most of the breeds that were studied based on breeding, it is possible to track the animal's genealogy back to the fourth or fifth generation (depending on when the breeders' association's started), so that genetic evaluations in this aspect are efficient.

- Economic important traits can generally be measured in both genders.

In most of the breeds, especially the ones for some kinds of performance, the possibility of measuring both genders generates a greater information and knowledge volume about the behavior of traits in the population, what enables more efficient genetic evaluations. Racing, jumping, dressage, barrel, performances[2] - are examples of this kind.

On the other hand, even in breeds (or strains within a breed) where the breeding economic interest is the production of animals only for the conformation, not for performance, both genders can be evaluated.

- Economic important traits can usually be measured repeatedly in short time periods.

In horses it is possible to get repeated performance in relatively short time periods in a large portion of the population, while that doesn't happen with some domestic species, where economic important traits aren't repeated in the animal life (weight at weaning, weight at the year), or require a relatively long time to repeat (milk production, weight of the fleece).

In this sense, considering that the average number of starts per horse in Thoroughbred race season 2007/2008 was 6.5 [6], in a year is possible to get reasonable performance information (position, awards, time) about the horses. Although this average represents a new performance every two months, there are animals that race every 15 days, or even during the same week.

On the other hand, certain difficulties related to equine species and others due to various issues have been some of the major problems when research on horse breeding is done.

3.1. Inherent to the species

- Low reproductive rates

Since the early ancestors emerged from 55 to 60 million years ago, horses are adapting in order to develop a reproduction model that ensures survivability in the wild, adopting different reproductive strategies to ensure that their progeny are born in the appropriate time of year [7]. However, domestication has strongly influenced reproductive performance, with selection pressure on fertility being either small or null, and mating usually dictated by the functional performance of the animals [8].

Thus, considering that reproductive traits usually have low heritability estimate [8], and have been selected on horses by an indirect way, genetic alterations in order to enhance characteristics of this nature are evidently neither fast nor simple, especially in a species with a long generation interval such as the equine's (see item "High ranges of generation and delivery").

Hence, horses have low reproductive performance when compared to other farm animal species. The birth rate ranges from 59% [9, 8] to 74% [10, 7] the higher percentages being usually found in tests involving a small number of mares. Table 1 illustrates the result of 42,750 matings done with 7,278 Thoroughbred mares. It was observed that birthrates for males and females were 49.26% and 50.74%, respectively, whereas abortion and stillborn foals were 1.41% and 2.02%, respectively. It was also recorded that 9.07% of coverings were classified as empty, whereas 23.7% of matings did not show any latter records to track success.

Table 1. Occurrences after mating

Occurrence	Number of observations and %
Male	(29.35%)
Female	12,927 (30.23%)
Abortion	603 (1.41%)
Empty	3,878 (9.07%)
Without information	9,864 (7.23%)
Stillborn	866 (2.02%)
Not mated	1,945 (4.54%)
Mated with another breed	135 (0.31%)

This relatively low fertility is described in different races, countries and purposes, and may be related to hormonal dysfunction, genital infections in mares, parasitic infestations and inadequate handling practices before the breeding season [11]. These factors are even more imposing in the case of animals used in sports, since they have very different handling from those that are exclusively used in breeding [12]. Furthermore, maintenance of older mares and stallions due to their progeny's superior sporting performance can decrease the rate of conception in the herd, as the rate decreases progressively with increasing age [13]. Table 2 describes this aspect, illustrating the conception and apparent fertility rates (defined by [14] as the ratio of the total number of mares that conceived by the total number of mated mares and as mares that delivered living foals by the number of mated mares, respectively) according to parturition order.

Table 2. Number of observations (N), conception rates and apparent fertility according to the parturition order of Thoroughbred mares.

Parturition Order	N	Conception rate (%)	Apparent fertility rate (%)
1	5.531	75	71
2	4.534	73	68
3	3.655	71	68
4	2.794	69	67
5	2.155	67	65
6	1.560	64	61
7	1.070	62	59
8	777	59	56
9	513	56	54
10	329	54	52
11	171	48	47
12	92	48	46
"/> 13	79	41	38

Additionally, in some breeds (Thoroughbred, Quarter Horse, Standardbred, etc.), the existence of the so-called Racing Year[3] - also contributes to the decrease in reproductive rates. This is because, in certain competitions, animals are usually grouped according to, among other parameters (awards, past performance, gender, etc.), the horse's age defined by the racing year. Thus, breeders try to get the products born closer to the beginning of the equestrian year (July 1st - southern hemisphere or January 1st - northern hemisphere) in order to seize a competitive advantage (better developed, mature and trained horses) in relation to the animals born later that year [15]. Figure 3 shows the concentration of births of Thoroughbred foals in Brazil, a country located in the southern hemisphere.

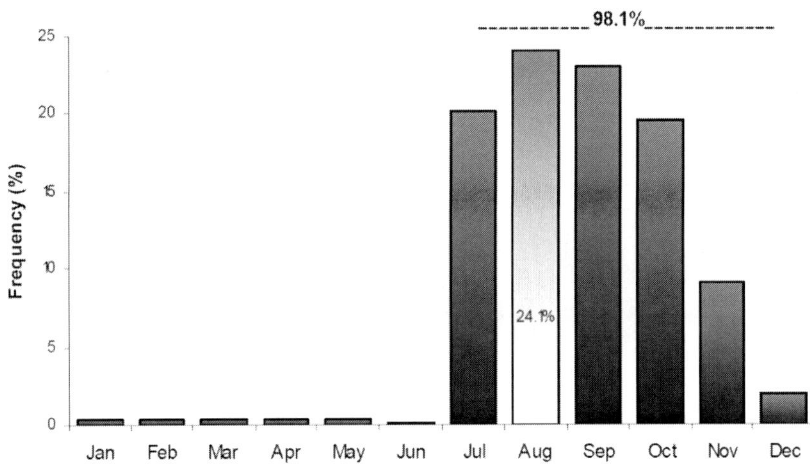

Figure 3. Percentage of Thoroughbred foal births by month in Brazil

With that objective, the breeding season, that usually lasts for 4 to 5 months, usually starts in August 15th in the southern hemisphere (or February 15th in the northern hemisphere). However, the percentage of mares that naturally ovulate in this period is quite low, since mares are seasonal poliestric, with the onset of the natural breeding season in spring, associated with increases in photoperiod, temperature and food availability [7]

In this sense, [15] found in Thoroughbred horses raised in Ireland (Northern Hemisphere), that the change of the beginning of the breeding season from February 15th to April 15th (hence delaying the beginning of the racing year by the same amount - January 1 to March 1), would better accommodate the natural reproductive cycle of females and could potentially increase the pregnancy rate by approximately 10% (Figure 4). Similar gains were likely to occur in countries in the southern hemisphere if the beginning of the breeding season changed from August 15th to October

15th, consecutively postponing the beginning of the racing year from July 1st to September 1st. Following this guidance, Australia postponed the beginning of the national racing year to August 1st (breeding season between September and December), mitigating the problem, although [13] considers the best breeding season in that country to be between November and February.

Figure 5 helps to understand the idea suggested by [13], as it represents the photoperiod in the southern hemisphere according to the latitude. It is observed that photoperiods are longer between October and February and higher percentages of ovulating mares are expected, favoring successful coverings.

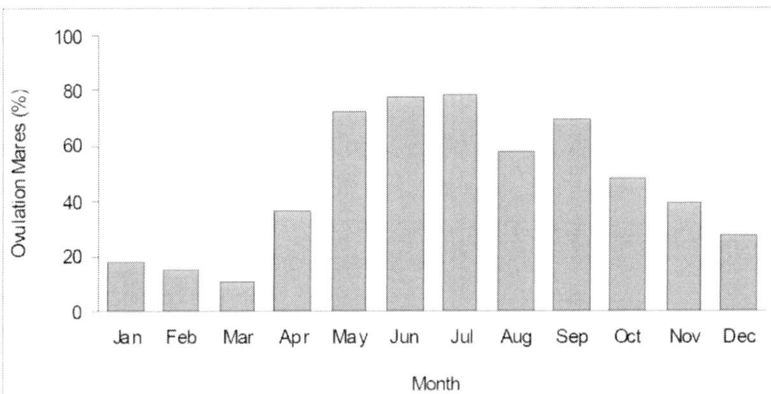

Figure 4. Percentage of mares ovulating by month in the northern hemisphere (adapted from [15])

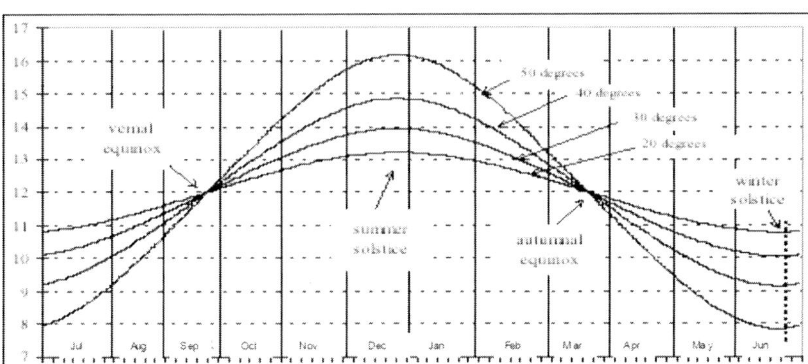

Figure 5. Photoperiods for different degrees south latitude (adapted from [16])

- high generation and parturition intervals

Generation interval represents the time needed to replace the next generation, and the shorter it is, the greater the expected annual gene change rate. Thus, considering that in horses this interval varies from 8 to 12 years, overall genetic changes due to selection tend to be slower when compared to cattle (4-6 years), sheep (3-5 years) pigs (1.5 to 2 years) and birds (1 to 1.5 years). Sportive breeds, in which reproductive technologies are not permitted, typically show higher ranges as the superior performance animals, especially females, usually start into reproduction after the end of their competitive life.

The generation intervals of some equine breeds are described in Table 3.

Table 3. Generation intervals from a few equine breeds

Breed	**Generation Interval (years)**	**Reference**
Andalusian	10.1	[17]
Thoroughbred	7.10	[18]
Icelandic Toeler	9.7	[14]
Friesian	9.6	[19]
Hanoverian	8.4	[20]
Mangalarga	9.5	[21]
Arabian	9.7	[22]
Lusitano	10.5	[23]
Mangalarga Marchador	8.9	[24]

The parturition interval is the amount of time between two consecutive parturitions, including the time from the parturition until the appearance of the first heat, from the first heat to the conception and finally the duration of pregnancy. It is an important component when estimating the herds' reproductive efficiency, with great influence on the economic return and breeding, due to its effects on the generation interval and selection intensity to be applied.

Among the domestic species, the mare has the capacity to provide fertile estrus a few days after birth, the so called foal heat. The main advantage of this phenomenon seems to be the maintenance of a 12 months foaling interval [25]. According to these authors, due to the enrollment of horses in sport activities, there is great pressure in order to have as many pregnant mares as possible in the breeding season. In this sense, efforts are made to cover mares in the foal heat. Considering the average pregnancy lasting around 11 months (see topics below) and the possibility of new pregnancy in the close postpartum days, a 12 months foaling interval would be obtained [25]. Thus, most breeders seek to take the opportunity of the foal heat, being aware that during this heat mares ovulate quickly, conception rates are lower and early embryonic mortality rate is higher [8].

There is a great variation among the equine foaling intervals, depending on the breed and breeding purposes, although most studies point values greater than 365 days (Table 4).

Table 4. Foaling intervals in some equine breeds

Breed	Foaling interval (days)	Reference
Halflinger	468	[26]
Thoroughbred	490	[8]
Marwari	535	[27]
Kathiawari	567	[27]
Mangalarga Marchador	548	[24]
Arabian	387	[7]

- Relatively long gestation period and low number of offspring per parturition

Although the duration of a pregnancy isn't directly associated with a breeding farm's production costs, its study may be extremely important in the preparation of breeding plans.

The gestation period can be defined as the time between fertilization of egg and the fetus' delivery. According to [28], the average duration of a mare's pregnancy is typically 340 days, ranging from 300 to 400 days. This wide time range until the birth of the foals indicates that mares may be highly susceptible to both internal and external factors afecting the duration of the pregnancy [29]. The ages of mare and stallion, year and month of birth, breeding season, foal sex, breed and nutritional status are factors that should be considered in the study of the pregnancy's duration [28].

Studies focusing the stallion used on coverings also deserve special attention when studying the pregnancy period in mares. The pregnancy's duration for females mated with specific stallions may be a criterion when choosing the stallion. This is because when a mare is bred late in a breeding season, yet the owner wants to mate her during this season, choosing a stallion associated with shorter pregnancy durations may be profitable [28].

Working with horses in the northern hemisphere [30] observed that the mating season was the most important factor affecting the duration of pregnancy in mares. According to these authors, the pregnancies that derived from mating during the period from December to May were 10.4 days longer than those derived from mating from June to November.

Studying Arab mares in Egypt, [31] observed that pregnancies with longer durations were the ones that ended in the winter, suggesting that the mares seem to be able to adapt the length of the gestation so the births happen in spring, which may be important for the survival of the species in the wild.

Table 5. Gestation periods in some equine breeds

Breed	Gestation period (days)	Reference
Carthusian	332	[32]
Thoroughbred/Quarter Horse	341	[33]
Andalusian	336	[34]
Standardbred	349	[35]
Mangalarga Marchador	327	[24]
Arabian	334	[7]
Criollo	335	[36]
Freiberger	336	[37]

Regardless of the discrepancy between studies and breeds, mares have a relatively long gestation period compared to other domestic species such as cattle (270-290 days), goats (145-151 days), sheep (144-152 days), pigs (112-115 days) and buffalo (298-317 days). Furthermore, as a uniparous species, twins (or multiple) are rare in mares, incidence varying from 0.5 to 1.6% of parturitions [38], so that the annual availability of animals for selection is comparatively small.

3.2. Other aspects

- Use of reproductive technologies

Although reproductive technologies (artificial insemination and embryo transfer) make possible different practical advantages such as lower disease and injury transmission, long-term storage of genetic material, easier transportation, earlier onset of reproduction in females and, in the case of embryo transfer, reproduction during the sports career, in the context of animal breeding they are usually considered additional tools to optimize breeding programs [5].

The commercial impact of these techniques in horses varies greatly. For example, in Thoroughbred horses meant for racing, the use of artificial insemination[4] - "The fecundation of mares can only be made by direct sexual contact, not admitting artificial insemination, but may exceptionally be authorized by the Brazilian Stud Book, by virtue of proven physical impairment of the player, the use of immediate reinforcement with fresh semen collected during the coverage. This procedure, when authorized, will be held only by a veterinarian authorized in advance by the Brazilian Stud Book " is officially banned worldwide, while in several sportive breeds, especially in Europe, the percentage of inseminated animals exceeds 80% (French Saddle Horse - 84%, Hanoverian - 91% Holsteiner - 95%, Belgian and Dutch Warmblood - over 95%) [5].

On the other hand, according to the International Embryo Transfer Society (IETS) the higher number of embryo transfers occur in the United States, Argentina and Brazil, in order of importance, countries which together represent over 90% of the activities in this area worldwide. In 2005, 5,700 embryos were transferred in Brazil, none frozen, while across Europe only 711 transfers were done in horses [39].

In this context, it is observed that, although several countries present a prominent role in the worldwide scenario with respect to the use of these reproductive biotechniques, especially embryo transfer, there are few studies about their impact (generation interval, accuracy, selective intensity, inbreeding) on horses breeding programs. Research in this subject would be important to understand the direction that has been given to these tools in different countries and monitor their actual benefit for the horse population.

- Knowledge of interesting traits' economic values

A fundamental requirement for any breeding scheme aiming the improvement of quantitative traits is the establishment of the breeding objectives, involving the relative values of genetic change for all desirable features included in the breeding program. Typically, these values are expressed in monetary terms as weights to be applied to each feature of economic importance [40].

However, few scientific studies have been performed in order to obtain economic weights for traits involved in horse selection programs, were a combination of empirical experience, some biological factors and intuition of designers prevail. There are a few reasons for that.

It's often very difficult to determine, in horses, the value of one unit of expression for a given trait in relation to the animal's total value. Economically quantifying units for traits such as speed, dressage, jumping, etc., is far more complex than attributing values to liters of milk, kilos of meat or wool.

The long period of time between mating and expressig the traits of interest in the progeny, besides the difficulties in determining an appropriate function for profitability, provide part of this deficiency. Moreover, according to recent authors, another problem arises from the fact that not always the relative economic weights are linear in a breeding program. Thus, the amount of increase in the genotype for certain character can be strongly dependent on the values of other genetic traits. For example, in horses with outstanding ability to jump, the additional genetic values affecting their training capability is almost neglected, whereas in animals with a low ability to jump, a genotype corresponding to training characteristics can greatly increase its value.

- Diversity of goals within certain breeds

There are equine breeds that are commonly selected by breeder in only one direction, as is the case of the Thoroughbred, where the objective is basically to obtain animals with superior performance in races. In Quarter horses, in addition to races, performance in work tests and conformation may also be targets of breeders since the race is subdivided amongst these strains.

Moreover, in breeds in which animals are involved in a wide variety of uses (work on farms, horseback riding, trekking, exhibitions, equine therapy, equestrian tourism, unskilled riding sports, etc.) breeders seek very different traits, depending on the purpose of raising the animal, hindering the implementation of breeding programs that cover all segments. Brazilian breeds such as the Mangalarga and the Mangalarga Marchador fall into this category.

In these cases, studies involving quantitative and molecular aspects of traits that can meet the desire for the greater proportion of breeders can result in meaningful contributions to selection of reproductive, behavioral, immunological and other traits.

REFERENCES

1. GHILPA (2011) The Global Livestock Production and Health Atlas. Available: http://kids.fao.org/glipha/. Acessed 2011 November 20.
2. FAO (2011) Food and Agriculture Organization of the United Nations. Available: http://faostat.fao.org/site/569/default.aspx#ancor. Acessed 2011 November 22.
3. Mulliez J (1983) Les chevaux du royaume. Ed Montalba, 385p.
4. Langlois B (1996) A consideration of the genetic aspects of some current practices in Thoroughbred horse breeding. Ann. Zootech. 45:41-51.
5. Bruns EW (2008) Horse Breeding in Germany. In: Horse breeding and genetics course, class notes. Graduation Program on Animal Breeding and Genetics, Unesp/FCAV, Jaboticabal, 45 p.
6. IFHA (2012) International Federation of Horseracing Authorities, Statistical Information, Available: www.IFHAOnline.org. Acessed 2012 May 5.
7. Cilek S (2009) The Survey of reproductive success in Arabian horse breeding from 1976-2007 at Anadolu State Farm in Turkey. Journal of Animal and Veterinary Advances 8:389-396.
8. Taveira R Z, Mota MDS (2007) Genetic and quantitative evaluation of breeding traits in Thoroughbred mares. Revista Eletrónica de Veterinaria 8:1-11.
9. Mota MDS (2005) Avaliação populacional da raça Mangalarga. In: Reunião do Conselho Técnico da Associação Brasileira de Criadores do Cavalo Mangalarga, Ribeirão Preto 63p.
10. Zúccari CESN, Nunes DB, Costa e Silva EV (2004) Harém Pantaneiro sob monta natural a campo na Região do Rio Negro, Pantanal, MS. In: Anais, IV Simpósio sobre Recursos Naturais e Sócio-econômicos do Pantanal. Corumbá, 1-4.
11. Sullivan J J, Turner PC, Self LC, Gutteridge HB, Bartlett DE (1975) Survey of reproductive efficiency in the Quarter-Horse and Thoroughbred. J. Reprod. Fert. Suppl. 23: 315-318.
12. Jackson RS (1971) Equine infertility. Preliminary report on a survey of courses taken by the A.A.E.P. Newsletter Am. Ass. Equine Pract.
13. Bailey CJ (1998) Wastage in the Australian Thoroughbred Racing Industry. Rural Industries Research & Development Corporation, No 98/52, 67p.
14. Hugason K, Arnason TH, Jónmundsson JV (1985) A note on the fertility and some demographical parameters of Icelandic toelter horses. Livest. Prod. Sci. 12: 161-167.
15. Cunningham EP (1991) The Genetics of Thoroughbred Horses. Scientific American 264: 92-98.
16. Diven D (2000) Keys to the low cost/calf program. Missouri forage and grassland council Annual meeting, Lake Ozark, Missouri, 2000. Available: http://agebb.missouri.edu/mfgc/2000mtg/lowcost.htm. Acessed 2011 November 22.
17. Valera M, Molina A, Gutiérrez JP, Gómez J, Goyache F (2005) Pedigree analysis in the Andalusian horse: Population structure, genetic variability and influence of the Carthusian strain. Livestock Production Science: 95, 57–66.
18. Taveira RZ, Mota MDS, Oliveira HN (2004) Population parameter in Brazilian Thoroughbred. J. Anim. Breed. Genet. 121: 384-391.
19. Sevinga M, Vrijenhoek T, Hesselink JW, Barkema HW, Groen AF (2004) Effect of inbreeding on the incidence of retained placenta in Friesian horses. Journal of Animal Science, 82:982–986.
20. Hamann H, Distl O (2008) Genetic variability in Hanoverian Warmblood horses using pedigree analysis. Journal of Animal Science 86:1503-13.
21. Mota MDS, Almeida Prado RS, Sobreiro J (2006) Caracterização da população de cavalos Mangalarga no Brasil. Archivos de Zootecnia, 55: 31-37.
22. Moureaux S, Verrier E, Ricard A, Meriaux JC (1996) Genetic variability within French race and riding horse breeds from genealogical data and blood marker polymorphism. Genetics Selection and Evolution 28: 83–102.
23. Vicente AA, Carolino N, Gama LT (2012) Genetic diversity in the Lusitano horse breed assessed by pedigree analysis. Livestock Science – in press.

24. Gonçalves RW, Costa MD, Rocha Júnior VR, Costa MR, Silva ESP, Ribeiro AMF (2011) Inbreeding effect on reproductive traits in a herd of Mangalarga Marchador Brazilian horses. Rev. Bras. Saúde Prod. Anim., Salvador 12: 641-649.
25. Lowis TC, Hyland JH (1991) Analysis of post-partum fertility in mares on a Thoroughbred stud in southern Victoria. Australian Veterinary Journal 68, n.9, September.
26. Meregalli A, Valzania C (1984) Reproductive parameters of Halflingers kept on Tuscan farms. Rivista di Zootecnia e Veterinaria 12: 190-98.
27. Singh MK, Yadav MP, Mehta NT (2002) Breed characteristics of Marwari and Kathiawari horses. Indian Journal of Animal Sciences 72: 319-323.
28. Marteniuk JV, Carleton CL, Lloyd JW, Shea ME (1998) Association of sex of fetus, sire, month of conception, or year of foaling with duration of gestation in Standarbred mares. Javma 212: No. 11.
29. Davies Morel MCG, Newcombe JR, Holland SJ(2002) Factors affecting gestation length in the Thoroughbred mare. Animal Reproduction Science 74: 175-185.
30. Rollins WC, Howell CE (1951) Genetic sources of variation in the gestation length of the horse. Journal Animal Science 10: 797.
31. Hura V, Hajurka J, Kacmárik J, Csicsai G, Valocky I (1997) The effect of some factors on gestation length in Nonius breed mares in Slovakia. UVL, 04101 Kosice, Komenského 73, Slovak Republic.
32. Satué K, Felipe M, Mota J, Munoz A (2011) Gestational length in Carthusian broodmares:effects of breeding season, foal gender, age of mare, year of parturition, parity and sire. Polish Journal of Veterinary Sciences 14: 173-180.
33. McCue PM, Ferris RA (2012) Parturition, dystocia and foal survival: A retrospective study of 1047 births. Equine Veterinary Journal 44, Suppl. 41: 22–25.
34. Valera M, Blesa F, Dos Santos R, Molina A (2006) Genetic study of gestation length in Andalusian and Arabian mares. Animal Reproduction Science 95:75-96.
35. Dicken M, Gee EK, Rogers CW, Mayhew IG (2012) Gestation length and occurrence of daytime foaling of Standardbred mares on two stud farms in New Zealand. New Zealand Veterinary Journal 60: 42-46.
36. Winter GHZ, Rubin MIB, De La Corte FD, Silva CAM (2007) Gestational Length and First Postpartum Ovulation of Criollo Mares on a Stud Farm in Southern Brazil. Journal of Equine Veterinary Science 27: 531-534.
37. Giger R, Meier HP, Küpfer U (1997) Gestation lengths of Freiberger» mares with mule and horse foals. Schweizer Archiv fur Tierheilkunde, 139: 303-307.
38. Basrur PK (1983) Genetics in Veterinary Medicine. University of Guelph, Guelph, Canada, 261p.
39. Thibier M (2006) Transfers of both *in vivo* derived and *in vitro* produced embryos in cattle still on the rise and contrasted trends in other species in 2005. International Embryo Transfer Society Newsletter, 24: 12-18, 2006. Available: http://www.iets.org/pdf/data_retrieval/ december2006.pdf. Acessed: 2010 August 2.
40. Arnason T, Van Vleck LD (2000) Genetic Improvement of the Horse. In: The Genetics of the horse. Cab Publising. pp. 473-497.

CHAPTER 18

An interpretive review of selective sweep studies in Bos taurus cattle populations: identification of unique and shared selection signals across breeds

Beatriz Gutiérrez-Gil[1*], Juan J. Arranz[1] and Pamela Wiener[2]

[1]*Departamento de Producción Animal, Universidad de León, León, Spain*

[2]*Division of Genetics and Genomics, Roslin Institute and R(D)SVS, University of Edinburgh, Midlothian, UK*

This review compiles the results of 21 genomic studies of European Bos taurus breeds and thus provides a general picture of the selection signatures in taurine cattle identified by genome-wide selection-mapping scans. By performing a comprehensive summary of the results reported in the literature, we compiled a list of 1049 selection sweeps described across 37 cattle breeds (17 beef breeds, 14 dairy breeds, and 6 dual-purpose breeds), and four different beef-vs.-dairy comparisons, which we subsequently grouped into core selective sweep (CSS) regions, defined as consecutive signals within 1 Mb of each other. We defined a total of 409 CSSs across the 29 bovine autosomes, 232 (57%) of which were associated with a single-breed (Single-breed CSSs), 134 CSSs (33%) were associated with a limited number of breeds (Two-to-Four-breed CSSs) and 39 CSSs (9%) were associated with five or more breeds (Multi-breed CSSs). For each CSS, we performed a candidate gene survey that identified 291 genes within the CSS intervals (from the total list of 5183 BioMart-extracted genes) linked to dairy and meat production, stature, and coat color traits. A complementary functional enrichment analysis of the CSS positional candidates highlighted other genes related to pathways underlying behavior, immune response, and reproductive traits. The Single-breed CSSs revealed an over-representation of genes related to dairy and beef production, this was further supported by over-representation of production-related pathway terms in these regions based on a functional enrichment analysis. Overall, this review provides a comparative map of the selection sweeps reported in European cattle breeds and presents for the first time a characterization of the selection sweeps that are found in individual breeds. Based on their uniqueness, these breed-specific signals could be considered as "divergence signals," which may be useful in characterizing and protecting livestock genetic diversity.

Keywords: cattle, breeds, selection signals, candidate genes, diversity, selective sweep, domestication

1. INTRODUCTION

The genetic diversity of livestock species is an economical and cultural inheritance from our ancestors, and an indispensable resource to meet the unpredictable needs of our future (Larson et al., 1992). The history of this diversity involves the spread of livestock populations from their centers of domestication as small samples of the original domesticated populations. Under new environments and the effects of genetic drift and natural selection, the different groups developed into distinct local populations (FAO, 2013). Associated with later advances in animal husbandry and breeding, more specialized breeds and breeding lines were developed. During the past 250 years, there has been a development of individually uniform but collectively highly diverse and distinguishable populations, which are known as "standardized breeds" (FAO, 2013).

In livestock populations, approximately half of the genetic diversity is shared across breeds while the other half is observed within single breeds (Sponenberg and Bixby, 2007). Hence, the substantial loss of biodiversity associated with the loss of a breed means that effective management of breeds is essential to managing the overall biodiversity of domesticated species.

During the establishment of modern livestock breeds, the genomes of domestic animal species have been subjected to multiple human-imposed selection events influencing traits of concern to agriculturists. In comparison to natural selection, artificial selection has the ability to rapidly change the genome. Selection not only affects the favored mutation but it produces a "hitchhiking" effect on the frequency of neutral alleles at linked loci (Maynard Smith and Haigh, 1974; Kaplan et al., 1989). Selection-mapping or hitchhiking mapping approaches exploit this phenomenon by searching for genomic regions of reduced variability as signatures of strong positive selection, with the aim of identifying causal mutations controlling selected phenotypes (e.g., Kohn et al., 2000; Harr et al., 2002; Storz et al., 2004; Pollinger et al., 2005; Voight et al., 2006). The different methods developed for detection of selection signatures through the analysis of genetic markers are based either on the distribution of allelic frequencies or the properties of haplotypes segregating within populations, or on the distribution of genetic differentiation between populations (reviewed by Hohenlohe et al., 2010).

In recent years, the availability of high-density, genome-wide single nucleotide polymorphism (SNP) arrays and parallel progress in statistical techniques have allowed the identification of genomic regions that have been subjected to positive artificial selection in livestock species ("selection scans"). While identifying a selection signature in the same region in different breeds gives support to the hypothesis that a particular genomic region has undergone selection for a given trait, many selection signatures appear to be breed-specific. By comparing the results of the studies that have searched for selection signatures in different cattle breeds, this review provides a map of selection footprints that could be considered a source of genetic diversity in these domestic populations and therefore represent a valuable resource that may be worth protecting independently of the productive ability of the breed(s) involved.

2. GENETIC DIVERSITY AND SELECTION SIGNATURE STUDIES IN *BOS TAURUS* CATTLE

Present day cattle breeds are the result of years of human selection, adaptation to different environments and cross-breeding, as well as demographic effects such as bottlenecks and migration, all of which contribute to the current patterns of genetic diversity (Bruford et al., 2003; Laloe et al., 2010). Human-mediated selective processes include those related to domestication, breed formation, and ongoing selection to enhance performance and productivity. In 2009, the Bovine HapMap Consortium presented the first detailed genome-wide characterization of the genetic variability of 19 geographically and phylogenetically diverse bovine breeds, based on the analysis of 37,470 SNPs. This study showed that taurine breeds (*Bos taurus*) showed a lower genetic diversity than indicine breeds (*Bos indicus*), probably due to a lower diversity within the pre-domestication ancestral population and/or post-domestication effects of stronger bottlenecks at breed formation and stronger selection for docility and productivity (Bovine HapMap Consortium, 2009). The authors concluded that despite the decline in effective population size (N_e) of some breeds, overall genetic diversity in cattle was "not low" and the between-breed differences in diversity were due to events at and before breed formation rather than differences in the intensity of natural or artificial selection post-domestication. This study was the first to perform a high-resolution, genome-wide examination of the structure of the cattle genome in different breeds and reported selection signatures in regions involving genes known to harbor causal mutations related to production traits (e.g., *GDF-8* and *ABCG2*, in relation to muscle conformation and milk composition, respectively) and genes associated with food conversion efficiency (e.g., *R3HDM1*). Since this initial analysis, many studies have followed, with the common aim of identifying specific genomic regions influenced by artificial selection in cattle breeds.

This review compiles the results of 21 genomic studies of European-related Bos taurus populations and thus provides a general picture of the selection signatures in taurine cattle identified by genome-wide selection-mapping scans. By performing a systematic comparison of the results reported in the literature, we have identified those regions that are found in several breeds showing the same production characteristics, and that therefore are very likely to harbor mutations with significant effects on production traits. In general, these are the regions that have already been highlighted by the different authors, as they show the highest statistical support for the presence of a positive selection signature, and because in many cases they contain genes related to the shared production characteristics that can be viewed as selection candidates. We also show that in many cases selection signatures are also shared by breeds showing different production characteristics. These may be regions of interest in relation to metabolic homeostasis or other general traits such as disease resistance and behavior. But one of the main objectives of the interpretive survey presented herein is to highlight those regions that have been reported in a single breed. In general, results of this type are not discussed in detail by the authors, and in some cases are not presented to the reader, such that a large portion of the biological information generated through these genomic studies is never interpreted. However, we hypothesize herein that these single-breed sweeps may indicate genomic sources of unique phenotypic characteristics of the target breed for which the selection signal has been detected. Although determining the phenotype associated with these single-breed sweeps may be particularly difficult, the identification and characterization of these regions as "divergence signals" may be of value as an initial step to protect, from a genomic point of view, the wealth of livestock diversity.

3. GENERAL OVERVIEW OF THE REVIEWED STUDIES

As an initial attempt to perform a systematic review of the available literature on cattle selection signals, this review targets the genome-wide selective sweep scans described in Bos taurus breeds of European origin and mainly focuses on the interpretation of selection sweeps associated with dairy and beef production specialization. Hence, some studies have not been considered, including studies limited to specific chromosomes (e.g., Hayes et al., 2008; Prasad et al., 2008), or studies mainly addressing Bos indicus (Somavilla et al., 2014), African taurine cattle breeds (Gautier et al., 2009) or cross-bred cattle (Flori et al., 2012) or studies focusing on Bos taurus-Bos indicus comparisons (Porto-Neto et al., 2013; Utsunomiya et al., 2013a).

Exceptions were four studies that included in their larger-scale analysis some Bos indicus and hybrid breeds (Bovine HapMap Consortium, 2009; Qanbari et al., 2011; Ramey et al., 2013; Porto-Neto et al., 2014), although we have only considered the results reported for the European Bos taurus breeds. Details of the 21 studies compiled in this review are provided in Table 1, including information about the breeds analyzed and their production characteristics, the statistical method(s) used for the identification of selection signatures, the SNP-chip or dataset analyzed, and other technical details such as the version of the reference genome on which the study was based.

Depending on the number of breeds analyzed, we classify the studies as those that focus on: (i) a single breed (Qanbari et al., 2010, 2014; Glick et al., 2012; Boitard and Rocha, 2013; Lee et al., 2013b; Lim et al., 2013; Pan et al., 2013); (ii) a pair-wise comparison of closely-related populations with divergent production characteristics (mostly beef vs. dairy breeds, e.g., Hayes et al., 2009; Wiener et al., 2011; Hosokawa et al., 2012; Pintus et al., 2014) and (iii) several breeds, from three (Flori et al., 2009) to 19 breeds (Bovine HapMap Consortium, 2009), of the same or different production characteristics, and for which both across- and within-population analyses are performed.

Overall, the selection sweeps considered in this review involved 37 breeds (including 17 beef breeds, 14 dairy breeds, and six dual-purpose breeds), and four different beef-vs.-dairy comparisons (Australian Holstein vs. Australian Angus, Charolais vs. Holstein, Japanese Black vs. Japanese Holstein, Piedmontese vs. Italian Brown) (Supplementary Table S1 in Supplementary Material).

Table 1. Summary of the genome-wide selection mapping studies included in this review.

Study	Original reference genome	Genotyping platform[1]	Number of analyzed SNPs	Number of regions considered in this review[2]	Number of breeds	Breeds[3]	Production category	Analysis method[4]	Origin of data from original study
Boitard and Rocha, 2013	UMD3.1	50K-chip	35,564	3	1	BLO	Beef	HMM-allele frequency	Table 1
Bovine HapMap Consortium, 2009	Btau4.0	AfC-10K 1536-chip	33,326	67	19	BMAS*, ANG, REDA, HER, NORW, HOL, LIM, CHA, BRSW, JER, GUE, PIE, ROM NDAM*, SHE*, NEL*, BRA*, GIR*, SGER*	Dairy, Beef, Dual-purpose	F_{ST}, IHS	Table 1; Table S8
Druet et al., 2013	UMD3.1	HD-chip	725,293	147[a]	12	BB, DBB, HOL, JER, LIM, HER, ANG, CHA, GUE, PIE, ROM, BRSW	Dairy, Beef, Dual-purpose	HMM-average heterozygosity	Tables S1–S12
Flori et al., 2009	Btau4.0	50K-chip	42,486	6	3	HOL, NORM, MONT	Dairy	F_{ST}	Table 2
Glick et al., 2012	Btau4.0	50K-chip	41,814	25	1	ISR-HOL	Dairy	REHH	Table 3
Hayes et al., 2009	UMD3.1	TM-9323	9,323	15	2	AUSHOL-vs.-AUSANG	Beef vs. Dairy	Differences in allele frequencies	Table 2
Hosokawa et al., 2012	Btau4.0	50K-chip	40,635	11	2	JAPBL-vs.-JAHOL	Beef vs. Dairy	Differences in allele frequencies	Table 2
Kemper et al., 2014	UMD3.1	HD-chip (real and imputed)	610,123	30	8	HOL, JER, ANG, CHA, HER, LIM, MUG, SHOR	Dairy, Beef	HAPH, IHS, F_{ST}	Table 2, Table 6
Lee et al., 2013b	UMD3.1	WGS	15,125,420	15	1	HAN	Beef (high intramuscular fat content)	LD-ω, CLR	Table 1
Lim et al., 2013	UMD3.1	AfM-10K	4,522	2		HAN	Beef (high intramuscular fat content)	EHH-IES	Table 1
Mancini et al., 2014	Btau4.0	50K-chip	29,848	5	5	IT-BR, IT-HOL, PIE, MAR, IT-PEZZ	Dairy, Beef, Dual-purpose	F_{ST}	Table 1
Pan et al., 2013	Btau4.0	50K-chip	40,130	16	1	CHI-HOL	Dairy	EHH	Table 3
Pintus et al., 2014	Btau4.0	50K-chip	42,514	53	2	PIE-vs.-ITBR	Beef vs. Dairy	F_{ST} (LOWESS regression correction)	Table 1
Porto-Neto et al., 2014	UMD3.1	HD-chip	680,000	55	9	HAN (against six European breeds, NEL, NDAM)	Beef (high intramuscular fat content)	F_{ST}	Table 2
Qanbari et al., 2010	Btau4.0	50K-chip	41,398	12	1	HOL	Dairy	REHH	Table 4
Qanbari et al., 2011	Btau4.0	50K-chip	40,595	14	10	HOL, BRSW, SIM, AUS-ANG, BELR*, HER, MUG, SGER, SHOR, BRA*	Dairy, Beef, Dual-purpose	IHS, F_{ST}	Table 2
Qanbari et al., 2014	Btau4.0	WGS-based imputation	Sequenced data (15,182,131 SNPs), medium density panel (39,304), high density panel (645,189).	140	1	FLE	Dual-purpose	IHS, CLR	Tables S3, S4

Ramey et al., 2013	UMD3.1	50K-chip AFFXB1P-chip-2787037	52,942 (50Kchip dataset); 2,575,339 SNPs (AFFXB1P)	132	15	ANG, BRA, CHA, HAN, HER, LIM, SAL, SHOR, SIM, BRSW, FINA, HOL, JER, WAG, BRA*	Dairy, Beef, Dual-purpose	Extended low diversity haplotypes	Tables 2, 3, 4
Rothammer et al., 2013	UMD3.1	50K-chip	47,651	61[b]	8	GAL, BB, RED-HOL, OBRA, MUR, FRGE, FLE, BRA*	Beef	XP-EEH	Tables S4–S13
Stella et al., 2010	Btau4.0	HMap-data	32,689	215[c]	5	BRSW, GUE, HOL, NORW, JER	Dairy	CLL	Within-breed results[5]
Wiener et al., 2011	Btau4.0	HMap-data	31,312	30[d]	2	CHA-vs.-HOL	Beef vs. Dairy	Difference in allele frequencies	Table S2

[1] *High-throughput genotyping arrays used for genotyping in the different studies 50k chip, Illumina's Bovine SNP50 Genotyping BeadChip; 1536-chip, Illumina 1536 BeadArray assays; HD-chip, Illumina Bovine HD chip genotyping assay; AfC-10K, Affymetrix Custom 10K; TM-9323, 9323 SNPs (Parallele TM/Affymetrix TM); HMap-data, data from Bovine HapMap Consortium; WGS, Whole genome sequence; AfM-10K, Affymetrix MegAllele GeneChip Bovine Mapping 10K SNP array; AFFXB1P-chip-2787037, Affymetrix Axiom Genome-wide BOS 1 assay = 2,787,037 SNPs.*

[2] *Where indicated the initial significant positions reported by the original authors were filtered in this review as described below (number of raw reported sweeps):*

[a] *From an initial list of 398 positions, only those showing an average heterozygosity < 0.003 were considered herein.*

[b] *From an initial list of 1251 breed-specific significant positions, only those with consecutive XP-EHH values (gap < 1 Mb) higher than 4 were considered herein.*

[c] *From the initial results from the within-breed analysis results kindly provided by the authors (a total of 13,077 significant positions at $P < 0.01$, genome-wide), we selected, for each breed, the top 10% of the positions showing a significant CLL, and then we grouped the positions in signatures regions. Following the authors' criteria, adjacent signature regions were considered "distinct" if they were separated by at least 3 consecutive windows with non-significant CLL ($P < 0.01$ genome-wide).*

[d] *From the initial list of 121 significant positions of top 1% MA_d values, the positions were grouped in regions of subsequent significant regions with no more than 1 Mb gaps.*

[3] *Breed abbreviations: ANG, Angus; AUS-ANG, Australian Angus; BB, Belgian Blue; BLO, Blonde d'Aquitaine; CHA, Charolais; GAL, Galloway; HER, Hereford; HAN, Korean Hanwoo; LIM, Limousin; MAR, Marchigiana; MUG, Murray Gray; PIE, Piedmontese; REDA, Red Angus; ROM, Romagnola; SAL, Salers; SHOR, Shorthorn; WAG, Wagyu; AUSHOL-vs.-AUSANG, Australian Holstein vs. Australian Angus; CHA-vs.-HOL, Charolais vs. Holstein; JAPBL-vs.-JAHOL, Japanese Black vs. Japanese Holstein; PIE-vs.-ITBR, Piedmontese vs. Italian Brown; BRA, Braunvieh; BRSW, Brown Swiss; CHI-HOL, Chinese Holstein; FINA, Finnish Ayrshire; GUE, Guernsey; HOL, Holstein; ISR-HOL, Israeli Holstein; IT-BR, Italian Brown; IT-HOL, Italian Holstein; IT-PEZZ, Italian Pezzeta rosa; JER, Jersey; MONT, Montbeliarde; NORM, Normande; NORW, Norwegian Red; RED-HOL, Red Holstein; DBB, Dual-purpose Belgian blue; FLE, Fleckvieh; FRGE, Franken Gelbvieh; MUR, Murnau-Werdenfelser; OBRA, Original Braunvieh; SIM, Simmental.*

[4] *Statistical methods used in the original studies reviewed in this work: CLL, Parametric composite log likelihood of the differences in allelic frequencies; CLR, Composite likelihood ratio of a model considering a section sweep as defined by Nielsen et al. (2005); EHH, Extended haplotype homozygosity; F_{ST}, Population differentiation statistic; HAPH, Haplotype homozygosity measurement; HMM allele frequency, Hidden Markov model-based test, local variations in the allele frequency; iHS, Integrated haplotype homozygosity score; LD ω, LD-based omega (ω) statistics; REHH, Relative extended haplotype homozygosity; XP-EEH, Cross population extended haplotype homozygosity.*

[5] *Boettcher and Stella, personal communication.*

* *Additional breeds analyzed in the original studies but not considered in this review: BMAS, Beefmaster; BELR, Belmont Red; NDAM, N'Dama; SHE, Sheko; NEL, Nelore; BRA, Brahman; GIR, Gir; SGER, Santa Gertrudis.*

In addition, we have considered those selection sweeps reported for Holstein populations from specific geographic regions, such as Italian, Israeli or Chinese Holstein cattle, for Angus and Australian Angus cattle, and for Simmental and its German strain Fleckvieh, as related to "distinct" breeds in order to investigate whether there is evidence for geographical region-specific sweeps for the same breed.

The genotyping platforms used in the considered studies demonstrate the rapid development of livestock genomic tools over the last few years (**Table 1**). The earliest study included, that of Hayes et al. (2009), involved the analysis of 9323 SNPs genotyped by Parallele TM or Affymetrix TM and the Bovine HapMap study (2009) generated data using a custom Affymetrix 10K genotyping chip and Illumina 1536 BeadArray assays (Taylor, personal communication). Additional analyses of the original Bovine HapMap dataset were reported later by Stella et al. (2010) and Wiener et al. (2011). But most of the studies compiled in this review (10 out of 17) are based on the medium density SNP-array platform (~50K SNPs) provided by the Illumina Bovine SNP50 Genotyping BeadChip (Matukumalli et al., 2009). This SNP-array provides an initial dataset of 54,001 SNPs of which quality control filtering left between 29,848 (Mancini et al., 2014) and 47,651 (Rothammer et al., 2013) markers available for analysis in the 10 studies considered (**Table 1**). The studies of Druet et al. (2013) and Porto-Neto et al. (2014) involved genotyping with the Illumina BovineHD genotyping assay (>770K SNPs), which, after quality control filtering, resulted in 680,000 and 725,293 markers, respectively. In Kemper et al. (2014) the genotypes obtained with the Illumina BovineHD chip were used to perform imputation of a second dataset generated with the Illumina Bovine SNP50v2.0 (Erbe et al., 2012), yielding a total of 616,350 and 692,527 SNPs for analysis within the groups of dairy and beef breeds, respectively. Ramey et al. (2013) used the Illumina's Bovine SNP50 Genotyping BeadChip and a pre-screening assay comprising almost 2.8 million SNPs that were used as an initial marker panel in the design of the Affymetrix Axiom Genome-wide BOS 1 assay (AFFXB1P). Finally, some of the most recent studies have used data generated by large-scale sequencing. Lee et al. (2013b) analyzed more than 15 million SNPs identified through the sequencing of 12 genomes of Hanwoo cattle, whereas Qanbari et al. (2014) performed a sequence-based imputation, from a 50K SNP panel bridged by a high-density panel to the full genome sequence of Fleckvieh individuals.

The reports reviewed here have applied different but complementary statistics to detect selection signatures (Table 1). We classify the studies in the following categories: (i) studies that have estimated differences in allele frequencies by contrasting pair of breeds through FST (or related statistics) or by differences in allelic frequencies (Flori et al., 2009; Hayes et al., 2009; Wiener et al., 2011; Hosokawa et al., 2012; Mancini et al., 2014; Pintus et al., 2014; Porto-Neto et al., 2014; the across-breed results of Stella et al., 2010);

(ii) studies based on extended regions of low diversity or the calculation of extended haplotype homozygosity (EHH) or variants of this statistic such as Relative Extended Haplotype Homozygosity (REHH), the long-range haplotype (LRH) test, and integrated Haplotype Homozygosity Score (iHS) (Qanbari et al., 2010; Glick et al., 2012; Lim et al., 2013; Pan et al., 2013; Ramey et al., 2013; Rothammer et al., 2013); and (iii) studies based on the allele frequency spectrum, in which regions with outlying allele frequency patterns within a single population are identified through various tests (e.g., the CLR, Composite likelihood-ratio test; CLL, parametric composite log likelihood; and HMM, Hidden Markov Model-based test) (Boitard and Rocha, 2013; Druet et al., 2013; the within-breed results of Stella et al., 2010). The studies based on FST and related statistics (category i) detect diversifying selection between breeds. Of within-breed studies, those based on differences in allele frequency patterns (category iii) have greatest power to detect completed selection (fixation of alleles) whereas the haplotype-based procedures (category ii) have greatest power to detect ongoing selection, as they explore the structure of haplotypes and essentially identify unusually long haplotypes carrying the ancestral and derived alleles (Qanbari et al., 2014). Some of the studies implement two or three different selective sweep mapping methods that fall into multiple categories (Bovine HapMap Consortium, 2009; Qanbari et al., 2011, 2014; Lee et al., 2013b; Kemper et al., 2014) (**Table 1**).

4. FILTERING CRITERIA AND COMPARATIVE APPROACH

In order to look for independent identification of the same regions and to identify those single-breed sweeps that could be uniquely associated with individual breeds, we compiled all the selection signals as reported in the different studies. For some studies reporting both regions identified in specific breeds and also across-breed analyses (Flori et al., 2009; Stella et al., 2010; Qanbari et al., 2011), we only considered the regions reported for specific breeds. The only exception to this criterion was the inclusion in our reviewed dataset of the 12 autosomal regions with extreme F_{ST} value across all populations reported by the Bovine HapMap Consortium (2009). In all cases except one, the details of the selection signatures (Start-End of the region; candidate genes included) were obtained from the original publications (tables in the main text or Supplementary Material); the only exception was the results reported for Stella et al. (2010). In this case, we compiled the genomic positions of the 13,000 significant positions ($P < 0.001$) identified for the five individual breeds (kindly provided by the authors, Boettcher and Stella, personal communication).

For four of the studies for which the original list of significant regions/positions included the results of all the positions/windows exceeding the significance threshold (Stella et al., 2010; Wiener et al., 2011; Druet et al., 2013; Rothammer et al., 2013), we applied additional filtering criteria by selecting the most significant regions or those on the top/bottom of the distribution and/or by grouping close significant positions (within 1 Mb of distance or a distance criteria previously applied by the authors) under the same sweep signals (see Table 1 for details about the additional filtering applied to these four studies).

An important issue when comparing the results of genomic studies in cattle is related to the use of different versions of the bovine genome assembly. Nine studies were based on the UMD_3.1 reference sequence, the version currently available at Ensembl and the UCSC browser (genome.ucsc.edu/). Eleven out of the remaining studies provided results with reference to the previous Btau_4.0 version of the assembly whereas Qanbari et al. (2014) referred to the Btau_4.6.1 version. To make the genomic positions reported by the different studies comparable across studies, we used LiftOver to translate all genomic positions to the UMD_3.1 assembly. Using default parameters, we automatically obtained the correspondence between Btau4.0/Btau4.6.1 (hereafter referred to as Btau_4) and UMD_3.1 coordinates for 403 out of the 612 regions. For the 209 other Btau_4-based regions for which the LiftOver analysis did not yield appropriate UMD_3.1 coordinates, we performed a manual search to provide approximate UMD_3.1 genomic positions (using the closest genes to the positions flanking the selection signal in the Btau_4 region as markers to localize the region in the UMD_3.1 reference genome).

Finally, the list of all reported selection sweeps across the 21 studies, which included a total of 1049 selection sweep regions, was sorted by UMD_3.1 genomic position. With the aim of generating an interpretable set of results, the initial 1049 selection signals were subsequently grouped into core selective sweep (CSS) regions, which were defined as signals within 1 Mb of each other. This criterion was established following a detailed analysis of the regions harboring genes such as *GDF-8*, *MC1R*, and *DGAT1*, with large phenotypic effect and previously identified as being subjected to positive selection. The flanking intervals of the defined CSSs were based on the most proximal and most distal positions of the individual selection signals included in each CSS; the breeds for which individual selection signals were included in each CSS were also noted.

5. INTERPRETATIVE ANALYSIS OF SELECTION SWEEPS REPORTED IN CATTLE

The number of detected selective sweeps varied across the 21 studies reviewed here (Table 1). Of the 1049 selection sweeps identified, the greatest number of regions, 215 (~20%), were obtained from the filtered data of the within-breed analysis provided by Stella et al. (2010) for five specialized dairy cattle breeds. The study from which the next highest number of regions was obtained was Druet et al. (2013) (147 regions; ~14%), who studied 12 breeds with different production characteristics (dairy, beef, dual-purpose). In contrast, four studies contributed fewer than 10 selective sweeps each to the total list (16 regions in total, ~1.5% of signals all together). These were based on breeds without wide distributions, such as French dairy breeds (Flori et al., 2009), Blonde d'Aquitaine (Boitard and Rocha, 2013), Hanwoo cattle (Lim et al., 2013), and Italian breeds (Mancini et al., 2014).

By grouping the consecutive selection sweeps reported by the different authors, (allowing gaps no greater than 1 Mb), we defined a total of 409 CSSs across the 29 bovine autosomes (Supplementary Table S2 in Supplementary Material; Figure 1), 232 (57%) of which were associated with a single-breed (Single-breed CSSs). For the remaining CSSs, we distinguished between 134 CSSs (33%) associated with a limited number (from 2 to 4) of breeds (76 two-breed CSSs, 42 three-breed CSSs, and 16 four-breed CSSs) and 39 CSSs (9.5%) that were associated with five or more breeds (from 5 to 19 breeds) (Supplementary Table S2 in Supplementary Material). We will refer to these two categories as Two-to-Four-breed CSSs and Multi-breed CSSs, respectively. In addition, four identified CSS regions were only detected in the across-breed FST analyses reported by the Bovine HapMap Consortium (2009), and will henceforth be referred to as HapMap-Unique CSSs. These four groups of CSSs are indicated by different cell color backgrounds in Supplementary Table S2, which also includes the genes that were highlighted by the original studies as possible candidate targets of the identified selection sweep.

We have also performed a thorough search of plausible candidate genes for the defined CSSs. This involved a systematic extraction of the annotated genes included in the corresponding interval of the UMD_3.1 bovine assembly using BioMart. Subsequently, a systematic search for functional candidate genes was conducted by searching within CSSs for genes from four lists of genes related to phenotypes for which cattle breeds have been subjected to strong positive selection (a total of 1255 genes). These lists comprised: (i) the database of 449 genes (considering only unique genes) related to milk production and mastitis provided by Ogorevc et al. (2009);

(ii) a list of 519 candidate genes for meat production and meat quality derived from the EU funded GemQual project (QLK5 – CT2000-0147; Williams et al., 2009; Sevane et al., 2013, 2014); (iii) a list of 176 genes related to coat color in cattle and other mammals (http://homepage.usask.ca/~schmutz/colors.html; Olson, 1999; Montoliu et al., 2014), and (iv) a list of 111 genes associated with stature and body size in humans and cattle (Pryce et al., 2011; Guo et al., 2012; Kemper et al., 2012) (See Supplementary Table S3 in Supplementary Material for a complete list of the candidate genes associated with the four phenotype groups). Note that some of the genes appear in more than one of the candidate gene lists (e.g., PPARGC1A, *MC1R*).

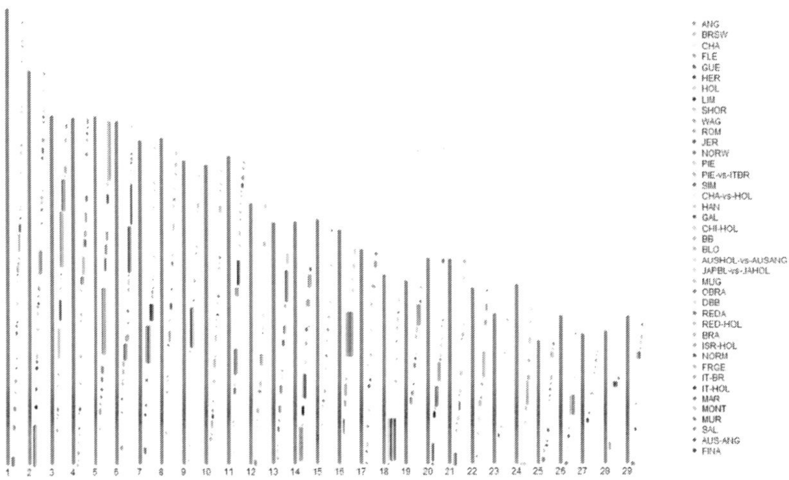

Figure 1. Graphical representation of the Core Selection Sweep (CSS) regions defined in the present work across the 29 bovine autosomes, based on the selection sweeps reported by the 21 genome-wide selection-mapping scans reviewed in the present study. Note that the proximal end of each chromosome (centromere) is represented at the bottom of the plot. ANG, Angus; AUS-ANG, Australian Angus; AUSHOL-vs.-AUSANG, Australian Holstein vs. Australian Angus; BB, Belgian Blue; BLO, Blonde d'Aquitaine; BRA, Braunvieh; BRSW, Brown Swiss; CHA, Charolais; CHA-vs.-HOL, Charolais vs. Holstein; CHI-HOL, Chinese Holstein; DBB, Dual-purpose Belgian blue; FINA, Finnish Ayrshire; FLE, Fleckvieh; FRGE, Franken Gelbvieh; GAL, Galloway; GUE, Guernsey; HER, Hereford; HOL, Holstein; ISR-HOL, Israeli Holstein; IT-BR, Italian Brown; IT-HOL, Italian Holstein; JAPBL-vs.-JAHOL, Japanese Black vs. Japanese Holstein; JER, Jersey; HAN, Korean Hanwoo; LIM, Limousin; MAR, Marchigiana; MONT, Montbéliarde; MUR, Murnau-Werdenfelser; MUG, Murray Gray; NORM, Normande; NORW, Norwegian Red; OBRA, Original Braunvieh; PIE, Piedmontese; PIE-vs.-ITBR, Piedmontese vs. Italian Brown; REDA, Red Angus; RED-HOL, Red Holstein; ROM, Romagnola; SAL, Salers; SHOR, Shorthorn; SIM, Simmental; WAG, Wagyu.

In addition, for the genes extracted for the three main CSS categories (Single-breed, Two-to-Four-breed, and Multi-breed CSSs), we carried out a functional enrichment analysis with WebGestalt (http://bioinfo.vanderbilt.edu/webgestalt/; Wang et al., 2013), using pathways defined by WikyPathways, and selecting "hsapiens_entrezgene_ protein-coding" as the reference set and hypergeometric *p*-value and the "top10" options.

6. PLAUSIBLE CANDIDATE GENES UNDERLYING CSS REGIONS

6.1 Overall Results

The BioMart analysis extracted a total of 5182 genes from the 409 CSSs (Supplementary Table S4 in Supplementary Material). The number of genes extracted for the CSSs was proportional to the length of the genomic intervals involved in the CSSs (Table 2). Hence, a larger number of genes was extracted for the Multi-breed CSSs (2440 genes), which spanned a total of 264.05 Mb across 20 out of the 29 bovine autosomes. From the Two-to-Four-breed CSSs, which involved 202.95 Mb across all the bovine autosomes except BTA23, we extracted 1886 genes (Supplementary Table S4 in Supplementary Material).

Table 2. Characterization of the four categories of core selection sweeps (CSSs) defined in this review based on the number of associated breeds.

	Single-breed CSS	Two-to-Four-breed-CSS	Multibreed-CSS	HapMap-Unique-CSS	Total
Number of CSS	232	134	39	4	409
Average CSS-interval length (Mb)	73.01	202.95	264.05	1.21	541.22
Total number of extracted genes[a]	839	1886	2440	17	5182
Number of candidate genes[b]	67	83	139	2	291
Color	3	19	18	-	40
Dairy-Mastitis	36	38	54	1	129
Beef	21	19	55	1	96
Stature/Body-size	7	7	12	-	26

[a] Genes extracted with the web-based BioMart tool available at ensembl.org and based on the UMD_3.1 bovine genome assembly.
[b] Candidate genes identified through the candidate gene survey performed in this study for four phenotype classes under putative selection in modern cattle breeds.

Although more than half of the defined CSSs were associated with a single breed, and these were located across all the autosomes, the breed-specific selection sweeps spanned a shorter genomic length (73.01 Mb) and thus included a smaller number of genes (839). Seventeen of the extracted genes were located within the four regions uniquely detected in the HapMap project (**Table 2**). Of the 5183 genes found within CSSs, 291 of them were included in the four lists of phenotype-related candidate genes. The number of candidate genes mapping within the CSS categories defined were two (HapMap-unique CSSs), 67 (Single-breed CSSs), 83 (Two-to-Four-breed CSSs), and 139 (Multi-breed CSSs) (**Table 2**). The number of candidate genes for dairy, beef or body-size related traits was similar among the Single-, Two-to-Four-, and Multi-breed CSS categories, whereas coat color genes were mainly detected in the CSSs involving more than one breed (18 in the Multi-breed-CSS group and 19 in the Two-to-Four-breed category) (**Table 2**). Considering the three main CSS categories (Single-breed, Two-to-Four-breed, and Multi-breed CSSs), the candidate genes were not over-represented in the genes located within CSSs although the subset of dairy-related genes was slightly over-represented (but not significantly, based on Fisher's Exact Test). When the same analysis was done separately at the Single-breed CSSs and other CSSs (merging the Two-to-Four-breed and the Multi-breed CSSs together), the Single-breed category was significantly enriched (Fisher's Exact Test, $p = 0.006$) for production genes (beef and dairy) (0.07 of genes are production genes versus 0.05 for the genome overall), whereas the CSSs involving more than one breed were not significantly enriched for production genes (only 0.04 of genes are production genes).

The candidate genes highlighted by this survey are detailed in Supplementary Table S2 within the corresponding CSS where they are included. The gene symbol is indicated with different font color depending on the database of candidates from which it was identified (blue = "dairy-related," red = beef-related," green = "coat-color-related," and pink = "stature/body-size-related").

6.2 Single-Breed CSSs

The 232 single-breed CSSs identified corresponded to selection signals reported in beef (54), dairy (87) and dual-purpose (64) cattle breeds, and 28 of them were reported in a beef vs. dairy pair-wise comparison (Supplementary Table S2 in Supplementary Material). Fleckvieh showed the largest number (49) of these breed-specific selection sweeps, followed by Holstein (33 CSSs), Korean Hanwoo cattle (22), Jersey (14), Guernsey (13), and Simmental (10). Most of the Fleckvieh-specific CSSs breed were reported by Qanbari et al. (2014), and the Korean Hanwoo-related ones were reported by Porto-Neto et al. (2014). The uniqueness of these regions may be biased due to the higher marker density of these studies, which were based on whole genome sequence and the HD-chip dataset, respectively, compared with studies performed in other breeds, which were based on the lower-density SNP panels. The 33 Holstein-specific regions, however, were also extracted from studies based on lower density panels (Qanbari et al., 2010; Stella et al., 2010) and thus their abundance does not appear to be an artifact of the methodology. The large number of such CSSs may be directly related to the very strong selection and resulting high level of dairy specialization in this breed. It is not possible to present a detailed discussion of the each of the single-breed regions and associated candidate genes, but we discuss below some of the regions for which plausible candidate genes could be identified.

A number of dairy-related candidate genes were identified in Holstein-specific regions. For example, several genes related to the immune response were located in Holstein-specific regions, including *IL12B* (Subunit beta of interleukin 12), (CSS-149), which is a cytokine expressed by activated macrophages that has been found to be expressed in milk somatic cells during intra-mammary infections (Lee et al., 2006), and *TLR4* (CSS-171), for which polymorphisms have been associated with mastitis (Wang et al., 2007; de Mesquita et al., 2012) and somatic cell score in cattle (Li et al., 2014b; Wang et al., 2014a). Finally, CSS-298 includes two genes expressed in the mammary gland, *G0S2* and *LAMB3* (Ron et al., 2007). These two genes are also associated with fat metabolism in cattle (Lee et al., 2013a; Ahn et al., 2014) and therefore may be linked to the fat mobilization related to high dairy production.

For the Chinese and Israeli Holsteins, four and seven population-specific CSSs were observed, respectively. None of the Single-breed CSSs were linked to Italian Holsteins, whose selection sweeps were shared with the general Holstein population. Despite the world-wide spread of the Holstein breed, the different conditions in which the animals are reared in some of the countries e.g., resistance to heat stress in Israeli Holstein (Flamenbaum and Galon, 2010), may underlie some of these population-specific CSSs. Apart from the candidate genes suggested by Pan et al. (2013) in the Chinese study, our candidate gene survey did not detect any additional genes associated with known cattle phenotypes. Regarding this point, it should be noted that the different Holstein subpopulations shared CSSs involving major dairy candidate genes, such as the *DGAT1* (CSS-251), *ABCG2* (CSS-123), and *PLAG1* (CSS-254) genes, all of them classified as Multibreed-CSSs. For the other CSSs involving more than one breed, the Chinese Holstein was found independently of other Holstein populations in two cases included in the Two-to-Four-breed category: CSS-110 (shared with Brown Swiss, Fleckvieh, Simmental) and CSS-34 (shared with Jersey), and the Multibreed CSS-67 region (shared with Guernsey, Jersey, Korean Hanwoo, and Angus).

For two other dairy breeds, Jersey and Guernsey, several breed-specific selection sweeps were identified (14 and 13, respectively). One Guersey-related sweep (CSS-118) includes the *NFKB1* gene, whose liver expression is altered in response to pre-partum energy intake and post-partum intramammary inflammatory challenge in dairy cows (Graugnard et al.,

2013). A Jersey-related sweep (CSS-383) includes *PTEN* (phosphatase and tensin homolog) which encodes a tumor suppressor gene regulating many cellular processes, including growth, adhesion, and apoptosis. *PTEN* has also recently been shown to function as an inhibitor during mammary gland development and lactation in dairy cows (Wang et al., 2014b). At the pathway level, the PTEN-AKT pathway is required for the initiation of lactation through the induction of autocrine prolactin (Chen et al., 2012). In addition, *PTEN* has been shown also to play a vital role in regulating fatty acid metabolism (Fu et al., 2012).

A number of Single-breed CSSs were identified in beef cattle breeds. There were several beef-related candidate genes located in the Angus-associated CSS-63, including *CTSK*, *CTSS* (cathepsin K and S), and *TMOD4* (tropomodulin 4). CTSS (cathepsin S) is known to be involved in antigen presentation and also cleaves some extracellular matrix proteins. Through its physiological role, which is to degrade type I collagen, CTSK appears to regulate adipocyte differentiation in adipose tissues of obese patients and animal models (Xiao et al., 2006; Han et al., 2009).

Beef-related candidate genes are located in several of the Korean Hanwoo-specific selection sweeps, including ITGB3 (β3 integrin; CSS-313), which is involved in cytoskeletal organization and plays a role in the adhesion between the cell cytoskeleton and cell extracellular matrix. During postmortem aging, degradation of integrin has been found to be associated with increased drip loss in pork (Lawson, 2004), suggesting it may also be related to meat quality traits in cattle. Furthermore, *MC2R* (adrenocorticotropin receptor) and *MC5R* (melanocortin 5 receptor) are located in the Hanwoo-specific CSS-365. *MC2R* encodes a receptor for the adrenocorticotropic hormone which plays a crucial role in the regulation of glucocorticoid secretion, while *MC5R* is involved in lipid metabolism, exocrine function, and proinflammatory activity (reviewed by Switonski et al., 2013). In addition, *MC5R* expression down-regulates leptin secretion in cultured adipocytes and in humans *MC5R* polymorphisms were reported to be associated with obesity (Switonski et al., 2013). In pigs, *MC2R* is located within a QTL region for intramuscular fat content and back fat thickness (Jacobs et al., 2002) and *MC5R* is close to a QTL influencing fatness and meat quality. Several reports have confirmed an association between porcine back fat thickness or feed intake and variants of the *MC5R* gene (Emnett et al., 2001; Kováčik et al., 2012).

Several Fleckvieh-specific CSSs also include functional candidate genes. The *MFGE8* (milk fat globule-EGF factor 8 protein), located in CSS-334, has been reported to be associated with an index assessing productivity and functional and conformation traits (Fontanesi et al., 2014), which may be relevant to the dual-purpose production characteristics of this breed. The same CSS also includes *ISG20* (interferon stimulated exonuclease gene 20 kDa), which is involved in cumulus oocyte growth and may be related to fertility (Puglisi et al., 2013).

Another Fleckvieh-specific region (CSS-352) includes the *ATP2B2* gene (plasma membrane Ca(2+)-ATPase). The protein encoded by this gene is involved in the transport of calcium across the mammary cell apical membrane. This protein is related to calcium-mediated cell death and has been suggested to play a part in early signaling of mammary gland involution (Reinhardt and Lippolis, 2009).

6.3 Two-to-Four-Breed CSSs

Mapping within the Two-to-Four-breed CSS intervals, we found a high proportion of coat-color related genes (22% of the 93 candidate genes associated with these regions) including *KITLG* (*KIT*-ligand, also known as mast cell growth factor) (CSS-103; identified in Hereford, Holstein, Normande, and a Piedmontese vs. Italian Brown comparison) and *MITF* (microphthalmia-associated transcription factor) (CSS-350; identified in Fleckvieh and Murray Gray), both of which are known to be associated with coat color in cattle (Seitz et al., 1999; Hayes et al., 2010).

Several other genes that have been associated with coat color phenotypes in species other than cattle fall in Two-to-Four breed CSSs, including *HS2ST1* (CSS-68; identified in Fleckvieh, Guernsey, Japanese Black vs. Japanese Holstein, Jersey), *AP3B1* (CSS-183; Guernsey, Piedmontese), *MAP2K1* (CSS-185; Holstein, Piedmontese, Romagnola), *MYC* (CSS-252; identified in Holstein, Piedmontese vs. Italian Brown), PTS (CSS-263; identified in Guernsey, Piedmontese vs. Italian Brown), *PDPK1* (CSS-368; identified in Brown Swiss, Jersey, Norwegian Red), and *ERCC2* (CSS-303; Angus, Simmental).

An interesting region (CSS-131) identified within the Two-to-Four-breed CSS category is that harboring the bovine casein gene cluster on BTA6 (84.66–97.99 Mb). The selection sweeps included in this CSS were identified in three dairy breeds: Braunvieh, Israeli Holstein, and Jersey. Caseins (CSN1S1, CSN1S2, CSN2, CSN3, etc.) represent the primary protein constituents of cow's milk (approximately 80%). The amount and allelic variants of caseins are associated with clotting properties and cheese yield (Wedholm et al., 2006). Due to the importance of caseins in milk production, it is intriguing that only three out of 14 dairy breeds included in this study show a selective sweep near the casein cluster. Nevertheless, this observation agrees with the discordant results reported in the 1980s and 1990s regarding the association of specific casein alleles with production traits, which appear to be breed-specific (reviewed by Caroli et al., 2009).

The *LEP* (leptin) gene appears as a strong candidate gene underlying the selection sweeps reported for one dairy (Guernsey) and two beef breeds (Piedmontese, Red Angus) (CSS-96). Leptin regulates feed intake and energy balance in mammals (Houseknecht et al., 1998) and is involved in the regulation of nutritional status and reproductive functions. Polymorphisms in the bovine *LEP* gene are associated with feed intake (Lagonigro et al., 2003) as well as production traits in both beef (Woronuk et al., 2012) and dairy cattle (Liefers et al., 2002).

6.4 Multi-breed CSSs

Only 39 of the 409 CSSs defined herein involved at least five breeds. As was observed for the Two-to-Four-breed CSSs, the number of CSSs generally decreased as the number of breeds associated with the CSS increased. Hence, we found 12 five-breed CSSs, six six-breed CSSs, three seven-breed CSSs, three eight-breed CSS, and five nine-breed CSSs, and 10 CSSs involving 10–19 breeds.

The two CSSs involving the largest number of breeds were located on BTA6 (CSS-123) and BTA16 (CSS-278), and included 18 and 19 breeds (or pair of breeds), respectively, out of the 41 breed/breed pairs considered in this study (Supplementary Table S2 in Supplementary Material). CSS-123 involved selection sweeps reported in a large number of dairy breeds (Brown Swiss, Chinese Holstein, Guernsey, Holstein, Italian Brown, Italian Holstein, Jersey, Norwegian Red, Montbéliarde) but also beef production (Angus, Hereford, Romagnola, Piedmontese, Marchigiana) and dual-purpose (Fleckieh, MurnauWerdenfelser, OriginalBraunvieh) breeds (a selective signal was also identified for Piedmontese vs. Italian Brown, a beef-dairy comparison). This CSS includes the *ABCG2* (ATP-Binding Cassette, Sub-Family G Member 2) gene, which harbors a QTN for milk composition previously reported in cattle (Olsen et al., 2007). The precise role this gene plays in milk compositions was not initially understood but a later study suggested that ABCG2 plays a role in mammary epithelial cell proliferation and that functional polymorphisms in this gene may influence the cellular compartment of the mammary gland and potentially milk production (Wei et al., 2012). This interval also includes the SPP1 (osteopontin) gene, which has been shown to have significant role in the modulation of milk protein gene expression (Sheehy et al., 2009) and whose allelic variants have also been shown to be associated with variation in milk compositions (Leonard et al., 2005; Khatib et al., 2007). Possibly due to its role as a cytokine, osteopontin has been shown to be beneficial for reducing the incidence of infection during the transition period in lactating cows (Dudemaine et al., 2014). As mentioned above, CSS-123 was also identified in major beef production breeds. In this regard, the NCAPG gene, also located in this genomic region, harbors a causal mutation (I442M) related to fetal growth, carcass performance, and body frame size in cattle (Eberlein et al., 2009; Setoguchi et al., 2009, 2011). Interestingly, a later study has also shown a possible association of this polymorphism on milk production traits (Weikard et al., 2012). NCAPG overlaps with the *LCORL* (ligand dependent nuclear receptor corepressor-like) gene and in many cases these two genes are jointly referred to as *LCORL/NCAPG*. The LCORL/NCAPG locus influences feed intake, gain, meat and carcass traits in beef cattle (Lindholm-Perry et al., 2011) and has been associated with human height (Soranzo et al., 2009; Lango-Allen et al., 2010) and withers height in horses (Tetens et al., 2013). Another notable gene located within CSS-123 is LAP3 (leucine aminopeptidase 3), which has been associated with milk production traits (Zheng et al., 2011). The region involving NCAPG, LCORL, and LAP3 genes has been associated with calving ease in Norwegian Red dairy cows (Olsen et al., 2008) and in Piedmontese beef cattle (Bongiorni et al., 2012). The results in the latter breed suggest that selection on LAP3 for better calving ease is driving the selection signature in this region. Therefore, the large number of breeds included in CSS-123 probably results from the presence of multiple genes influencing various traits of economic interest in cattle.

The region associated with 19 different breeds was CSS-278 on BTA16 (38.500–53.307 Mb). Although it involves selection sweeps reported in a large number of beef-related breeds (Angus, Australian Angus, Charolais, Hereford, Korean Hanwoo, Limousin, Piedmontese, Red Angus, Salers, Shorthorn), it was also related to dairy (Brown Swiss, Guernsey, Holstein, Jersey) and dual-purpose (Simmental, Fleckvieh, FrankenGelbvieh, Braunvieh) breeds. The BioMart extraction for this CSS interval included 253 annotated genes among which we did not identify any gene with known major effects. The genes suggested by the corresponding authors for the selection sweeps included in this CSS involve several genes related to different biological functions: immune response (*PIK3CD, SPSB1, ISG15, TNFRSF9*), development (*RERE*), lipid transportation (*GLTPD1*) muscle physiology (*AGRN*), and apoptosis (programmed cell death) (*FASLG, TNFRSF1B, DFFB, TNFRSF25, DFFA, CASP9*). Among these, CASP9 (caspase 9) is the strongest candidate as it belongs to a subgroup of proteases involved in the phase of apoptosis initiation that occurs in the post-mortem conditioning period and that, together with the calpain system, influences the ultimate meat tenderness (Ouali et al., 2006).

Our candidate gene survey in relation to CSS-278 also identified one dairy-related gene (*PEX14,* peroxisomal biogenesis factor 14), genes related to muscle physiology within the beef-candidate list (SLC2A5, solute carrier family 2 member 5; *TNNT2*, troponin T type 2, cardiac; *TNNI1*, troponin I type 1, skeletal, slow; *SKI*, v-ski avian sarcoma viral oncogene homolog: and CTRC, caldecrin), and one gene related to coat color (*ZBTB17*, Zinc Finger And BTB Domain Containing 17). *ZBTB17* is required for hair follicle structure and hair morphogenesis, and mutations in the murine gene are associated with darkened coat, dark skin, dark dermis around hairs, and abnormal follicles.

Two Multi-breed CSSs regions on BTA14 were identified in 13 and 14 breeds. One of these regions was located on the proximal end of the chromosome (CSS-251, 1.657–12.713 Mb) and involved both dairy and beef breeds (Angus, Australian Holstein vs. Australian Angus, Charolais, Charolais vs. Holstein, Chinese Holstein, Guernsey, HapMap project, Hereford, Holstein, Jersey, Korean Hanwoo, Limousin, Norwegian Red, Piedmontese, Wagyu). It is highly likely that CSS-251 incorporates the selection sweep reported in relation to the *DGAT1* (diacylglycerol O-acyltransferase 1) gene for many dairy cattle breeds, based on the causal role of the mutation K232A on milk composition (Grisart et al., 2002). In addition, the *DGAT1* gene has also been associated with carcass and meat quality traits in beef cattle (Thaller et al., 2003; Wu et al., 2012; Avilés et al., 2013). However, the *DGAT1* gene is located at the very proximal end of the chromosome (1.795–1.805 Mb), indicating that the large CSS-251 interval incorporates selection sweeps related to other genes, such as that reported near the TG (thyroglobulin; located at 9.262–9.509 Mb) gene. TG is known to influence carcass and meat quality traits in beef cattle (Gan et al., 2008; Bennett et al., 2013). Another gene highlighted by our candidate gene query in this region is *CYP11B1* (Cytochrome P450, Family 11, Subfamily B, Polypeptide 1), which influences energy metabolism. A study in German Holstein cattle has shown that SNPs in this gene are associated with milk production traits and somatic cell score independently of the *DGAT1* genotype (Kaupe et al., 2007).

The other CSS on BTA14 associated with a large number of breeds was CSS-254 (23.885–31.847 Mb). The related breeds included both dairy (Brown Swiss, Chinese Holstein, Holstein, Jersey, Montbéliarde, Normande, Norwegian Red) and beef (Angus, Braunvieh, Charolais, Limousin, Piedmontese, Red Angus) breeds as well as the dual-purpose Fleckvieh. Within this region we found the PLAG1 (pleiomorphic adenoma gene 1) gene, which has been shown to be associated with stature in Jersey × Holstein crosses (Karim et al., 2011) but also shows pleiotropic effects on fertility such that the PLAG1 allele associated with increased height and weight was also associated with reduced fat, greater feed intake, less residual feed intake, later puberty in both sexes, and longer post-partum interval before reconceiving in cows (Fortes et al., 2013). This region also encompasses a cluster of genes, including *CHCHD7* (coiled-coil-helix-coiled-coil-helix domain containing 7), *SDR16C5* (short chain dehydrogenase/reductase family 16C, member 5), MOS (v-mos Moloney murine sarcoma viral oncogene homolog), *LYN* (v-yes-1 Yamaguchi sarcoma viral related oncogene homolog), *PENK* (proenkephalin), and RPS20 (ribosomal protein S20), that have been associated with stature in cattle and humans (Utsunomiya et al., 2013b). In particular, a polymorphism ablating a polyadenylation signal of RPS20 has been proposed as the candidate causal mutation of a QTL influencing calving ease and stillbirth incidence in the Fleckvieh breed (Pausch et al., 2011). Another possible candidate for that CSS is *NCOA2* (nuclear receptor coactivator 2), which encodes a transcriptional coactivator for steroid receptors and nuclear receptor and has been found to influence puberty in tropical breeds of beef cattle (Fortes et al., 2011).

Also on BTA6, CSS-130 (67.850–83.375 Mb) involved selection signatures identified in 13 different breeds, involving dairy, beef, and dual-purpose cattle breeds. This region includes a cluster of tyrosine kinase receptor genes (*PDGFRA, KIT*, and *KDR*). The *KIT* (the Hardy-Zuckerman 4 feline sarcoma viral oncogene homolog) gene, which is centered in the CSS interval (71.796–71.917 Mb), explains a considerable proportion of the variation in patterned pigmentation (Hayes et al., 2010), such as the characteristic spotting phenotype of Holstein and other dairy breeds. Close to the *KIT* gene, at 71.374–71.421 Mb, *PDGFRA* (platelet-derived growth factor alpha receptor), has recently been identified as the strongest positional candidate for the non-MC1R-related reddening phenotype in an F2 Nellore-Angus population (Hanna et al., 2014). Other coat color genes were also located in Multi-breed CSSs. CSS-297, which contains the *MC1R* gene, was identified in 15 different breeds. Polymorphisms in this gene are related to the production of eumelanin and phaeomelanin pigments and determine the red-black axis in cattle coat color (Robbins et al., 1993). In addition, *MC1R*, through a competitive relationship for alpha-melanocyte stimulating hormone (α-MSH) with the Melanocortin 4 Receptor (appetite suppressing receptor), has been associated with growth and carcass traits (McLean and Schmutz, 2009). Some of the selection signals of the large CSS-297 interval are likely to be related to other genes. A large list of genes has been suggested by the corresponding authors, and our candidate gene survey also highlighted candidates for dairy traits (SLC7A5, solute carrier family 7 member 5), meat production (*CTRB1*; chymotrypsinogen B1; *FOXC2*, forkhead box C2; *CDH15*, cadherin 15, type 1, M-cadherin), and stature (*GALNS*; galactosamine (N-acetyl)-6-sulfate sulfatase).

Another coat color gene, *SILV* (silver), also known as *PMEL* (premelanosome protein), is located within CSS-109, related to 11 breeds (breed pairs). This gene has been associated with the white coat color characteristic of the Charolais breed (Gutiérrez-Gil et al., 2007; Kuehn and Weikard, 2007a) and pale Highland cattle (Schmutz and Dreger, 2013). It also may be involved in the gray coat phenotype of the Murray Gray breed. In addition, it has been suggested that this gene, based on its multiple splice variants expressed in a variety of tissues independent of pigmentation, could have functions other than melanosome development (Kuehn and Weikard, 2007b). The proximal section of CSS-109 also includes the *IFNG* (interferon, gamma) gene, of interest due to its relationship with the immune response.

The CSS-248 region, which was identified in seven breeds including beef breeds (Korean Hanwoo, Marchigiana, Piedmontese, Shorthorn, Simmental, Wagyu), one dairy breed (Holstein) and a beef vs. dairy comparison (Japanese Black vs. Japanese Holstein), includes the *ASIP* (agouti signaling protein) gene. Although this is a color-related gene in many species, mutations in the ASIP coding region have not been found to play an important role in coat color variation in cattle (Royo et al., 2005). However, a transcript variant of *ASIP* has been assumed to be the causal variant for the brindle coat color of Normande cattle (Girardot et al., 2006) and due to the expression of this gene in adipocytes and its implication in the obese yellow mouse, this transcript has also been suggested to be related to the milk composition traits in this dairy breed (Girardot et al., 2006) and intramuscular fat content in other breeds (Albrecht et al., 2012).

In addition to coat color and patterning, the presence or absence of horns is a breed hallmark in European B. taurus. The locus controlling the polled phenotype, *POLLED*, is located within CSS-1, on BTA1 (0.198–2.60 Mb) (Brenneman et al., 1996), which was identified in 10 breeds. The molecular basis of this phenotype has proven to be complex and the existence of allelic heterogeneity has been suggested for this locus, with the candidate causal mutations located outside known genes or regulatory regions (Drögemüller et al., 2005; Medugorac et al., 2012; Allais-Bonnet et al., 2013). Recently, a long intergenic non-coding RNA has been suggested as the most probable cause of horn bud agenesis for one of the defined allelic variants (Allais-Bonnet et al., 2013).

Another Multi-breed CSS including a gene with a major effect on a bovine qualitative phenotype is CSS-32, located at the proximal end of BTA2, and including selection sweeps described in Belgian Blue, Blonde d'Aquitaine, Limousin, Piedmontese, and a Piedmontese vs. Italian Brown breed-comparison. These are all breeds known to show disruptive or missense mutations in the myostatin (*GDF8* or *MSTN*) gene, associated with muscle conformation and in extreme cases, "double muscling" (Grobet et al., 1997; McPherron and Lee, 1997; Smith et al., 2000; Boitard and Rocha, 2013). CSS-32 encompasses the myostatin gene (2: 6.213–6.220 Mb) but also extends over a large region of BTA2 (0–13.850 Mb), due to the long selective sweep reported by Kemper et al. (2014) for Limousin, known to show very high frequency (~94.2%) of the *GDF8*-F94L mutation (Vankan et al., 2010), whereas for the other breeds, the selective sweep was closer to the *GDF-8* location. As a result of the large size of this CSS, selective sweeps that originally did not include the GDF-8 gene have been incorporated under the CSS-32 label; these include two sweeps described in a Piedmontese vs. Italian Brown comparison (Pintus et al., 2014) 6.717–

9.760 Mb, including the SLC40A1 (solute carrier family 40, member 1), COL5A2 (collagen, type V, alpha 2), COL3A1 (collagen, type III, alpha 1), *CALCRL* (calcitonin receptor-like) and *ITGAV* (integrin, alpha V) genes, and a Fleckvieh selective sweep located in a gene-desert region. Polymorphisms in the *SLC40A1* gene have been related to beef iron content (Duan et al., 2012).

Other Multi-breed regions include genes associated with production traits. These include the short CSS-124 region, which was identified in eight breeds and includes the *PPARGC1A* (peroxysome proliferator-activated receptor-γ coactivator-1α) gene, which mediates expression of genes involved in oxidative metabolism, adipogenesis, and gluconeogenesis (Puigserver and Spiegelman, 2003). Expression of this gene has been suggested to be required for the initiation and development of lactation in dairy cattle (Weikard et al., 2005). *PPARGC1A* has also been shown to be associated with milk composition (Weikard et al., 2005; Khatib et al., 2007; Schennink et al., 2009), reproduction (Komisarek and Walendowska, 2012), growth (Li et al., 2014a), carcass traits (Shin and Chung, 2013; Ramayo-Caldas et al., 2014), and meat quality (Sevane et al., 2013).

CSS-72 (identified in eight breeds) includes *LEPR* (leptin receptor), which due to its interaction with leptin, may be a target of selection in relation to a wide range of economically relevant traits, including growth (Guo et al., 2008), milk production (Suchocki et al., 2010), and calving interval (Trakovická et al., 2013). Finally, CSS-314 (identified in six breeds) includes the FASN gene, which has been associated with milk and beef fatty acid composition (Roy et al., 2006; Morris et al., 2007; Zhang et al., 2008). CSS-322 (identified in five breeds) includes *GHR* (growth hormone receptor), which has been shown to harbor a causal mutation of a QTL influencing milk yield and composition (Blott et al., 2003) and *FST* (follistatin), which encodes a protein related to ovary function and has also been suggested to play a key role in regulating bovine mammary branching morphogenesis and epithelial differentiation (Bloise et al., 2010).

We acknowledge that our candidate gene survey did not take into account genes related to the immune response and behavior, which are found in various CSSs, as reported by many of the original studies reviewed here. For example, the list of Biomart-extracted genes from all the CSS defined in this study (Supplementary Table S4 in Supplementary Material) includes genes directly associated with the immune response. Hence, the list includes 34 genes encoding proteins related to interferon and interleukin responses (16, 12, and five of them belonging to the Multi-breed, Two-to-Four-breed, and Single-breed CSSs, respectively). The CSSs also include 128 genes encoding olfactory receptors (106 of them within Multi-breed CSSs, and 11 in each of the two other categories) and 28 encoding olfactory receptor-like proteins (22 in the Multi-breed CSSs, five in the Single-breed category and one in the Two-to-Four CSSs), which are proposed to be associated with behavioral traits modified through domestication in cattle (Ramey et al., 2013) and other livestock species (Bovine HapMap Consortium, 2009).

Although our survey and the original papers have identified clear candidates for some of the Multibreed CSS regions (mainly genes influencing morphological traits but also some genes with large effects on production traits), it is worth noting that due to our method of merging multiple selection signals at similar positions under the same label, some of these CSSs involve a much larger region than that directly related to the gene with the major gene effect (as discussed above regarding the myostatin region).

6.5 Enrichment Analysis

In an attempt to highlight genes influencing traits other than those considered in our candidate gene survey and to identify the functional biological pathways that are over-represented in the genes included in the CSSs, we performed a complementary functional enrichment analysis for the genes extracted from the Single-breed, Two-to-Four-breed and Multi-breed CSS regions (Supplementary Table S5 in Supplementary Material).

Among the top 10 significant pathway terms in the Single-breed CSSs, five terms were related to the immune response [regulation of Toll-like receptor signaling, IL-1 and IL-4 signaling, and leucocyte-related validated miRNA (defined by tarBase database) pathways], and the others were linked to global metabolism (leptin signaling pathway), bone and muscle physiology (RNAKL-RANK signaling, osteopontin signaling pathways) and to one of the most important intracellular signal transduction pathways (MAPK signaling pathway).

To further explore these results and assess whether single-breed CSSs are linked to genes underlying the physiology of the production specialization for which they have been selected, we performed the functional enrichment analysis of the Single-breed CSSs, separately for the beef and dairy breeds and also for the dual-purpose breeds (Table 3). Whereas pathway terms related to the general immune response were found at similar proportions within the 10 top terms of the three subcategories (although Toll-like receptor signaling pathways terms were only identified in the analysis of the dairy CSSs), other pathway terms appeared to be subcategory-specific. For example, bone and muscle physiology-related terms constituted the majority (5/10) of the top 10 significant terms for beef breed CSSs (i.e., osteopontin signaling, RANKL-RANK signaling pathway, endochondral ossification, osteoclast signaling, striated muscle contraction) whereas those related to major metabolic pathways (leptin signaling pathway, vitamin D synthesis, insulin signaling) were found within the top 10 significant terms only in the dairy-related CSSs. For the Single-breed CSSs associated with dual-purpose breeds, the top 10 significant terms were mainly related to cell signaling pathways involving two important mitogen-activated protein kinases (MAP2K2, MAPK3), which are linked to pathways involving receptors of serotonin and histamine (see **Table 3**). Whereas serotonin is a local regulator in the mammary gland that regulates lactation and initiates the transition into the earliest phases of the involution process related to the return of the mammary gland to morphologically near pre-pregnant state (Horseman and Collier, 2014), the histamine receptors may, in addition to their involvement in local immune responses, also show central effects on modulation of behavior related to the biological function of histamine as a neurotransmitter in the central nervous system (Schneider et al., 2014). The analysis of the dual-purpose-related Single-breed CSSs also revealed over-representation of genes involved in myometrial relaxation and contraction pathways, which could be related to the selection of females that are good dairy cows and can also give birth to calves with meat-production characteristics (e.g., large size).

In the Two-to-Four-Breed CSSs the functional gene enrichment analysis (Supplementary Table S5 in Supplementary Material) highlighted three pathway terms related to global metabolism (insulin signaling, glucuronidation, and metapathway biotransformation; the latter term involves several enzymes from the cytochrome P450 superfamily of enzymes, sulfotransferases, and glucuronosyltransferases) and others were related to the immune response [regulation of the Toll-like receptor signaling pathway and lymphocyte-validated miRNAs (TarBase)], cell adhesion mechanisms (integrin-mediated cell adhesion, focal adhesion), and specific cell physiology pathways [MAPK signaling, epithelium-related validated miRNA (TarBase), and microRNAs in cardiomyocyte hypertrophy].

The enrichment analysis performed for the Multi-breed CSSs highlighted among the top 10 significant terms, two related to the immune response (complement activation, classical pathway, complement and coagulation cascades), two related to overall lipid-metabolism (adipogenesis, SREBF, and miR33 in cholesterol and lipid homeostasis), and others related to skeleton and reproductive physiology (regulation of actin cytoskeleton, and myometrial relaxation and contraction pathways).

Because of the large number of genes highlighted by this functional analysis, we do not present here a detailed discussion about the known effects of these genes in cattle. This could be the objective of future studies focusing on some of the CSS regions presented here.

Table 3. Results from the gene enrichment analysis performed using WikiPathway analysis (WebGestalt software; Wang et al., 2013) individually for the three production-based subcategories (beef, dairy, dual-purpose) of the Single-breed core selective sweep (CSSs).

Core selective sweep (CSS) subcategory (number of genes extracted with BioMart)	Pathway name	Number of genes	Gene names	Statistics[a]
Single-breed-CSSs involving BEEF breeds (304 genes)	Interleukin-11 signaling pathway	3	*TGFB1, RELA, IKBKB*	C = 49; O = 3; E = 0.30; *R* = 9.93; rawP = 0.0035; adjP = 0.0540
	Osteopontin signaling	2	*RELA, IKBKB*	C = 18; O = 2; E = 0.11; R = 18.01; rawP = 0.0054; adjP = 0.0540
	RANKL-RANK signaling pathway	3	*CTSK, RELA, IKBKB*	C = 66; O = 3; E = 0.41; R = 7.37; rawP = 0.0080; adjP = 0.0540
	Corticotropin-releasing hormone	4	*TGFB1, FOSL1, RELA, NCOA2*	C = 123; O = 4; E = 0.76; R = 5.27; rawP = 0.0072; adjP = 0.0540
	Endochondral ossification	3	*TGFB1, PTCH1, PLAT*	C = 69; O = 3; E = 0.42; R = 7.08; rawP = 0.0089; adjP = 0.0534
	Osteoclast signaling	2	*CTSK, ITGB3*	C = 19; O = 2; E = 0.12; R = 17.07; rawP = 0.0060; adjP = 0.0540
	B Cell receptor signaling pathway	4	*PIP5K1A, GSK3A, RELA, IKBKB*	C = 114; O = 4; E = 0.70; R = 5.69; rawP = 0.0056; adjP = 0.0540
	NOD pathway	3	*RELA, IKBKB, NLRC4*	C = 39; O = 3; E = 0.24; R = 12.47; rawP = 0.0018; adjP = 0.0540
	Androgen receptor signaling pathway	4	*KAT5, RELA, KAT7, NCOA2*	C = 91; O = 4; E = 0.56; R = 7.13; rawP = 0.0025; adjP = 0.0540
	Striated muscle contraction	2	*MYL4, ACTN3*	C = 38; O = 2; E = 0.23; R = 8.53; rawP = 0.0230; adjP = 0.0690
Single-breed-CSSs involving DAIRY breeds (239 genes)	Regulation of toll-like receptor signaling pathway	6	*MAPK13, TLR4, MAPK14, PELI2, IL12B, NFKB1*	C = 154; O = 6; E = 0.69; R = 8.75; rawP = 7.00e-05;adjP = 0.0033
	Estrogen signaling pathway	3	*ESR1, MAPK14, NFKB1*	C = 30; O = 3; E = 0.13; R = 22.46; rawP = 0.0003; adjP = 0.0047
	Toll-like receptor signaling pathway	5	*MAPK13, TLR4, MAPK14, IL12B, NFKB1*	C = 116; O = 5; E = 0.52; R = 9.68; rawP = 0.0002; adjP = 0.0047
	Leptin signaling pathway	4	*ESR1, PDE3B, MAPK14, NFKB1*	C = 81; O = 4; E = 0.36; R = 11.09; rawP = 0.0005; adjP = 0.0059
	Vitamin D synthesis	2	*CYP27A1, CYP2R1*	C = 10; O = 2; E = 0.04; R = 44.92; rawP = 0.0009; adjP = 0.0070
	TFs Regulate miRNAs related to cardiac hypertrophy	2	*RASGRF1, NFKB1*	C = 10; O = 2; E = 0.04; R = 44.92; rawP = 0.0009; adjP = 0.0070
	Focal Adhesion	5	*COL6A2, SHC3, PTEN, LAMA5, LAMB3*	C = 185; O = 5; E = 0.82; R = 6.07; rawP = 0.0015; adjP = 0.0088
	Alpha 6 Beta 4 signaling pathway	3	*MAPK14, LAMA5, LAMB3*	C = 50; O = 3; E = 0.22; R = 13.48; rawP = 0.0015; adjP = 0.0088
	Complement activation, classical pathway	2	*C1QA, C1QC*	C = 17; O = 2; E = 0.08; R = 26.43; rawP = 0.0026; adjP = 0.0136
	Insulin signaling	4	*MAPK13, SHC3, PTEN, MAPK14*	C = 163; O = 4; E = 0.73; R = 5.51; rawP = 0.0062; adjP = 0.0291
Single-breed-CSSs involving DUAL-PURPOSE breeds (195 genes)	Serotonin receptor 4-6-7 and NR3C Signaling	3	*HTR4, MAP2K2, MAPK3*	C = 18; O = 3; E = 0.06; R = 46.67; rawP = 3.50e-05; adjP = 0.0014
	Calcium regulation in the cardiac cell	5	*RGS18, RGS2, RGS1, ATP2B2, PRKAR1B*	C = 151; O = 5; E = 0.54; R = 9.27; rawP = 0.0002; adjP = 0.0041
	Serotonin Receptor 2 and ELK-SRF-GATA4 signaling	2	*MAP2K2, MAPK3*	C = 16; O = 2; E = 0.06; R = 35.01; rawP = 0.0015; adjP = 0.0205
	IL-7 signaling pathway	2	*MAP2K2, MAPK3*	C = 25; O = 2; E = 0.09; R = 22.40; rawP = 0.0036; adjP = 0.0246
	IL-9 signaling pathway	2	*MAP2K2, MAPK3*	C = 25; O = 2; E = 0.09; R = 22.40; rawP = 0.0036; adjP = 0.0246

Myometrial relaxation and contraction pathways	4	*RGS18, RGS2, RGS1, PRKAR1B*	C = 162; O = 4; E = 0.58; R = 6.91; rawP = 0.0028; adjP = 0.0246
Monoamine GPCRs	2	*HTR4, HRH1*	C = 33; O = 2; E = 0.12; R = 16.97; rawP = 0.0062; adjP = 0.0261
Serotonin HTR1 group and FOS pathway	2	*MAP2K2, MAPK3*	C = 33; O = 2; E = 0.12; R = 16.97; rawP = 0.0062; adjP = 0.0261
Hypothetical network for drug addiction	2	*MAP2K2, MAPK3*	C = 35; O = 2; E = 0.12; R = 16.00; rawP = 0.0070; adjP = 0.0261
EPO receptor signaling	2	*MAP2K2, MAPK3*	C = 35; O = 2; E = 0.12; R = 16.00; rawP = 0.0070; adjP = 0.0261

[a] Wikipathway analysis Statistics: C, the number of reference genes in the category; O, the number of genes in the gene set and also in the category; E, the expected number in the category; R, ratio of enrichment; rawP, p-value from hypergeometric test; adjP, p-value adjusted by the multiple test adjustment.

7. OVERALL CONCLUSIONS

Compilation of the results from many selection sweep mapping studies in cattle provides an ideal opportunity to investigate how artificial selection has influenced the variability and architecture of the bovine genome. Selection is likely to have eroded the levels of genetic variation that existed in the original domesticated population. At the same time, selection on a livestock breed has tended to fix specific variants that have become distinctive genetic signals of that breed compared with others. Strong selection for improvement of productivity, such as milk or beef production traits, has led to specialization of cattle breeds. It might be expected that breeds that share the same production characteristics would show a similar picture of selection sweeps related to such specialization, and conversely, that divergently specialized breeds would share few selection sweeps. However, our review shows that in many cases selection signatures are also shared by breeds showing different production characteristics. These may include regions containing genes associated with metabolic homeostasis or other general traits such as disease resistance and behavior, but may also reflect the pleiotropic effects of genes on traits relevant to both beef and dairy production. Because of the large number of selective sweeps compared here, we have not performed a detailed analysis of all genes included within the CSSs, although in a number of cases, it was possible to speculate as to which gene or genes could be the targets of selection.

This review presents an initial comparative map of the selection sweeps reported in European Bos taurus cattle breeds and provides an integrated dataset that can also incorporate results from future studies and thus allow the researchers to perform systematic comparisons of selection sweeps reported in cattle. This type of comparative tool is essential to properly interpret the results of individual studies for such a complex topic as selection sweeps across different breeds of the same species.

Considering the three main CSS categories defined here, the Single-breed and the Two-to-Four-breed CSS groups together accounted for about 90% of the CSSs, whereas only 9.5% of the CSSs were identified in five or more breeds (**Figure 1**). These Multi-breed CSSs appear to encompass the sweeps involving the limited number of genes that have large phenotypic effects across different breeds and also, in part due to the long Multi-breed CSS intervals resulting from our CSS-labeling approach, other putatively selected genes with small effect sizes, some of which are breed-specific. Regarding the large phenotypic effects linked to the Multi-breed CSSs, many of them appear to relate to physical rather than production traits, consistent with a simpler genetic architecture (i.e., fewer genes involved in determination of the phenotype) for the former. The putatively strongly selected phenotypes include physical hallmarks that define a breed, such as coat color and patterning (*MC1R, KIT*) or obvious morphological traits such as lack of horns (*POLL* locus) and stature (*PLAG1*). The strong signals of selection in relation to morphological traits (e.g., body size and color-patterning traits) are consistent with the theory of the "domestication syndrome" in mammals, which suggests that selective pressure for tameness during the initial stages of domestication involved a developmental reduction in neural crest cell populations and led to multiple phenotypic changes shared by various domesticated animals species (e.g., depigmentation, floppy and reduced ears, shorter muzzles, docility, smaller brain, or cranial capacity) (Wilkins et al., 2014).

In addition to the Multi-breed CSS regions including genes that influence physical traits, there are also several genomic regions that show evidence of selection across many breeds and appear to be driven by selection on production-related genes such as *ABCG2, DGAT1, NCAPG*, and *GHR*. For the CSSs including genes with large effects, there was a correspondence between the production profiles of the breeds associated with these CSSs and the known effects of the putative target gene. It is interesting that some of these genes for which the initial major effect was related to a specific specialization (e.g., *DGAT1* for milk and *NCAPG* for growth traits), latter studies have shown that they also have effects on traits of interest in the alternative production group (e.g., *DGAT1* for beef composition and *NCAPG* for milk traits). These observations for genes with known major effects provide insights into the complexity of the relationship between genes and phenotypes; this complexity may be even more pronounced for genes of small effect.

In addition to these genes with major effect, the Multi-breed CSS intervals also included other potential selection candidates related to production (dairy and beef) traits (Supplementary Table S2 in Supplementary Material), which could represent some of the small size effect genes underlying the complex genetic architecture of quantitative traits. The functional enrichment analysis for these genomic regions suggested that genes related to the immune response and reproduction traits may also have been selection targets shared by many breeds. We also found a significant over-representation of genes related to olfactory receptors (protein coding and pseudogenes) in the Multi-breed CSSs. The abundance of these genes within selection sweep intervals, which has previously been highlighted (Bovine HapMap Consortium, 2009; Ramey et al., 2013; Qanbari et al., 2014), suggests that these behavior-related loci may have played a role in cattle domestication, whereas newly evolving functions have been suggested for these genes based on their reported duplication in the cattle genome (Elsik et al., 2009). Regarding the large number of olfactory receptor genes included in the Multi-breed CSS regions, it should be taken into account that this gene family shows one of the highest frequency of somatic mutations in their coding regions due to low expression levels, late replication time during the cell cycle and high regional non-coding mutation rate (Lawrence et al., 2013). This observation may suggest these genes as false positive results in GWAS analyses, as pointed out by Lawrence et al. (2013), and may also be relevant in interpreting results from selection signature analyses.

As mentioned above, about 90% of the CSSs defined involved a single breed (57%) or a limited number of breeds (33%, Two-to-Four CSSs). The Single-breed CSSs included an over-representation of genes related to dairy and beef production; this observation was supported by the functional enrichment analysis, which highlighted production-related pathway terms associated with these regions (**Table 3**). Hence, the Single-breed CSS regions may include genes with small effects that influence quantitative traits of economic interest. This also suggests that similar selective pressures on different breeds, for example, for milk and meat production traits, can result in allele frequency changes in different genomic regions. This interpretation agrees with the hypothesis that many genes influence the complex traits under selection in cattle and that few of them show large phenotypic effects (Hayes et al., 2010). Alternatively, although within the same production category (dairy, beef, dual-purpose), the breeds may have been selected for subtly different production characteristics or have been subjected to differential natural (environmental) selection. In any case, each breed retains its own unique signature of its selection history. The functional enrichment analysis performed for the dual-purpose breeds, for which extremely strong selection has not been performed on either dairy or beef traits, primarily revealed genes related to reproduction traits and behavior-physiological pathways. Overall, the Single-breed CSSs pinpoint specific regions that appear to have been uniquely selected in the corresponding breeds. We propose these regions as potential markers of unique diversity and further studies focusing on the molecular basis of these selection sweeps are recommended. Furthermore, we acknowledge that a more comprehensive review also covering Bos indicus and African Bos taurus cattle would provide an enhanced overview of the impact of artificial and natural selection on the cattle genome. For example, for a selection sweep that appears to be related to short, slick hair coat (which in turn is associated with heat-stress tolerance) in tropical Senepol cattle

(Flori et al., 2012), a mutation in the PRLR (prolactin receptor) has been identified as the putative causal mutation (Littlejohn et al., 2014). The identification of this effect, associated with a gene of major importance in lactation, provides a clear example of pleiotropy and the complex genetic architecture of physiological traits and suggests that examining selection sweeps in a broader range of cattle breeds could help to dissect the genetic architecture of traits of economic relevance.

This large-scale review of selection sweeps in European cattle reveals the historical impacts of long-term selection pressures on a species of great importance in human history. This review also presents for the first time a characterization of the selection sweeps that are breed-specific, and suggests that based on their uniqueness, these could be considered as "divergence signals," which may be important for the management and prioritization of livestock diversity.

ACKNOWLEDGMENTS

The authors would like to acknowledge Milagros Sánchez for providing graphical assistance for the CSS representation. BGG is funded through the Spanish "Ramón y Cajal" Programme (RYC-2012-10230) from the Spanish Ministry of Economy and Competitiveness (State Secretariat for Research, Development and Innovation). The authors gratefully acknowledge the initial support for collaborative work between these two groups from the European Science Foundation through the GENOMIC-RESOURCES Exchange Grant awarded to Beatriz Gutierrez (EX/3723). This work is also included in the framework of the project AGL2012-34437 funded by the Spanish Ministry of Economy and Competitiveness (MINECO). The Roslin Institute receives core funding from the British Biotechnology and Biological Sciences Research Council (BBSRC).

REFERENCES

1. Ahn, J., Li, X., Choi, Y. M., Shin, S., Oh, S. A., Suh, Y., et al. (2014). Differential expressions of G0/G1 switch gene 2 and comparative gene identification-58 are associated with fat content in bovine muscle. Lipids 49, 1–14. doi: 10.1007/s11745-013-3866-3
2. Albrecht, E., Komolka, K., Kuzinski, J., and Maak, S. (2012). Agouti revisited: transcript quantification of the ASIP gene in bovine tissues related to protein expression and localization. PLoS ONE 7:e35282. doi: 10.1371/journal.pone.0035282
3. Allais-Bonnet, A., Grohs, C., Medugorac, I., Krebs, S., Djari, A., Graf, A., et al. (2013). Novel insights into the bovine polled phenotype and horn ontogenesis in Bovidae. PLoS ONE 8:e63512. doi: 10.1371/journal.pone.0063512
4. Avilés, C., Polvillo, O., Peña, F., Juárez, M., Martínez, A. L., and Molina, A. (2013). Associations between DGAT1, FABP4, LEP, RORC, and SCD1 gene polymorphisms and fat

deposition in Spanish commercial beef. J. Anim. Sci. 91, 4571–4577. doi: 10.2527/jas.2013-6402

5. Bennett, G. L., Shackelford, S. D., Wheeler, T. L., King, D. A., Casas, E., and Smith, T. P. (2013). Selection for genetic markers in beef cattle reveals complex associations of thyroglobulin and casein1-s1 with carcass and meat traits. J. Anim. Sci. 91, 565–571. doi: 10.2527/jas.2012-5454
6. Bloise, E., Cassali, G. D., Ferreira, M. C., Ciarmela, P., Petraglia, F., and Reis, F. M. (2010). Activin-related proteins in bovine mammary gland: localization and differential expression during gestational development and differentiation. J. Dairy Sci. 93, 4592–4601. doi: 10.3168/jds.jds.2009-2981
7. Blott, S., Kim, J. J., Moisio, S., Schmidt-Küntzel, A., Cornet, A., Berzi, P., et al. (2003). Molecular dissection of a quantitative trait locus: a phenylalanine-to-tyrosine substitution in the transmembrane domain of the bovine growth hormone receptor is associated with a major effect on milk yield and composition. Genetics 163, 253–266. Available online at: http://www.genetics.org/content/163/1/253.long
8. Boitard, S., and Rocha, D. (2013). Detection of signatures of selective sweeps in the Blonde d'Aquitaine cattle breed. Anim. Genet. 44, 579–583. doi: 10.1111/age.12042
9. Bongiorni, S., Mancini, G., Chillemi, G., Pariset, L., and Valentini, A. (2012). Identification of a short region on chromosome 6 affecting direct calving ease in Piedmontese cattle breed. PLoS ONE 7:e50137. doi: 10.1371/journal.pone.0050137
10. Brenneman, R. A., Davis, S. K., Sanders, J. O., Burns, B. M., Wheeler, T. C., Turner, J. W., et al. (1996). The polled locus maps to BTA1 in a Bos indicus × Bos taurus cross. J. Hered. 87, 156–161. doi: 10.1093/oxfordjournals.jhered.a022975
11. Bruford, M. W., Bradley, D. G., and Luikart, G. (2003). DNA markers reveal the complexity of livestock domestication. Nat. Rev. Genet. 4, 900–910. doi: 10.1038/nrg1203
12. Caroli, A. M., Chessa, S., and Erhardt, G. J. (2009). Invited review: milk protein polymorphisms in cattle: effect on animal breeding and human nutrition. J. Dairy Sci. 92, 5335–5352. doi: 10.3168/jds.2009-2461
13. Chen, C. C., Stairs, D. B., Boxer, R. B., Belka, G. K., Horseman, N. D., Alvarez, J. V., et al. (2012). Autocrine prolactin induced by the Pten-Akt pathway is required for lactation initiation and provides a direct link between the Akt and Stat5 pathways. Genes Dev. 26, 2154–2168. doi: 10.1101/gad.197343.112
14. Bovine HapMap Consortium. (2009). Genome-wide survey of SNP variation uncovers the genetic structure of cattle breeds. Science 324, 528–532. doi: 10.1126/science.1167936
15. de Mesquita, A. Q. E., Rezende, C. S., de Mesquita, A. J., Jardim, E. A., and Kipnis, A. P. (2012). Association of TLR4 polymorphisms with subclinical mastitis in Brazilian Holsteins. Braz. J. Microbiol. 43, 692–697. doi: 10.1590/S1517-83822012000200034
16. Drögemüller, C., Wöhlke, A., Mömke, S., and Distl, O. (2005). Fine mapping of the polled locus to a 1-Mb region on bovine chromosome 1q12. Mamm. Genome 16, 613–620. doi: 10.1007/s00335-005-0016-0
17. Druet, T., Pérez-Pardal, L., Charlier, C., and Gautier, M. (2013). Identification of large selective sweeps associated with major genes in cattle. Anim. Genet. 44, 758–762. doi: 10.1111/age.12073
18. Duan, Q., Tait, R. G. Jr., Mayes, M. S., Garrick, D. J., Liu, Q., Van Eenennaam, A. L., et al. (2012). Genetic polymorphisms in bovine transferrin receptor 2 (TFR2) and solute carrier family 40 (iron-regulated transporter), member 1 (SLC40A1) genes and their association with beef iron content. Anim. Genet. 43, 115–122. doi: 10.1111/j.1365-2052.2011.02224.x
19. Dudemaine, P. L., Thibault, C., Alain, K., and Bissonnette, N. (2014). Genetic variations in the SPP1 promoter affect gene expression and the level of osteopontin secretion into bovine milk. Anim. Genet. 45, 629–640. doi: 10.1111/age.12176
20. Eberlein, A., Takasuga, A., Setoguchi, K., Pfuhl, R., Flisikowski, K., Fries, R., et al. (2009). Dissection of genetic factors modulating fetal growth in cattle indicates a substantial role of the non-SMC condensin I complex, subunit G (NCAPG) gene. Genetics 183, 951–964. doi: 10.1534/genetics.109.106476

21. Elsik, C. G., Tellam, R. L., and Worley, K. C. (2009). The genome sequence of taurine cattle: a window to ruminant biology and evolution. Science 324, 522–528. doi: 10.1126/science.1169588
22. Emnett, R., Moeller, S., Irwin, K., Rothschild, M. F., Plastow, G., and Goodwin, R. (2001). "Association studies with leptin receptor, melanocortin-4 receptor, melanocortin-5 receptor, and peroxisome proliferator activated receptor-γ," in Research and Reviews: Swine 2001 OARDC Special Circular, eds M. L. Eastridge, W. L. Bacon, C. L. Knipe, D. L. Meeker, T. B. Turner, and D. L. Zartman (Columbus: The Ohio State University), 185, 57–63. Available online at: http://ohioline.osu.edu/sc185/index.html
23. Erbe, M., Hayes, B. J., Matukumalli, L. K., Goswami, S., Bowman, P. J., Reich, C. M., et al. (2012). Improving accuracy of genomic predictions within and between dairy cattle breeds with imputed high-density single nucleotide polymorphism panels. J. Dairy Sci. 95, 4114–4129. doi: 10.3168/jds.2011-5019
24. FAO. (2013). In Vivo Conservation of Animal Genetic Resources. Rome: FAO Animal Production and Health Guidelines. No. 14.
25. Flamenbaum, I., and Galon, N. (2010). Management of heat stress to improve fertility in dairy cows in Israel. Reprod. Dev. 56 Suppl. S36–S41. doi: 10.1262/jrd.1056S36
26. Flori, L., Fritz, S., Jaffrézic, F., Boussaha, M., Gut, I., Heath, S., et al. (2009). The genome response to artificial selection: a case study in dairy cattle. PLoS ONE 4:e6595. doi: 10.1371/journal.pone.0006595
27. Flori, L., Gonzatti, M. I., Thevenon, S., Chantal, I., Pinto, J., Berthier, D., et al. (2012). A quasi-exclusive European ancestry in the Senepol tropical cattle breed highlights the importance of the slick locus in tropical adaptation. PLoS ONE 7:e36133. doi: 10.1371/journal.pone.0036133
28. Fontanesi, L., Calò, D. G., Galimberti, G., Negrini, R., Marino, R., Nardone, A., et al. (2014). A candidate gene association study for nine economically important traits in Italian Holstein cattle. Anim. Genet. 45, 576–580. doi: 10.1111/age.12164
29. Fortes, M. R., Reverter, A., Nagaraj, S. H., Zhang, Y., Jonsson, N. N., Barris, W., et al. (2011). A single nucleotide polymorphism-derived regulatory gene network underlying puberty in 2 tropical breeds of beef cattle. J. Anim. Sci. 89, 1669–1683. doi: 10.2527/jas.2010-3681
30. Fortes, M. R. S., Kemper, K. E., Sasazaki, S., Reverter, A., Pryce, J. E., Barendse, W., et al. (2013). Evidence for pleiotropism and recent selection in the PLAG1 region in Australian beef cattle. Anim. Genet. 44, 636–647. doi: 10.1111/age.12075
31. Fu, S., Deng, Q., Yang, W., Ding, H., Wang, X., Li, P., et al. (2012). Increase of fatty acid oxidation and VLDL assembly and secretion overexpression of PTEN in cultured hepatocytes of newborn calf. Cell. Physiol. Biochem. 30, 1005–1013. doi: 10.1159/000341477
32. Gan, Q. F., Zhang, L. P., Li, J. Y., Hou, G. Y., Li, H. D., Gao, X., et al. (2008). Association analysis of thyroglobulin gene variants with carcass and meat quality traits in beef cattle. J. Appl. Genet. 49, 251–255. doi: 10.1007/BF03195621
33. Gautier, M., Flori, L., Riebler, A., Jaffrézic, F., Laloé, D., Gut, I., et al. (2009). A whole genome Bayesian scan for adaptive genetic divergence in West African cattle. BMC Genomics 10:550. doi: 10.1186/1471-2164-10-550
34. Girardot, M., Guibert, S., Laforet, M. P., Gallard, Y., Larroque, H., and Oulmouden, A. (2006). The insertion of a full-length Bos taurus LINE element is responsible for a transcriptional deregulation of the Normande Agouti gene. Pigment Cell Res. 19, 346–355. doi: 10.1111/j.1600-0749.2006.00312.x
35. Glick, G., Shirak, A., Uliel, S., Zeron, Y., Ezra, E., Seroussi, E., et al. (2012). Signatures of contemporary selection in the Israeli Holstein dairy cattle. Anim. Genet. 43 Suppl. 1, 45–55. doi: 10.1111/j.1365-2052.2012.02348.x
36. Graugnard, D. E., Moyes, K. M., Trevisi, E., Khan, M. J., Keisler, D., Drackley, J. K., et al. (2013). Liver lipid content and inflammometabolic indices in peripartal dairy cows are altered in response to prepartal energy intake and postpartal intramammary inflammatory challenge. J. Dairy Sci. 96, 918–935. doi: 10.3168/jds.2012-5676

37. Grisart, B., Coppieters, W., Farnir, F., Karim, L., Ford, C., Berzi, P., et al. (2002). Positional candidate cloning of a QTL in dairy cattle: identification of a missense mutation in the bovine DGAT1 gene with major effect on milk yield and composition. Genome Res. 12, 222–231. doi: 10.1101/gr.224202
38. Grobet, L., Martin, L. J., Poncelet, D., Pirottin, D., Brouwers, B., Riquet, J., et al. (1997). A deletion in the bovine myostatin gene causes the double-muscled phenotype in cattle. Nat. Genet. 17, 71–74. doi: 10.1038/ng0997-71
39. Guo, J., Jorjani, H., and Carlborg, Ö. (2012). A genome-wide association study using international breeding-evaluation data identifies major loci affecting production traits and stature in the Brown Swiss cattle breed. BMC Genet. 13:82. doi: 10.1186/1471-2156-13-82
40. Guo, Y., Chen, H., Lan, X., Zhang, B., Pan, C., Zhang, L., et al. (2008). Novel SNPs of the bovine LEPR gene and their association with growth traits. Biochem. Genet. 46, 828–834. doi: 10.1007/s10528-008-9197-z
41. Gutiérrez-Gil, B., Wiener, P., and Williams, J. L. (2007). Genetic effects on coat colour in cattle: dilution of eumelanin and phaeomelanin pigments in an F2-Backcross Charolais x Holstein population. BMC Genet. 8:56. doi: 10.1186/1471-2156-8-56
42. Han, J., Luo, T., Gu, Y., Li, G., Jia, W., and Luo, M. (2009). Cathepsin K regulates adipocyte differentiation: possible involvement of type I collagen degradation. Endocr. J. 56, 55–63. doi: 10.1507/endocrj.K08E-143
43. Hanna, L. L., Sanders, J. O., Riley, D. G., Abbey, C. A., and Gill, C. A. (2014). Identification of a major locus interacting with MC1R and modifying black coat color in an F2 Nellore-Angus population. Genet. Sel. Evol. 46:4. doi: 10.1186/1297-9686-46-4
44. Harr, B., Kauer, M., and Schlotterer, C. (2002). Hitchhiking mapping: a population-based fine-mapping strategy for adaptive mutations in Drosophila melanogaster. Proc. Natl. Acad. Sci. U.S.A. 99, 12949–12954. doi: 10.1073/pnas.202336899
45. Hayes, B. J., Chamberlain, A. J., Maceachern, S., Savin, K., McPartlan, H., MacLeod, I., et al. (2009). A genome map of divergent artificial selection between Bos taurus dairy cattle and Bos taurus beef cattle. Anim. Genet. 40, 176–184. doi: 10.1111/j.1365-2052.2008.01815.x
46. Hayes, B. J., Lien, S., Nilsen, H., Olsen, H. G., Berg, P., Maceachern, S., et al. (2008). The origin of selection signatures on bovine chromosome 6. Anim. Genet. 39, 105–111. doi: 10.1111/j.1365-2052.2007.01683.x
47. Hayes, B. J., Pryce, J., Chamberlain, A. J., Bowman, P. J., and Goddard, M. E. (2010). Genetic architecture of complex traits and accuracy of genomic prediction: coat colour, milk-fat percentage, and type in Holstein cattle as contrasting model traits. PLoS Genet. 6:e1001139. doi: 10.1371/journal.pgen.1001139
48. Hohenlohe, P. A., Phillips, P. C., and Cresko, W. A. (2010). Using population genomics to detect selection in natural populations: key concepts and methodological considerations. Int. J. Plant Sci. 171, 1059–1071. doi: 10.1086/656306
49. Horseman, N. D., and Collier, R. J. (2014). Serotonin: a local regulator in the mammary gland epithelium. Annu. Rev. Anim. Biosci. 2, 353–374. doi: 10.1146/annurev-animal-022513-114227
50. Hosokawa, D., Ishii, A., Yamaji, K., Sasazaki, S., Oyama, K., and Mannen, H. (2012). Identification of divergently selected regions between Japanese Black and Holstein cattle using bovine 50k SNP array. Anim. Sci. J. 83, 7–13. doi: 10.1111/j.1740-0929.2011.00927.x
51. Houseknecht, K. L., Baile, C. Z., Matteri, R. L., and Spurlock, M. C. (1998). The biology of leptin: a review. J. Anim. Sci. 76, 1405–1420.
52. Jacobs, K., Van Poucke, M., Mattheeuws, M., Chardon, P., Yerle, M., Peelman, L. J., et al. (2002). Characterization of the porcine melanocortin 2 receptor gene (MC2R). Anim. Genet. 33, 415–421. doi: 10.1046/j.1365-2052.2002.00899.x
53. Kaplan, N. L., Hudson, R. R., and Langley, C. H. (1989). The hitchhiking effect revisited. Genetics 123, 887–899.

54. Karim, L., Takeda, H., Lin, L., Druet, T., Arias, J. A. C., Baurain, D., et al. (2011). Variants modulating the expression of a chromosome domain encompassing PLAG1 influence bovine stature. Nat. Genet. 43, 405–413. doi: 10.1038/ng.814
55. Kaupe, B., Brandt, H., Prinzenberg, E. M., and Erhardt, G. (2007). Joint analysis of the influence of CYP11B1 and DGAT1 genetic variation on milk production, somatic cell score, conformation, reproduction, and productive lifespan in German Holstein cattle. J. Anim. Sci. 85, 11–21. doi: 10.2527/jas.2005-753
56. Kemper, K. E., Saxton, S. J., Bolormaa, S., Hayes, B. J., and Goddard, M. E. (2014). Selection for complex traits leaves little or no classic signatures of selection. BMC Genomics 15:246. doi: 10.1186/1471-2164-15-246
57. Kemper, K. E., Visscher, P. M., and Goddard, M. E. (2012). Genetic architecture of body size in mammals. Genome Biol. 13:244. doi: 10.1186/gb-2012-13-4-244
58. Khatib, H., Zaitoun, I., Wiebelhaus-Finger, J., Chang, Y. M., and Rosa, G. J. (2007). The association of bovine PPARGC1A and OPN genes with milk composition in two independent Holstein cattle populations. J. Dairy Sci. 90, 2966–2970. doi: 10.3168/jds.2006-812
59. Kohn, M. H., Pelz, H. J., and Wayne, R. K. (2000). Natural selection mapping of the warfarin-resistance gene. Proc. Natl. Acad. Sci. U.S.A. 97, 7911–7915. doi: 10.1073/pnas.97.14.7911
60. Komisarek, J., and Walendowska, A. (2012). Analysis of the PPARGC1A gene as a potential marker for productive and reproductive traits in cattle. Folia Biol. 60, 171–174. doi: 10.3409/fb60_3-4.171-174
61. Kováčik, A., Bulla, J., Trakovická, A., Žitný, J., and Rafayová, A. (2012). The effect of the porcine melanocortin-5 receptor (MC5R) gene associated with feed intake, carcass and physic chemical characteristics. J. Microbiol. Biotechnol. Food Sci. 1, 498–506. Available online at: http://www.jmbfs.org/wp-content/uploads/2012/01/Kovacik.pdf
62. Kuehn, C., and Weikard, R. (2007a). An investigation into the genetic background of coat colour dilution in a Charolais x German Holstein F2 resource population. Anim. Genet. 38, 109–113. doi: 10.1111/j.1365-2052.2007.01569.x
63. Kuehn, C., and Weikard, R. (2007b). Multiple splice variants within the bovine silver homologue (SILV) gene affecting coat color in cattle indicate a function additional to fibril formation in melanophores. BMC Genomics 8:335. doi: 10.1186/1471-2164-8-335
64. Lagonigro, R., Wiener, P., Pilla, F., Woolliams, J. A., and Williams, J. L. (2003). A new mutation in the coding region of the bovine leptin gene associated with feed intake. Anim. Genet. 34, 371–374. doi: 10.1046/j.1365-2052.2003.01028.x
65. Laloe, D., Moazami-Goudarzi, K., Lenstra, J. A., Ajmone Marsan, P., Azor, P., Baumung, R., et al. (2010). Spatial trends of genetic variation of domestic ruminants in Europe. Diversity 2, 932–945. doi: 10.3390/d2060932
66. Lango-Allen, H., Estrada, K., Lettre, G., Berndt, S. I., Weedon, M. N., Rivadeneira, F., et al. (2010). Hundreds of variants clustered in genomic loci and biological pathways affect human height. Nature 467, 832–838.
67. Larson, J., Swanson, J., Berry, D'. A., and Smith, C. (1992). Selected Readings on the History and Use of Old Livestock Breeds. Beltsville, MD: American Minor Breeds Conservancy.
68. Lawrence, M. S., Stojanov, P., Polak, P., Kryukov, G. V., Cibulskis, K., Sivachenko, A., et al. (2013). Mutational heterogeneity in cancer and the search for new cancer-associated genes. Nature 499, 214–218. doi: 10.1038/nature12213
69. Lawson, M. A. (2004). The role of integrin degradation in post-mortem drip loss in pork. Meat Sci. 68, 559–566. doi: 10.1016/j.meatsci.2004.05.019
70. Lee, H. J., Jang, M., Kim, H., Kwak, W., Park, W., Hwang, J. Y., et al. (2013a). Comparative transcriptome analysis of adipose tissues reveals that ecm-receptor interaction is involved in the depot-specific adipogenesis in cattle. PLoS ONE 8:e66267. doi: 10.1371/journal.pone.0066267
71. Lee, J. W., Bannerman, D. D., Paape, M. J., Huang, M. K., and Zhao, X. (2006). Characterization of cytokine expression in milk somatic cells during intramammary infections with Escherichia coli or Staphylococcus aureus by real-time PCR. Vet. Res. 37, 219–229. doi: 10.1051/vetres:2005051

72. Lee, T., Cho, S., Seo, K. S., Chang, J., Kim, H., and Yoon, D. (2013b). Genetic variants and signatures of selective sweep of Hanwoo population (Korean native cattle). BMB Rep. 46, 346–351. doi: 10.5483/BMBRep.2013.46.7.211
73. Leonard, S., Khatib, H., Schutzkus, V., Chang, Y. M., and Maltecca, C. (2005). Effects of the osteopontin gene variants on milk production traits in dairy cattle. J. Dairy Sci. 88, 4083–4086. doi: 10.3168/jds.S0022-0302(05)73092-7
74. Li, Q., Wang, Z., Zhang, B., Lu, Y., Yang, Y., Ban, D., et al. (2014a). Single nucleotide polymorphism scanning and expression of the pig PPARGC1A gene in different breeds. Lipids 49, 1047–1055. doi: 10.1007/s11745-014-3928-1
75. Li, Z., Zhang, H., Wang, H., Chen, L., Wang, L., Liu, X., et al. (2014b). Polymorphism in the promoter of TLR4 gene by PCR-RFLP and its association with somatic cell score in Chinese Holstein. Archiv. Tierzucht 57, 1–6. doi: 10.7482/0003-9438-57-006
76. Liefers, S. C., te Pas, M. F., Veerkamp, R. F., and van der Lende, T. (2002). Associations between leptin gene polymorphisms and production, live weight, energy balance, feed intake, and fertility in Holstein heifers. J. Dairy Sci. 85, 1633–1638. doi: 10.3168/jds.S0022-0302(02)74235-5
77. Lim, D., Gondro, C., Park, H. S., Cho, Y. M., Chai, H. H., Seong, H. H., et al. (2013). Identification of recently selected mutations driven by artificial selection in Hanwoo (Korean cattle). Asian-Australas J. Anim. Sci. 26, 603–608. doi: 10.5713/ajas.2012.12456
78. Lindholm-Perry, A. K., Sexten, A. K., Kuehn, L. A., Smith, T. P., King, D. A., Shackelford, S. D., et al. (2011). Association, effects and validation of polymorphisms within the NCAPG - LCORL locus located on BTA6 with feed intake, gain, meat and carcass traits in beef cattle. BMC Genet. 12:103. doi: 10.1186/1471-2156-12-103
79. Littlejohn, M. D., Henty, K. M., Tiplady, K., Johnson, T., Harland, C., Lopdell, T., et al. (2014). Functionally reciprocal mutations of the prolactin signalling pathway define hairy and slick cattle. Nat. Commun. 5:5861. doi: 10.1038/ncomms6861
80. Mancini, G., Gargani, M., Chillemi, G., Nicolazzi, E. L., Marsan, P. A., Valentini, A., et al. (2014). Signatures of selection in five Italian cattle breeds detected by a 54K SNP panel. Mol. Biol. Rep. 41, 957–965. doi: 10.1007/s11033-013-2940-5
81. Matukumalli, L. K., Lawley, C. T., Schnabel, R. D., Taylor, J. F., Allan, M. F., Van Tassell, C. P., et al. (2009). Development and characterization of a high density SNP genotyping assay for cattle. PLoS ONE 4:e5350. doi: 10.1371/journal.pone.0005350
82. Maynard Smith, J., and Haigh, J. (1974). The hitch-hiking effect of a favourable gene. Genet. Res. 23: 23–35. doi: 10.1017/S0016672300014634
83. McLean, K. L., and Schmutz, S. M. (2009). Associations of melanocortin 1 receptor genotype with growth and carcass traits in beef cattle. Can. J. Anim. Sci. 89, 295–300. doi: 10.4141/CJAS08094
84. McPherron, A. C., and Lee, S. J. (1997). Double muscling in cattle due to mutations in the myostatin gene. Proc. Natl. Acad. Sci. U.S.A. 94, 12457–12461. doi: 10.1073/pnas.94.23.12457
85. Medugorac, I., Seichter, D., Graf, A., Russ, I., Blum, H., Göpel, K. H., et al. (2012). Bovine polledness–an autosomal dominant trait with allelic heterogeneity. PLoS ONE 7:e39477. doi: 10.1371/journal.pone.0039477
86. Montoliu, L., Oetting, W. S., and Bennett, D. C. (2014). Color Genes. European Society for Pigment Cell Research. World Wide Web. Available online at: http://www.espcr.org/micemut.
87. Morris, C. A., Cullen, N. G., Glass, B. C., Hyndman, D. L., Manley, T. R., Hickey, S. M., et al. (2007). Fatty acid synthase effects on bovine adipose fat and milk fat. Mamm. Genome 18, 64–74. doi: 10.1007/s00335-006-0102-y
88. Nielsen, R., Williamson, S., Kim, Y., Hubisz, M. J., Clark, A. G., and Bustamante, C. (2005). Genomic scans for selective sweeps using SNP data. Genome Res. 15, 1566–1575. doi: 10.1101/gr.4252305
89. Ogorevc, J., Kunej, T., Razpet, A., and Dovc, P. (2009). Database of cattle candidate genes and genetic markers for milk production and mastitis. Anim. Genet. 40, 832–851. doi: 10.1111/j.1365-2052.2009.01921.x

90. Olsen, H. G., Meuwissen, T. H., Nilsen, H., Svendsen, M., and Lien, S. (2008). Fine mapping of quantitative trait Loci on bovine chromosome 6 affecting calving difficulty. J. Dairy Sci. 91, 4312–4322. doi: 10.3168/jds.2008-1000
91. Olsen, H. G., Nilsen, H., Hayes, B., Berg, P. R., Svendsen, M., Lien, S., et al. (2007). Genetic support for a quantitative trait nucleotide in the ABCG2 gene affecting milk composition of dairy cattle. BMC Genet. 8:32. doi: 10.1186/1471-2156-8-32
92. Olson, T. A. (1999). "Genetics of color variation," in The Genetics of Cattle, eds R. Fries and A. Ruvinsky (Wallingford: CABI), 33–53.
93. Ouali, A., Herrera-Mendez, C. H., Coulis, G., Becila, S., Boudjellal, A., Aubry, L., et al. (2006). Revisiting the conversion of muscle into meat and the underlying mechanisms. Meat Sci. 74, 44–58. doi: 10.1016/j.meatsci.2006.05.010
94. Pan, D., Zhang, S., Jiang, J., Jiang, L., Zhang, Q., and Liu, J. (2013). Genome-wide detection of selective signature in Chinese Holstein. PLoS ONE 8:e60440. doi: 10.1371/journal.pone.0060440
95. Pausch, H., Flisikowski, K., Jung, S., Emmerling, R., Edel, C., Götz, K. U., et al. (2011). Genome-wide association study identifies two major loci affecting calving ease and growth-related traits in cattle. Genetics 187, 289–297. doi: 10.1534/genetics.110.124057
96. Pintus, E., Sorbolini, S., Albera, A., Gaspa, G., Dimauro, C., Steri, R., et al. (2014). Use of locally weighted scatterplot smoothing (LOWESS) regression to study selection signatures in Piedmontese and Italian Brown cattle breeds. Anim. Genet. 45, 1–11. doi: 10.1111/age.12076
97. Pollinger, J. P., Bustamante, C. D., Fledel-Alon, A., Schmutz, S., Gray, M. M., and Wayne, R. K. (2005). Selective sweep mapping of genes with large phenotypic effects. Genome Res. 15, 1809–1819. doi: 10.1101/gr.4374505
98. Porto-Neto, L. R., Lee, S. H., Sonstegard, T. S., Van Tassell, C. P., Lee, H. K., Gondro, C., et al. (2014). Genome-wide detection of signatures of selection in Korean Hanwoo cattle. Anim. Genet. 45, 180–190. doi: 10.1111/age.12119
99. Porto-Neto, L. R., Sonstegard, T. S., Liu, G. E., Bickhart, D. M., Da Silva, M. V. B., Machado, M., et al. (2013). Genomic divergence of zebu and taurine cattle identified through high-density SNP genotyping. BMC Genomics 14:876. doi: 10.1186/1471-2164-14-876
100. Prasad, A., Schnabel, R. D., McKay, S. D., Murdoch, B., Stothard, P., Kolbehdari, D., et al. (2008). Linkage disequilibrium and signatures of selection on chromosomes 19 and 29 in beef and dairy cattle. Anim. Genet. 39, 597–605. doi: 10.1111/j.1365-2052.2008.01772.x
101. Pryce, J. E., Hayes, B. J., Bolormaa, S., and Goddard, M. E. (2011). Polymorphic regions affecting human height also control stature in cattle. Genetics 187, 981–984. doi: 10.1534/genetics.110.123943
102. Puglisi, R., Cambuli, C., Capoferri, R., Giannino, L., Lukaj, A., Duchi, R., et al. (2013). Differential gene expression in cumulus oocyte complexes collected by ovum pick up from repeat breeder and normally fertile Holstein Friesian heifers. Anim. Reprod. Sci. 141, 26–33. doi: 10.1016/j.anireprosci.2013.07.003
103. Puigserver, P., and Spiegelman, B. M. (2003). Peroxisome proliferator-activated receptor-gamma coactivator 1 alpha (PGC-1 alpha): transcriptional coactivator and metabolic regulator. Endocr. Rev. 24, 78–90. doi: 10.1210/er.2002-0012
104. Qanbari, S., Gianola, D., Hayes, B., Schenkel, F., Miller, S., Moore, S., et al. (2011). Application of site and haplotype-frequency based approaches for detecting selection signatures in cattle. BMC Genomics 12:318. doi: 10.1186/1471-2164-12-318
105. Qanbari, S., Pausch, H., Jansen, S., Somel, M., Strom, T. M., Fries, R., et al. (2014). Classic selective sweeps revealed by massive sequencing in cattle. PLoS Genet. 10:e1004148. doi: 10.1371/journal.pgen.1004148
106. Qanbari, S., Pimentel, E. C., Tetens, J., Thaller, G., Lichtner, P., Sharifi, A. R., et al. (2010). A genome-wide scan for signatures of recent selection in Holstein cattle. Anim. Genet. 41, 377–389. doi: 10.1111/j.1365-2052.2009.02016.x
107. Ramayo-Caldas, Y., Fortes, M. R., Hudson, N. J., Porto-Neto, L. R., Bolormaa, S., Barendse, W., et al. (2014). A marker-derived gene network reveals the regulatory role of PPARGC1A, HNF4G, and FOXP3 in intramuscular fat deposition of beef cattle. J. Anim. Sci. 92, 2832–2845. doi: 10.2527/jas.2013-7484

108. Ramey, H. R., Decker, J. E., McKay, S. D., Rolf, M. M., Schnabel, R. D., and Taylor, J. F. (2013). Detection of selective sweeps in cattle using genome-wide SNP data. BMC Genomics 14:382. doi: 10.1186/1471-2164-14-382

109. Reinhardt, T. A., and Lippolis, J. D. (2009). Mammary gland involution is associated with rapid down regulation of major mammary Ca2+-ATPases. Biochem. Biophys. Res. Commun. 378, 99–102. doi: 10.1016/j.bbrc.2008.11.004

110. Robbins, L. S., Nadeau, J. H., Johnson, K. R., Kelly, M. A., Roselli-rehfuss, L., Baack, E., et al. (1993). Pigmentation phenotypes of variant extension locus alleles result from point mutations that alter MSH receptor function. Cell 72, 827–834. doi: 10.1016/0092-8674(93)90572-8

111. Ron, M., Israeli, G., Seroussi, E., Weller, J. I., Gregg, J. P., Shani, M., et al. (2007). Combining mouse mammary gland gene expression and comparative mapping for the identification of candidate genes for QTL of milk production traits in cattle. BMC Genomics 8:183. doi: 10.1186/1471-2164-8-183

112. Rothammer, S., Seichter, D., Förster, M., and Medugorac, I. (2013). A genome-wide scan for signatures of differential artificial selection in ten cattle breeds. BMC Genomics 14:908. doi: 10.1186/1471-2164-14-908

113. Roy, R., Ordovas, L., Zaragoza, P., Romero, A., Moreno, C., Altarriba, J., et al. (2006). Association of polymorphisms in the bovine FASN gene with milk-fat content. Anim. Genet. 37, 215–218. doi: 10.1111/j.1365-2052.2006.01434.x

114. Royo, L. J., Alvarez, I., Fernández, I., Arranz, J. J., Gómez, E., and Goyache, F. (2005). The coding sequence of the ASIP gene is identical in nine wild-type coloured cattle breeds. J. Anim. Breed. Genet. 122, 357–360. doi: 10.1111/j.1439-0388.2005.00541.x

115. Schennink, A., Bovenhuis, H., Léon-Kloosterziel, K. M., van, Arendonk, J. A., and Visker, M. H. (2009). Effect of polymorphisms in the FASN, OLR1, PPARGC1A, PRL and STAT5A genes on bovine milk-fat composition. Anim. Genet. 40, 909–916. doi: 10.1111/j.1365-2052.2009.01940.x

116. Schmutz, S. M., and Dreger, D. L. (2013). Interaction of MC1R and PMEL alleles on solid coat colors in Highland cattle. Anim. Genet. 44, 9–13. doi: 10.1111/j.1365-2052.2012.02361.x

117. Schneider, E. H., Neumann, D., and Seifert, R. (2014). Modulation of behaviour by the histaminergic system: lessons from HDC-, H3R- and H4R-deficient mice. Neurosci. Biobehav. Rev. 47C, 101–121. doi: 10.1016/j.neubiorev.2014.07.020

118. Seitz, J. J., S., Schmutz, M., Thue, T. D., and Buchanan, F.C. (1999). A missense mutation in the bovine MGF gene is associated with the roan phenotype in Belgian blue and Shorthorn cattle. Mamm. Genome 10, 710–712. doi: 10.1007/s003359901076

119. Setoguchi, K., Furuta, M., Hirano, T., Nagao, T., Watanabe, T., Sugimoto, Y., et al. (2009). Cross-breed comparisons identified a critical 591-kb region for bovine carcass weight QTL (CW-2) on chromosome 6 and the Ile-442-Met substitution in NCAPG as a positional candidate. BMC Genet. 10:43. doi: 10.1186/1471-2156-10-43

120. Setoguchi, K., Watanabe, T., Weikard, R., Albrecht, E., Kühn, C., Kinoshita, A., et al. (2011). The SNP c.1326T<G in the non-SMC condensin I complex, subunit G (NCAPG) gene encoding a p.Ile442Met variant is associated with an increase in body frame size at puberty in cattle. Anim. Genet. 42, 650–655. doi: 10.1111/j.1365-2052.2011.02196.x

121. Sevane, N., Armstrong, E., Cortés, O., Wiener, P., Wong, R. P., Dunner, S., et al. (2013). Association of bovine meat quality traits with genes included in the PPARG and PPARGC1A networks. Meat Sci. 94, 328–335. doi: 10.1016/j.meatsci.2013.02.014

122. Sevane, N., Armstrong, E., Wiener, P., Pong, W. R., Dunner, S., and GemQual Consortium (2014). Polymorphisms in twelve candidate genes are associated with growth, muscle lipid profile and meat quality traits in eleven European cattle breeds. Mol. Biol. Rep. 41, 4721–4731. doi: 10.1007/s11033-014-3343-y

123. Sheehy, P. A., Riley, L. G., Raadsma, H. W., Williamson, P., and Wynn, P. C. (2009). A functional genomics approach to evaluate candidate genes located in a QTL interval for milk production traits on BTA6. Anim. Genet. 40, 492–498. doi: 10.1111/j.1365-2052.2009.01862.x

124. Shin, S., and Chung, E. (2013). Novel SNPs in the bovine ADIPOQ and PPARGC1A genes are associated with carcass traits in Hanwoo (Korean cattle). Mol. Biol. Rep. 40, 4651–4660. doi: 10.1007/s11033-013-2560-0
125. Smith, J. A., Lewis, A. M., Wiener, P., and Williams, J. L. (2000). Genetic variation in the bovine myostatin gene in UK beef cattle: allele frequencies and haplotype analysis in the South Devon. Anim. Genet. 31, 306–309. doi: 10.1046/j.1365-2052.2000.00521.x
126. Somavilla, A. L., Sonstegard, T. S., Higa, R. H., Rosa, A. N., Siqueira, F., Silva, L. O., et al. (2014). A genome-wide scan for selection signatures in Nellore cattle. Anim. Genet. 45, 771–781. doi: 10.1111/age.12210
127. Soranzo, N., Rivadeneira, F., Chinappen-Horsley, U., Malkina, I., Richards, J. B., Hammond, N., et al. (2009). Meta-analysis of genome-wide scans for human adult stature identifies novel loci and associations with measures of skeletal frame size. PLoS Genet. 5:e1000445. doi: 10.1371/journal.pgen.1000445
128. Sponenberg, D. P., and Bixby, D. E. (2007). Managing Breeds for a Secure Future: Strategies for Breeders and Breed Associations. Pittsboro, NC: American Livestock Breeds Conservancy.
129. Stella, A., Ajmone-Marsan, P., Lazzari, B., and Boettcher, P. (2010). Identification of selection signatures in cattle breeds selected for dairy production. Genetics 185, 1451–1461. doi: 10.1534/genetics.110.116111
130. Storz, J. F., Payseur, B. A., and Nachman, M. W. (2004). Genome scans of DNA variability in humans reveal evidence for selective sweeps outside of Africa. Mol. Biol. Evol. 21, 1800–1811. doi: 10.1093/molbev/msh192
131. Suchocki, T., Komisarek, J., and Szyda, J. (2010). Testing candidate gene effects on milk production traits in dairy cattle under various parameterizations and modes of inheritance. J. Dairy Sci. 93, 2703–2717. doi: 10.3168/jds.2009-2550
132. Switonski, M., Mankowska, M., and Salamon, S. (2013). Family of melanocortin receptor (MCR) genes in mammals-mutations, polymorphisms and phenotypic effects. J. Appl. Genet. 54, 461–472. doi: 10.1007/s13353-013-0163-z
133. Tetens, J., Widmann, P., Kühn, C., and Thaller, G. (2013). A genome-wide association study indicates LCORL/NCAPG as a candidate locus for withers height in German Warmblood horses. Anim. Genet. 44, 467–471. doi: 10.1111/age.12031
134. Thaller, G., Kühn, C., Winter, A., Ewald, G., Bellmann, O., Wegner, J., et al. (2003). DGAT1, a new positional and functional candidate gene for intramuscular fat deposition in cattle. Anim. Genet. 34, 354–357. doi: 10.1046/j.1365-2052.2003.01011.x
135. Trakovická, A., Moravčíková, N., and Kasarda, R. (2013). Genetic polymorphisms of leptin and leptin receptor genes in relation with production and reproduction traits in cattle. Acta Biochim. Pol. 60, 783–787. http://www.actabp.pl/pdf/4_2013/783.pdf
136. Utsunomiya, Y. T., do Carmo, A. S., Carvalheiro, R., Neves, H. H., Matos, M. C., Zavarez, L. B., et al. (2013b). Genome-wide association study for birth weight in Nellore cattle points to previously described orthologous genes affecting human and bovine height. BMC Genet. 14:52. doi: 10.1186/1471-2156-14-52
137. Utsunomiya, Y. T., Pérez O'Brien, A. M., Sonstegard, T. S., Van Tassell, C. P., do Carmo, A. S., Mészáros, G., et al. (2013a). Detecting loci under recent positive selection in dairy and beef cattle by combining different genome-wide scan methods. PLoS ONE 8:e64280. doi: 10.1371/journal.pone.0064280
138. Vankan, D. M., Waine, D. R., and Fortes, M. R. (2010). Real-time PCR genotyping and frequency of the myostatin F94L mutation in beef cattle breeds. Animal 4, 530–534. doi: 10.1017/S175173110999139X
139. Voight, B. F., Kudaravalli, S., Wen, X., and Pritchard, J. K. (2006). A map of recent positive selection in the human genome. PLoS Biol. 4: 446–458. doi: 10.1371/journal.pbio.0040154
140. Wang, J., Duncan, D., Shi, Z., and Zhang, B. (2013). WEB-based GEne SeT AnaLysis Toolkit (WebGestalt): update 2013. Nucleic Acids Res. 41, W77–W83. doi: 10.1093/nar/gkt439
141. Wang, X. P., Luoreng, Z. M., Gao, S. X., Guo, D. S., Li, J. Y., Gao, X., et al. (2014a). Haplotype analysis of TLR4 gene and its effects on milk somatic cell score in Chinese commercial cattle. Mol. Biol. Rep. 41, 2345–2351. doi: 10.1007/s11033-014-3088-7

Wang, X., Xu, S., Gao, X., Ren, H., and Chen, J. B. (2007). Genetic polymorphism of TLR4 gene and correlation with mastitis in cattle. J. Genet. Genomics 34, 406–412. doi: 10.1016/S1673-8527(07)60044-7

142. Wang, Z., Hou, X., Qu, B., Wang, J., Gao, X., and Li, Q. (2014b). Pten regulates development and lactation in the mammary glands of dairy cows. PLoS ONE 9:e102118. doi: 10.1371/journal.pone.0102118
143. Wedholm, A., Larsen, L. B., Lindmark-Månsson, H., Karlsson, A. H., and Andrén, A. (2006). Effect of protein composition on the cheese-making properties of milk from individual dairy cows. J. Dairy Sci. 89, 3296–3305. doi: 10.3168/jds.S0022-0302(06)72366-9
144. Wei, J., Geale, P. F., Sheehy, P. A., and Williamson, P. (2012). The impact of ABCG2 on bovine mammary epithelial cell proliferation. Anim. Biotechnol. 23, 221–224. doi: 10.1080/10495398.2012.696567
145. Weikard, R., Kuehn, C., Goldammer, T., Freyer, G., and Schwerin, M. (2005). The bovine PPARGC1A gene: molecular characterization and association of an SNP with variation of milk fat synthesis. Physiol. Genomics 21, 1–13. doi: 10.1152/physiolgenomics.00103.2004
146. Weikard, R., Widmann, P., Buitkamp, J., Emmerling, R., and Kuehn, C. (2012). Revisiting the quantitative trait loci for milk production traits on BTA6. Anim. Genet. 43, 318–323. doi: 10.1111/j.1365-2052.2011.02258.x
147. Wiener, P., Edriss, M. A., Williams, J. L., Waddington, D., Law, A., Woolliams, J. A., et al. (2011). Information content in genome-wide scans: concordance between patterns of genetic differentiation and linkage mapping associations. BMC Genomics 12:65. doi: 10.1186/1471-2164-12-65
148. Wilkins, A. S., Wrangham, R. W., and Fitch, W. T. (2014). The "domestication syndrome" in mammals: a unified explanation based on neural crest cell behaviour and genetics. Genetics 197, 795–808. doi: 10.1534/genetics.114.165423
149. Williams, J. L., Dunner, S., Valentini, A., Mazza, R., Amarger, V., Checa, M. L., et al. (2009). Discovery, characterization and validation of single nucleotide polymorphisms within 206 bovine genes that may be considered as candidate genes for beef production and quality. Anim. Genet. 40, 486–491. doi: 10.1111/j.1365-2052.2009.01874.x
150. Woronuk, G. N., Marquess, F. L., James, S. T., Palmer, J., Berryere, T., Deobald, H., et al. (2012). Association of leptin genotypes with beef cattle characteristics. Anim. Genet. 43, 608–610. doi: 10.1111/j.1365-2052.2012.02320.x
151. Wu, X. X., Yang, Z. P., Shi, X. K., Li, J. Y., Ji, D. J., Mao, Y. J., et al. (2012). Association of SCD1 and DGAT1 SNPs with the intramuscular fat traits in Chinese Simmental cattle and their distribution in eight Chinese cattle breeds. Mol. Biol. Rep. 39, 1065–1071. doi: 10.1007/s11033-011-0832-0
152. Xiao, Y., Junfeng, H., Tianhong, L., Lu, W., Shulin, C., Yu, Z., et al. (2006). Cathepsin K in adipocyte differentiation and its potential role in the pathogenesis of obesity. J. Clin. Endocrinol. Metab. 91, 4520–4527. doi: 10.1210/jc.2005-2486
153. Zhang, S., Knight, T. J., Reecy, J. M., and Beitz, D. C. (2008). DNA polymorphisms in bovine fatty acid synthase are associated with beef fatty acid composition. Anim. Genet. 39, 62–70. doi: 10.1111/j.1365-2052.2007.01681.x
154. Zheng, X., Ju, Z., Wang, J., Li, Q., Huang, J., Zhang, A., et al. (2011). Single nucleotide polymorphisms, haplotypes and combined genotypes of LAP3 gene in bovine and their association with milk production traits. Mol. Biol. Rep. 38, 4053–4061. doi: 10.1007/s11033-010-0524-1

INDEX